# SPACE, TIME, and SPACETIME

# SPACE, TIME, and SPACETIME

## Lawrence Sklar

UNIVERSITY OF CALIFORNIA PRESS
BERKELEY • LOS ANGELES • LONDON

### TO LISSY

*"Did I ever tell you my theory of Time and Space?"*
*They ignored him.*

John Gardner,
*The Wreckage of Agathon.*

University of California Press
Berkeley and Los Angeles, California
University of California Press, Ltd.
London, England
Copyright © 1974, 1976 by The Regents of the University of California
All rights reserved.
First Paperback Edition, 1977
ISBN: 0-520-03174-1
Library of Congress Catalog Card Number: 73-76096
Printed in the United States of America

6 7 8 9 0

The paper used in this publication meets the minimum
requirements of American National Standard for Information
Sciences—Permanence of Paper for Printed Library Materials,
ANSI Z39.48–1984. ⊗

# ACKNOWLEDGMENTS

My greatest intellectual debt is to Hans Reichenbach, who, inspired by the scientific and philosophical work of Einstein, Poincaré, Eddington, and Schlick, is most responsible for the philosophical study of space and time in its current form.

My deepest source of inspiration from contemporary philosophers has been the work of John Earman, from whose published and unpublished works I have taken ideas without stint. Substantial portions of this book can be looked upon as explications of and expansions on his contributions to the philosophy of space and time.

I have also greatly profited from the work on space and time of Adolph Grünbaum, Hilary Putnam, Howard Stein, Clifford Hooker, Hugh Lacey, John Graves, and Bas van Fraassen.

I am very deeply indebted to Peter Hempel and Hilary Putnam who, in their very different ways, inspired my interest in the philosophy of science. Whatever correct perspectives I now have on how to do it, I owe to them.

For detailed comments on the text of this book while it was in preparation, I wish to thank John Earman. The mathematical and physical errors which remain, and especially the philosophical confusions present are, of course, my own.

Thanks are also due to Robert Zachary and James Kubeck of the University of California Press for their help in publication, and especially to David Dexter for his enormous help in preparing the manuscript for publication.

Finally, I wish to express my very deepest thanks to Elizabeth Sklar and Moses Frohlich, without whose special help this book would never have been written.

L.S.

# PREFACE FOR
# THE PAPERBACK EDITION

Mechanical and economic considerations are such that the text of this paperback edition differs from the original only in the correction of a number of typographical errors and other minor mistakes redeemable by changing a few words.

Were a new edition to appear, though, I would make a number of substantial revisions and additions. I will use the opportunity of this preface just to indicate the direction in which some of these more important changes might go.

1. In (IV, E, 1) a "causal" definition of temporal betweenness is given. In (IV, E, 2) it is suggested that this definition breaks down in spacetimes which are causally pathological. As David Malament argues (in his review of this book in *Journal of Philosophy*) the account here is misleading. There exist definitions of temporal betweenness, framed in terms of the same primitives I use, which will hold up in any general relativistic spacetime.

The situation is, rather, like this. In appropriately non-pathological spacetimes a full definition of the topology in terms of the relation of *causal connectibility* among events is possible. In pathological spacetimes such a definition breaks down. But the full topology is always at least implicitly defined, in a sense which can be made precise, by the class of *causal curves* of the spacetime.

Nonetheless, the basic argument of (IV, E, 3) still goes through I believe. The causal theory of time as it is usually meant philosophically still appears to me to be an allegation that from a primitive knowledge of certain causal relationships, without presupposing any "direct awareness" of spatio-temporal relationships, one could reconstruct the spatio-temporal order of the world. But, I would still allege, the kind of causal relationships one would need for such a "reconstruction by definition" of the spacetime topology would be those (knowing the class of continuous causal curves, for example) which we could determine only if we had already the ability to discriminate topological features of the spacetime (the class of continuous timelike or null curves, for example).

Still, there may be something philosophically important to the fact that knowing a proper subset of the topological features is enough to fix the rest. For one might then argue that it is just those features of the topology which we discover, upon epistemological analysis, to be "immediately apprehendible" that are sufficient to fully determine the topology. So now a kind of "reductionism" seems to have some plausibility. But it is not a "reduction" of spatio-temporal to (prima facie) non-spatio-temporal, but of all the spatio-temporal concepts to an (apparently) proper subset of them.

These considerations are relevant in another way as well. If these topological aspects of spacetime are really the directly ascertainable root facts upon which our theories are to be constructed, many of Reichenbach's allegations about the conventionality of topology require further scrutiny. In particular one might make a case for certain global features being conventional (depending upon our choices in individuating events) but that certain local aspects (what classes of spacetime points are continuous) are among the non-conventional "brute facts" of experience.

Despite the skepticism evidenced throughout the book about "necessary" or "analytic" propositions so common in philosophical theories of space and time, it would seem that further exploration of the connections between such concepts as spatio-temporal continuity, genidentity and causal connection is in order. Not in the hopes of reducing all spatio-temporal concepts to those prima facie non-spatio-temporal but, rather, in the hopes of sorting out the spatio-temporal components of the other concepts. Finally, we need to be much clearer about the very fundamental problem of connecting the "immediately apprehendible" features of our "subjective" spatio-temporal experience with our hypothesized physical spatio-temporal structure of the world.

2. In Chapter V I believe a greatly expanded treatment of the problem of the direction of time is necessary. First of all the discussion of the physical aspects of the asymmetry of the world in time requires much supplementation to be adequate. More importantly I think much more needs to be said about just what the content of a philosophical theory of the direction of time is supposed to be and whether the assertions of such a theory could be justified. In particular I think it would be very worthwhile to explore the question as to why a "gravitational theory of up and down" seems obviously correct, why a "weak interaction theory of left and right" seems obviously wrong, and, most importantly, why an entropic theory of the direction of time seems so persistently controversial. I believe that this would require a much more thorough exploration of the notion of our "subjective" sense of time direction and its relation to the "physical direction of time" and of the alleged inter-connections between such varied asymmetries as the subjective the entropic, the causal, the epistemic, etc. In particular I now think the present treatment is unfair to the subtlety of the Boltzmann-Reichenbach approach, although I am still inclined to believe that that approach leaves many questions to be resolved.

3. (III, D, 3) should be supplemented by a treatment of a curved neo-Newtonian spacetime in which the gravitational field is assimilated to the geometry in a completely non-relativistic context just as it is in the relativistic context by the theory of general relativity.

4. The treatment of abstractions from Riemannian geometry in (II, B, 6) should be supplemented by a treatment of projective and conformal geometries which lie, in their abstractness, between affine geometry and the theory of differential manifolds in general.

5. The treatment of global aspects of spacetime in (IV, D) should be supplemented by a discussion of the hierarchy of notions of causal non-pathology, including the notion of stable causality, and by a discussion of the notion of a global time function for a spacetime. There should also be treatments of so-called non-predictive and indistinguishable spacetimes in which interesting consequences of the relativistic limits on the physical possibility of acquiring information are applied to world models for general relativity.

6. (II, D, 2) should be supplemented by a discussion of alleged limitations on the principle of equivalence.

7. Finally, there are numerous minor errors of mathematics and physics throughout the text which it has been impossible to repair in the

present addition. The author would be delighted to have these pointed out to him by readers in the hopes that at some future date a more accurate version may appear.

Since the manuscript was submitted for publication numerous books and articles of interest have appeared.

Some recent works in physics of interest are:

> Misner, C., Thorne, K., and Wheeler, J., *Gravitation*

which, among other things, gives an extensive account of the application of modern mathematics to general relativity;

> Hawking, S., and Ellis, G., *The Large Scale Structure of Space-Time,*

a brilliant and comprehensive study of global properties of spacetime in general relativity; and

> Davies, P., *The Physics of Time Asymmetry,*

which is an important study of the physics relevant to the problem of the "direction of time."

A sample of recent philosophical work in space and time can be obtained from the forthcoming:

> *Minnesota Studies in the Philosophy of Science,* Vol. VIII.

Lawrence Sklar<br>February 1976

*6 ₂*
*3 ₂ ₂*

# CONTENTS

# CHAPTER I

# INTRODUCTION

Philosophical concern with the nature of space and time is as old as philosophy itself. The mathematical and physical developments of the scientific progress since the seventeenth-century scientific revolution have made this concern take on quite new aspects, however. The philosopher can no longer take the nature of space and time as a pretty much self-evident given, assuming that this nature and our knowledge of it requires philosophical clarification, perhaps, but is in itself a definite, unchanging given to him. For with the advent of non-Euclidean geometries, their application to physical theories, the relativistic revisions in our concept of time, etc., the nature of space and time, or better, spacetime, has become itself a matter of scientific and philosophical dispute.

A good deal of philosophical speculation about space and time continues to take place in a manner that suggests that, for at least some questions, the results of contemporary mathematics and physics are irrelevant. Without necessarily disparaging such investigations as totally misguided from the start, I disavow any attempt in this book to deal with these issues in this "purely philosophical" way. The problems I will deal with are all those to which the results of physical theorizing are prima facie relevant. Indeed, some of them are the kinds of issues philosophers would never even consider as problematic, were it not for the developments of science.

The primary aim of this book, in fact, is to convince the reader of the interdependence of science and philosophy. I choose the term "interdependence" deliberately here, since I will argue that it is not simply that one cannot do good philosophy without relying upon the results of scientific theorizing, but that a careful examination of science shows that the acceptance or rejection of particular scientific theories depends as much upon the adoption of specific philosophical presuppositions as it does upon the evidence of observation and experimentation. In one sense, the major aim of this book is to cast as much doubt as possible on the view that science and philosophy are independent pursuits that can be carried out in total ignorance of each other.

I will deal with four overall issues in this book. Although they are treated chapter by chapter, they are not, of course, totally independent problems. They intertwine with one another, both in matters of detail and in the more important general way that, as we shall see, certain common underlying philosophical concerns are at the heart of all of these issues.

Chapter II is concerned with the epistemology of geometry. Given that mathematics provides us with more than one consistent description of possible spaces or spacetimes, and given that some physical theories utilize these possibilities to characterize worlds with quite different geometric structures, to what extent can we empirically determine just what the geometric structure of the space or spacetime of the world is? Is this an "empirical" matter? Can we, instead, fix upon a structure for the world independently of any observation or experiment? Or is the situation more appropriately described as one in which the geometric structure of the world we posit is one we can *choose* to please ourselves, "as a matter of convention"?

Chapter III is concerned with a closely related issue. When we deal with the spatiotemporal structure of the world, must we view ourselves as attempting to discover the nature of an entity of the world, space or spacetime itself, or should we instead view ourselves as attempting merely to determine certain general truths about the structure of a set of relations holding among concrete material happenings?

Chapter IV is devoted to discussing the relation between the temporal order of the world and its causal order. The first part of the chapter discusses the revolution in our views about time brought about by the adoption of the special theory of relativity and deals with such questions as: the empirical basis for this conceptual change, the "philosophical" ingredients in the adoption of special relativity, the problem of whether the special-relativistic theory of time is a matter of "convention," etc. The later parts of the chapter examine the claim that temporal relations among events can be "reduced" to causal relations among events in some plausible way—the so-called causal theory of time.

In Chapter V I discuss the philosophical issue sometimes called the question of the direction of time. The primary focus of the chapter is on the question of whether the intuitive notion of the direction of time receives any clarification by the study of various physical processes that we might intuitively describe as behaving asymmetrically in time.

The problem of setting an appropriate level for a work in the philosophy of physics is in some ways insoluble, so naturally I don't claim to have solved it. It is my hope that there is material in this book of interest to professional scientists and philosophers and, at the same time, that all of the book will be accessible to undergraduate students of both science and philosophy. Aiming for such a spread of readers leads, of course, to difficulties. But for better or for worse the philosophy of physics is not such a highly populated profession of specialists that one can usefully write an extensive work directed solely to specialists in the field. This book is aimed at physicists and students of science interested in the philosophical issues at the foundations of their discipline, and at philosophers and students of philosophy who are willing to accept my assurances that some knowledge of the results of physical theory is very relevant indeed to philosophical concerns.

The material of the book splits up, more or less, into popular science and philosophical analysis. The treatment of scientific results is skeletal, to say the least, but it is my hope that most of my assertions about "what science shows us" will be both intelligible to the reader not possessing a sophisticated mathematical or physical background and, at the same time, correct. If one is going to bring science to bear on philosophical issues for the benefit of philosophically trained readers, the science must be presented in a sketchy and popular manner. But this need not lead to popular science that is unsophisticated or, worse, incorrect. I hope that the scientific reader will have the patience and tolerance to put up with some fairly extended verbal presentations of the results of physical theory which could, of course, be handled far more concisely and elegantly using the sophisticated conceptual apparatus of mathematical physics. And I hope that the philosophical reader will find the popular science both intelligible and interesting enough to pursue the expansion of his scientific knowledge through the utilization of some of the resources noted in the bibliographies.

Just as with the science, frequently it will be necessary to present philosophical arguments, familiar in detail to some readers versed in philosophy, in a very abbreviated and sketchy manner. Again, I hope that as brief as these treatments are, they are not grossly oversimplified or naïve. The philosophy of physics has frequently suffered from wrongly or partially understood science. It has suffered even more, I believe, and at the hands of philosophers as well as scientists, from an extraordinarily myopic view of possible philosophical positions and of the arguments for and against them on the part of some of those who have attempted to bring philosophy to bear on physics, and vice versa.

In the main body of the text, little is done in the way of attributing various positions and arguments to particular philosophers. This is a deliberate attempt not to impede the reader's comprehension of the course of an argument by being forced to provide detailed and qualified presentations of positions and arguments in the various forms they have taken at the hands of various authors. One beneficial result of this is that wrong, but plausible, positions can be critically examined without worrying too much about whether one is doing full justice to some particular version of a position at the hands of some particular proponent of it. Hopefully, this has not resulted on my part in a mass assault upon straw men. A disadvantage of this procedure is that little credit is given in the text to sources of arguments and insights which I have drawn upon freely. I hope that in the annotated bibliographies fuller justice is done to the originators of interesting views and arguments.

The reader will note that each chapter is followed by a fairly brief annotated bibliography. These lists of resources make no claim whatever to completeness. I have attempted to make as accurate an assessment as I could of what directions for further reading would appeal to readers with various backgrounds, and to select what I have found to be the one or two most useful references for each particular question at each level of difficulty. Much more extensive and all-encompassing bibliographies can be found in the literature, and these are occasionally cited, but I hope that what the bibliographies here lack in extensiveness they make up for in the annotation accompanying each reference. I have aimed for a bibliographic apparatus that is far from encyclopedic, but which for that very reason may prove more useful than a mere listing of resources.

Items otherwise unmarked in the bibliography should be accessible to the reader whose background in mathematics and physics is sufficient to understand the body of the text of this book. Items marked with a single asterisk require somewhat more in the way of mathematical and physical background. Items marked with two asterisks (**) require a rather advanced level of comprehension of technical material for their full utilization.

# BIBLIOGRAPHY FOR CHAPTER I

Three general philosophical treatments of problems in space and time in the tradition that this book continues are:

Reichenbach, H., *The Philosophy of Space and Time,*

which sets the direction for much later discussion.

Grünbaum, A., *Philosophical Problems of Space and Time,*

which is a continuation of (and debate with) Reichenbach; and

Van Fraassen, B., *An Introduction to the Philosophy of Time and Space,*

which treats the issues debated by Reichenbach and Grünbaum at an elementary level and with a wealth of historical reference.

Two recent works that treat issues in space and time from alternative philosophical perspectives are:

Swinburne, R., *Space and Time,*

which has features in the British "ordinary language" tradition as well as parts more in the tradition this book follows; and

Whiteman, M., *Philosophy of Space and Time,*

which is an extensive treatment of the issues from a more idiosyncratic point of view.

A useful anthology of short selections, many extracted from works cited in the bibliographies to this book, is:

Smart, J., *Problems of Space and Time.*

CHAPTER **II.**

# THE EPISTEMOLOGY OF GEOMETRY

# SOME PRELIMINARY REMARKS

Epistemological questions about theories are frequent and familiar. Given that we *do* accept a certain theory as a correct description of the world, why *should* we hold to this belief? Is rational belief in the theory founded upon evidence, or is the theory one that is "self-justifying," requiring no previously-believed evidence for its support? If the theory is such that the rational man would believe it only upon evidence, what kind of evidence is requisite? And, finally, what must the relation between evidence and theory be for us to rationally agree that, on the basis of *that* evidence, belief in this theory is reasonable or correct?

It doesn't take long to discover that epistemological questions are inextricably entangled with questions of metaphysics (or ontology) and semantics—questions such as these: What is the theory about? What kinds of entities does it posit as existing "in the world," and what features does it attribute to them? And these: What are the meanings of the terms of the theory? What can the assertions of the theory be taken to mean? Our puzzles about "reality" and about "meaning" are not long kept on the shelf when puzzles about evidence and rational belief are brought to the fore. But epistemological issues provide as convenient an entry into the network of philosophical issues as any.

The epistemology of geometry has its own special interest. For several reasons, the question of the evidential basis for belief in a geometric hypothesis provides rather remarkable illumination of the question of the evidential basis of theories in general. There is the fact that geometry for many centuries provided the paradigm example of an allegedly a priori theory: a theory that, although deeply informative about the nature of the world, was one believable by the rational man on no evidence whatever, except *self-evidence*. Such allegedly a priori theories have, traditionally, provided a "touchstone" for the evaluation of general theories of knowledge. And then, geometry provides us with the clearest case of a theory whose evolution *qua* theory through time has strongly motivated an equally changing set of viewpoints about its evidential status. As our knowledge of geometry changed, so did our views about just what kind of knowledge geometric knowledge is. Last, and perhaps most important, contemporary views about geometric knowledge provide a lucid and "concrete" example of contemporary disputes about scientific knowledge in general.

The pursuit of the philosophical puzzles engendered by a study of geometry, from the viewpoint of the theory of knowledge, will have a number of stages. First, I will outline some important aspects in the historical development of geometry as a mathematical and physical theory. Second, I will discuss very briefly some traditional viewpoints about the evidential basis of geometric beliefs, viewpoints that are engendered jointly by presupposed philosophical positions and actual geometric practice. Third, I will outline the basic changes in the underlying view of the nature of space and time necessary to provide a framework for describing the one contemporary physical theory whose claims have had the most radical effect upon our epistemological attitude toward geometry: that is, general relativity. Fourth, I will try to present some basic features of general relativity. The treatment at this stage will be quite abbreviated, in that only enough will be said about the theory to allow us to use its claims for an object of study in my present epistemological investigation. Later I shall return to the theory, probing a little deeper into features of general-relativistic accounts of the nature of spacetime, where these additional aspects of the theory will become crucial for the discussion of questions more metaphysical than epistemological.

Finally, we shall be able to attack the ultimate goal of this section, utilizing the resources we have developed. I will examine current controversy over the epistemic status of geometry, controversy whose origin stems from Poincaré's brilliant and justly famous espousal of a *conventionalist* doctrine. I will try to expound Poincaré's original conventionalist claim and some replies to it, replies that, while taking Poincaré's results as deeply illuminating, attempt to show that the truths of Poincaré, properly interpreted, lead not to a genuine conventionalism but rather to a sophisticated *empiricist* view of geometry.

I will very briefly examine a number of attempts at elucidating the meaning of the conventionalist claim. I will argue that these attempts at "unpacking" the meaning of 'is conventional' when applied to geometry, although possibly illuminating in some wa᠈ s, fail to get at the heart of the most important epistemological issues.

Then, I shall attempt my own exegesis of the conventionalist claim, arguing that the debate over the question of the conventionality of geometry, properly understood, provides us with a very clear and concrete *example* of an overall connected group of issues about the status of physical theories in general. I will end by trying to show how various attempts at resolving the basic philosophical perplexities have their own idiosyncratic virtues and, at least apparently, debilitating defects as well.

Let me finish these preliminary remarks with two apologies and their accompanying excuses:

1. Superficiality will, I am afraid, be a pervasive feature of the material in this section. Superficial mathematics and physics, superficial history of science, and, worst of all, superficial philosophical analysis. Both detail and sometimes subtlety will be sacrificed at the altar of expediency. The history of geometry is a complex and fascinating study in the history of ideas; the mathematics of contemporary geometry is a lifetime pursuit in any one of its branches; general relativity is a full-time career for theoretical physicists; and the detailed treatment of the basic epistemological and metaphysical questions with which we shall end up would exhaust a significant portion of contemporary analytic philosophy.

But the understanding of the problem we have set ourselves requires *some* comprehension of material from all these areas, so we shall just have to pay for breadth with thinness of treatment. Hopefully, the resources for further study offered at the end of each chapter will guide the interested reader into deeper treatments of all of these component areas.

2. More disturbing to some readers than the superficiality of the treatment will be its inconclusiveness. I shall end not with a definitive statement of what I take to be the "truth" about the epistemic status of geometry, but rather with an indication of why, at least in my

opinion, all of the outstanding contenders for a philosophical resolution of the problem ultimately fail to be satisfactory.

I wish I had the answers to the questions I will end up with. I don't, and I here forestall claims of dishonesty on my part, if not of inadequacy. What I hope to show is the relevance of the discussion of the status of geometry to more general philosophical issues in metaphysics and epistemology. I don't pretend to solve the overriding philosophical questions.

# AN OUTLINE OF THE HISTORY
# OF GEOMETRY

## 1. The Development of Euclidean Geometry

The origins of geometric knowledge, which perhaps are in Egypt, are naturally enough clouded in obscurity. By the time of the great era of Greek philosophy, however, it is clear that there is quite systematic knowledge of a wide range of geometric truth. Many problems are treated by the Greek mathematicians (congruence of plane figures, areas of figures, theory of regular solids, division of angles into equal parts, etc.), but the greatest bulk of their systematic knowledge is in the study of plane figures bounded by segments of straight lines. This area of study contains some of their most monumental results—such as the fact that the sum of the interior angles of a triangle is equal to a straight angle, and that the square of the length of the hypotenuse of a right triangle is equal to the sum of the squares of the lengths of its sides—and is the primary material treated in the great systematization of Euclid (ca. 300 B.C.).

What is important for our purposes is the notion of *justification* of belief in geometric hypotheses, implicit in Greek geometric practice and explicitly discussed from the methodological point of view by the great philosophers—in particular by Plato and Aristotle. The assumption made is that one rationally believes in the truth of a geometric proposition only if the proposition can be shown to be inferable or deducible from other propositions whose truth is more "immediate." The structure of the great Euclidean synthesis of plane geometry, familiar from its perennial infliction upon the schoolboy, is well known. The basic elements are taken as definitions, axioms, and postulates, with the definitions as the weakest part of the edifice. The information that a line is that element with length but no breadth, etc., plays no role in the formal development that follows. At best, the definitions can be taken as heuristic devices indicating to the student what connections exist between the technical terms of the theory and the intuitive concepts of spatial figures and elements familiar to him from his pretheoretical discourse.

The axioms and postulates are more important. The class of axioms and the class of postulates are both classes of propositions. They are simply asserted to be true. From them the remaining truths of geometry are to be deduced, and the deduction is to make use of pure logical inference alone, without the postulation of additional nonlogical truths over and above those enunciated in the two classes of basic truths. It is the deducibility of these further truths from the axioms and postulates which rationalizes our belief in them.

The distinction between axioms and postulates is not all that clear. The axioms treat of "common notions," truths whose importance transcends their geometric application; the postulates deal with more specifically geometric matters. For concreteness here are the axioms and postulates, not in the original expression but like enough for our purposes:

AXIOMS:    1. *Things equal to the same thing are equal to each other.*
           2. *Equals added to equals yield equals.*
           3. *Equals removed from equals yield equals.*
           4. *Coincident figures are equal to one another in all re-respects.*
           5. *A whole is greater than any of its parts.*

POSTULATES: 1. *Two points determine a straight line.*
            2. *A straight line may be extended in a straight line in either direction.*
            3. *About any point a circle of a specified radius exists.*
            4. *All right angles are equal.*
            5. *If a straight line falling across two straight lines makes the sum of the interior angles on the same side less*

*than two right angles, then the two straight lines inter-*
*sect, if sufficiently extended, on that side.*

Postulate 5 will be particularly important for our discussion, so I might add here that, given the remaining axioms and postulates, it is logically equivalent to the proposition that through a point outside a given line one and only one line can be drawn which does not intersect the given line, no matter how far it is extended.

From this rather parsimonious basis the Greek mathematicians' famous results, such as that about the sum of the interior angles of a triangle and the Pythagorean theorem, can be deduced using logic alone.

But not really. The axiom-postulate basis is not complete. In proving the results of the theorems, the propositions deducible from the basis, tacit assumptions are made. Euclid assumes, for example, that a given straight line that passes through the center of a circle intersects the circle exactly twice. This is true, but you cannot deduce it from the ten basic assumptions. We needn't pursue this in any detail at this point, however, since we shall later (II,B,3) discuss Hilbert's complete axiomatization in some detail.

Even if the axiomatization were "complete" in the sense that all the appropriate theorems were genuinely derivable from the basis alone, it is important to notice that the full set of these theorems would *not* exhaust the body of geometric truth, even the geometric truth known to the Greeks. The geometric truths, some known to Archimedes, that depend upon various "limit" operations and which become fully accessible only after the seventeenth-century invention of the differential and integral calculus, are outside the range of the fully-completed Euclidean axiomatization, as are, for example, truths about the figures constructed on curved surfaces.

Nonetheless, the power of the theory is astounding. Even allowing for the "missing axioms" that make the theory incomplete, the range and variety of the theorems deducible from the extraordinarily parsimonious basis of ten fundamental truths is remarkable. Even more remarkable is the simplicity and obviousness of the basic truths. Of course, some are more obviously true than others. In particular, the Fifth Postulate, the parallel postulate, has a complexity and "remoteness from intuitive truth" which marks it out from the other nine propositions. We shall see below (II,B,2) the effect this differential status for the Fifth Postulate had on the evolution of geometry.

The believability of the theorems depends upon their inferability from the basic propositions. This immediately suggests to the philosophical mind two questions: Why should one believe the basic propositions to be true? and Why should one believe that truth is "preserved" by the inference from basic propositions to theorems? I will not consider the second question in depth at all; it is the question of the epistemic status of the laws of deductive logic. I shall consider the first in detail below (II,E), but will forgo this obvious entry for philosophy in order to continue the historical sketch.

Let me just note at this time the following points: (1) Euclid's systematization of geometry provides us with a theoretical systematization of a vast body of geometric truth. (2) It presents the Greeks, and many other scientific communities that follow, with the unique example of a science that can lay claim to completeness and to "perfect" veracity. (3) It is presented in the form of the specification of a number of propositions of extreme simplicity and "obvious" truth, with the remaining truths of the theory derivable from the basic truths by logic alone.

It is no wonder that the science of geometry provided generations of philosophers with both the paradigm problem for philosophical analysis and the paradigm example of what a scientific theory should be.

## 2. The Rise of Non-Euclidean Geometries

It isn't as though geometry didn't change between Euclid's *Elements* and the development of non-Euclidean geometry in the eighteenth and nineteenth centuries. In particular, the development of analytic geometry, beginning with Descartes in the seventeenth century, and the systematic exposition of the use of infinitesimal methods, anticipated by Archimedes and his school and finally developed into a full mathematical discipline by Leibniz, Newton, and their successors in the creation of the calculus, provided geometers with a range of methods and problems beyond the reach of the Euclidean systematization. But these mathematical achievements were viewed by all as expanding geometric methods, and suggested to none an inadequacy in the Euclidean theory or even a genuine alternative to it. The non-Euclidean geometries, however, were viewed, rightly or wrongly, as for the first time casting doubt upon the *correctness* of the Euclidean theory, and for this reason provided the first real impetus toward a radical revision in philosophical attitude toward the Euclidean theory's epistemic foundations.

The thrust toward non-Euclidean geometries came about from the peculiar status of the Fifth Postulate. As I have noted, to geometers it never seemed to possess the intuitive "self-evidentness" of the other axioms and postulates. But, as we shall see below (II,E), most philosophical accounts of the grounds for rational belief in the propositions of geometry rested very heavily on the claimed self-evident truth of the basic propositions. A natural reaction to this was to encourage attempts to show that the Fifth Postulate was, in fact, unnecessary as a part of the basis. In other words, to show that it was a *nonindependent* postulate and that it could be *derived* from the remaining nine propositions by logic alone. If this were true, belief in its truth, like belief in the truth of the other theorems of geometry, would rest upon its derivability from the remaining genuinely self-evidently true basic propositions.

The most interesting attempt at deducing the Fifth Postulate from the other basic propositions was that of the Italian mathematician Girolamo Saccheri in his *Euclid Freed From Every Flaw* (1733). We have seen that the Fifth Postulate is equivalent, given the truth of the remaining axioms and postulates, to the assertion that through a point outside a given straight line one and only one straight line can be drawn which fails to intersect the given line no matter how far extended. To demonstrate the deductibility of the Fifth Postulate, it would be sufficient to show that both the hypothesis that through a point outside a given line no such nonintersecting line could be drawn and the hypothesis that more than one nonintersecting line could be drawn would, in combination with the remaining nine basic propositions, lead to self-contradictions or logical absurdities.

Saccheri did not, in fact, use this version of the Fifth Postulate, in whose terms he was able to demonstrate that the hypothesis of *no* nonintersecting straight line through the point led to self-contradiction. To do this he had to read the Euclidean Postulate 2 as asserting that any straight line could be extended to a line of infinite length, or at least of length as great as one wished, not just that there were no end points to such an extension. It is clear that this was how the somewhat ambiguous wording of the postulate was normally intended and read, but we shall return to this distinction in the readings shortly. Given this reading of the Second Postulate, the part of the Fifth Postulate which asserts the existence of at least one nonintersecting line could be demonstrated from the remaining axioms and postulates.

Demonstrating the other half of the Fifth Postulate, the uniqueness of such a nonintersecting line, proved not so easy, however. Saccheri failed to deduce a logical contradiction from the assertion of more than one nonintersecting line. He showed that from this assertion and the remaining basic propositions many propositions followed that seemed to him patently false; but he could not discover a logical contradiction as a consequence. This is not too surprising, for, as was soon demonstrated, the postulate that through a point outside a given line more than one straight line can be drawn which fails to intersect the given line is logically consistent with the remaining nine basic propositions. And the postulate that no such nonintersecting lines can be drawn is also logically consistent with the remaining postulates if Postulate 2 is not read as asserting the infinite extendibility of any given line, but rather that all such lines are unbounded by endpoints, although possibly of finite length.

Let me call the hypothesis that more than one nonintersecting line can be drawn the Many-Parallel Postulate. The consistent geometry that results from replacing the Fifth Postulate with the Many-Parallel Postulate was developed independently and nearly simultaneously by the mathematicians Gauss, Schweikart, Bolyai, and Lobachevski. From now on we shall call it Lobachevskian geometry. The consistent geometry that results from replacing the Fifth Postulate by a No-Parallel Postulate, while simultaneously interpreting the Second Postulate *not* as asserting the infinitude of all extendible straight lines but only their unboundedness, was developed by Georg Riemann and is designated Riemannian geometry. ('Riemannian geometry' is also used to denote Riemann's extension of Gauss's analytic theory of surfaces, which I shall discuss in some detail below [II,B,5], but no confusion should result from this traditional terminological misfortune.)

We need not go into too much detail about the theorems of these geometries, but a few crucial results should be to hand.

In Lobachevskian geometry:

1. Through a point outside a given line an infinitude of nonintersecting lines may be drawn.

2. The sum of the interior angles of any triangle is less than a straight angle. As the area of the triangle approaches zero, the sum of the angles approaches a straight angle. All triangles have an area less than a maximum value, and as the maximum area is approached the sum of the interior angles of the specified triangle approaches zero.

3. The ratio of the circumference of a circle to its radius is greater than the Euclidean value of $2\pi$.

4. No similar figures of different areas exist—similar figures being those

having the *ratios* of their bounding line lengths constant—although the lengths of the bounding lines may differ.

5. Straight lines are, as in Euclidean geometry, extendible to infinite length.

In Riemannian geometry, on the other hand:

1. Through a point outside a given line no nonintersecting line can be drawn.

2. The sum of the interior angles of a triangle is greater than a straight angle. As the area of a triangle approaches zero, the sum of its interior angles approaches a straight angle. There is a maximum area to all triangles, and as the area of a triangle approaches the maximum, the sum of its interior angles approaches three times a straight angle.

3. The ratio of the circumference of a circle to its radius is less than $2\pi$.

4. Similar figures of different area do not exist.

5. All straight lines are of the same finite length when fully extended.

I have referred above to *consistent* Lobachevskian and *consistent* Riemannian geometries. But how do we know that these geometries are, in fact, consistent? Granted that Saccheri was unable to derive a logical contradiction from what amounted to the Lobachevskian basic premises, and that no one has yet found a logical contradiction in either Lobachevskian or Riemannian geometry, how do we know that this reflects not a genuine consistency in these theories, but rather a mere lack of cleverness on our part?

There are several ways of demonstrating the consistency of a theory. A "brute force" method is to treat the theory itself as an object of study, examining its logical form from the viewpoint of a "metalanguage" and attempting a direct demonstration that the logical forms of the basic propositions of the theory are such that no logical contradiction can follow from them.

Another technique is to offer instead a *relative consistency proof*. Such a technique relies upon the assumption that some distinct theory is consistent, to show consistency of the theory in question. Of course, given such a proof, one's confidence in the consistency of the theory under examination is only as great as one's confidence in the theory assumed consistent and one's confidence in the consistency of the means used to "transfer" belief in the consistency of the one theory to the other. I shall have more to say about the notion of consistency below (II,G,1), but for the moment I will outline some proofs of the consistency of Lobachevskian and Riemannian geometry relative to the assumed consistency of Euclidean geometry. Besides reassuring the skeptical reader as to the consistency of the two non-Euclidean geometries, we will gain a greater intuitive understanding of the content of these theories by inspecting these informal proofs.

First, two preliminary remarks: (1) We will be dealing at this point only with the two-dimensional non-Euclidean geometries. These are the "surrogates" of Euclidean plane geometry, not Euclidean solid geometry. Later, however, our concern will be with three-dimensional non-Euclidean theories, or even with four-dimensional theories (see [II,D,2] below). (2) We are concerned here with the consistency of these geometries, not with their truth. Consistency is a matter of "logical form" as abstracted from the content of a theory. It can be formally defined in a number of ways, but for our purposes it will simply mean the nondeducibility from the theory, by logic alone, of prima facie self-contradictions. It is not hard, of course, to construct perfectly consistent theories that are false. Indeed, later we shall be concerned not with the mere consistency of some non-Euclidean geometry, but with its truth.

The fact that we are dealing merely with consistency, and that consistency is only a matter of the logical form of a theory and is independent of its asserted content over and above the form of its assertions, allows us the following method of proving the relative consistency of the non-Euclidean to the Euclidean geometries. Take the non-Euclidean geometry whose consistency is in question. Abstract the logical form of its basic propositions. See if you can find a group of *theorems* of Euclidean geometry which, taken together, form a group of propositions of identical logical form. Remember, they need not be about the same things or say the same things about them as the basic propositions of the non-Euclidean theory. All that is required of the two sets of propositions is a common logical form. I will call such a relative consistency proof of non-Euclidean geometry relative to Euclidean geometry finding a Euclidean *model* for the non-Euclidean basic premises.

There is a particularly elegant and illuminating Euclidean model for two-dimensional Riemannian geometry. Consider the surface of a sphere in three-dimensional Euclidean geometry. Let this "stand in" for the Riemannian plane surface. Consider great circles on the sphere; that is, lines formed in three-space by the intersection of the surface of the sphere with plane surfaces that pass through the center of the sphere. These will stand in for the Riemannian straight lines. Points on the Riemannian plane will be taken as points on the spherical surface. Distances between points on the Riemannian plane will be taken as the Euclidean distances between the points on the sphere, as measured along the spherical surface.

The basic propositions of Riemannian two-geometry are then converted into true propositions of Euclidean spherical geometry. I won't prove this, but we might look at some of the consequences of Riemannian geometry I have listed and see why they are "true" in our model (Figs. 1 and 2). First, note that since all great circles intersect (at a

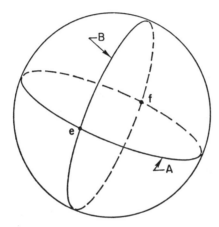

Fig. 1. "Straight lines" in the spherical model for Riemannian two-geometry. *A* and *B* are two great circles on the sphere. They are the "maximally extended straightest lines" of our model for Riemannian two-geometry. *e* and *f* are the two points of intersection of these great circles. Since all fully-extended great circles meet each other in two antipodal points, no two straightest lines are parallel to each other.

pair of antipodal points), the No-Parallel Postulate holds. Next, notice that while all great circles are unbounded (in the sense that they don't have endpoints), they are all of the same, finite length—the circumference of the sphere. To see the truth of the second consequence above, consider the typical spherical triangle. It is formed of arcs of great circles. Instead of the general case, consider the triangle such that one arc lies along the equator with the other two intersecting at the north pole. When the area of the triangle is zero, the two interior angles where the meridians of longitude meet the equator constitute the only nonzero interior angles, and since they are both right angles the

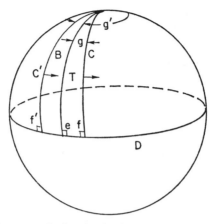

Fig. 2. Features of triangles in the spherical model for Riemannian two-geometry. The triangle $T$ is formed of arcs of
great circles, $B$, $C$, and $D$. Notice that when the angle $g$
is small, the area of the triangle is near zero and the sum of
its interior angles is the sum of the two right angles, $e$ and
$f$, and the small angle $g$. Hence the sum is only slightly
larger than 180°.

If we move the arc $C$ around the pole $p$ to obtain $C'$, we
end up with a triangle whose area approaches the area of
the hemisphere of the sphere, and whose sum of interior
angles is the two right angles, $e$ and $f'$, and the 360° angle
$g'$.

sum of the interior angles is a straight angle. Hold one longitude steady
and move the other around the globe. The largest area triangle one
can obtain has the area of the whole hemisphere. What is the sum of
its interior angles? Once again there are two right angles at the
equator, but there is now a full 360° angle at the pole; so the sum is,
as it should be, three straight angles. I leave it to the reader to examine
Riemannian consequences (3) and (4) in this model.

It is interesting also to notice the status of the Euclidean First
Postulate, i.e., two points determine a straight line. In the model, two
points usually determine a unique straight line (i.e., a great circle). In
the special case of antipodal points, however, an infinity of "straight
lines" (great circles) pass through the pair of points.

This model for Riemannian two-geometry is called "isometric," in
the sense that 'distance between two points' in the Riemannian plane
is mapped into 'Euclidean distance along the sphere between the corresponding points.' Further, the surrogate for the Riemannian plane

is embeddable in Euclidean three-space in a way that is intuitively clear and can be made mathematically precise. For Lobachevskian geometry we can't find quite as neat a Euclidean model. Beltrami showed that the geometry of shortest-distance curves on hyperboloids of revolution comes close to satisfying the Lobachevskian postulates. But all these models have singularities, boundary points or lines on the two-surface where the postulates break down. It can be shown, in fact, that there is no isometric model of Lobachevskian two-geometry which is embeddable without singularity in three-dimensional Euclidean space. But there is an illuminating nonisometric model, and it conclusively demonstrates the relative consistency of Lobachevskian to Euclidean geometry.

Take as the Lobachevskian plane a finite circular region of the Euclidean plane. We take as the points of the Lobachevskian plane the points *interior* to the bounding circle. Take as straight lines the chords of the circle. We "measure" the distance between points on this disk as illustrated, and we measure angles in a novel way as well. For example, to drop a perpendicular to a line we operate as illustrated below (Figs. 3, 4, and 5). This Euclidean structure provides a model

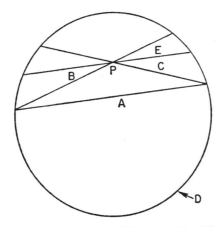

Fig. 3. Straight lines and parallels in the Euclidean model for Lobachevskian geometry. We take as the Lobachevskian plane the *interior* of the finite Euclidean disk *D*. The straight lines of Lobachevskian geometry are taken as chords of the bounding circle of the disk. Consider line *A* and point *P*. The chords through *P*, *B*, and *C*, which intersect *A* only on the boundary are "parallels" to *A* through *P*, as are all the chords between *B* and *C*, such as *E*.

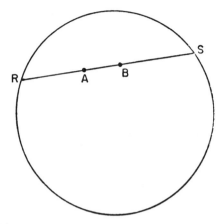

Fig. 4. Distances in the Euclidean model for Lobachevskian geometry. In our model for Lobachevskian two-geometry, the distance from $A$ to $B$ is taken as:

$$\text{dist } (A, B) = c/2 \left| \log \left( \frac{\text{dist } (A, R) \times \text{dist } (B, S)}{\text{dist } (B, R) \times \text{dist } (A, S)} \right) \right|$$

where $C$ is a constant.

If $A$ approaches $R$ and $B$ approaches $S$, this "distance" becomes infinite.

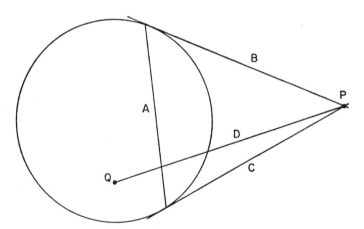

Fig. 5. Angles in the Euclidean model for Lobachevskian geometry. Suppose our problem is to find the line through point $Q$ which is perpendicular to a given line, $A$. Proceed as follows: Find the tangents to the bounding circle of the disk at the point where $A$ meets the circle. Where these tangents intersect, point $P$ is to be located. The line from $Q$ to $P$, line $D$, is the required "perpendicular."

for Lobachevskian two-geometry and hence reveals its consistency relative to Euclidean geometry.

Once again I will only look at one or two notable consequences of the Lobachevskian postulate to see how they are satisfied in the model. Consider a chord. In general, through a point outside the chord there will be an infinite number of drawable chords that do not intersect the given chord. So the Many-Parallel Postulate is satisfied. To see that all straight lines when fully extended have infinite length, just look at what happens to the "distance" between two points on the same chord as both points approach the bounding circle. I won't pursue this in any further detail, but it should give some idea of what the full relative consistency proof looks like.

So we know that the Lobachevskian and Riemannian two-dimensional geometries are as consistent as Euclidean geometry. The consistency of the three-geometries presents no real additional problem. When we come to discuss the possible *truth* of the non-Euclidean three-geometries (or, perhaps, four-geometries), however, the issues will become more philosophically interesting.


## 3. Hilbert's Study of Axiomatic Geometries

In 1898–99 the German mathematician David Hilbert offered an axiomatic treatment of geometries which we might look at briefly. By this time the analytic methods we shall look at below (II,B,4,5) had provided mathematicians with resources far outranging the axiomatic Euclidean and non-Euclidean geometries in scope and power. Nevertheless, Hilbert's work on axiomatics is useful both in completing the treatment of the axiomatic methods and in throwing light on its particular nature.

Hilbert sought a set of axioms for Euclidean geometry (1) which were complete, so that one did not require, as Euclid did, "hidden" assumptions in deriving the theorems, and (2) such that each axiom was independent of all the others. In Hilbert's system, in other words, all the theorems of Euclidean geometry can be logically deduced from the set of basic propositions, and none of the basic propositions can be deduced from the remaining axioms once the axiom in question is deleted.

Hilbert's axioms divide into five groups. Let me name the groups and offer one or two representative axioms from each class:

*I. Axioms of Connection (seven axioms):*

1. *Two distinct points, A and B, always determine completely a straight line, a. We write* AB = a.
2. *Any two distinct points determine a line they lie upon; that is, if* AB = a *and* BC = a, *then* AC = a.

*II. Axioms of Order (five axioms):*

1. *If A, B, and C are points of a line and if B lies between A and C, then B lies between C and A.*
2. *If A and C are two points of a line, there is at least one point, B, between A and C on the line and at least one point, D, on the line such that C is between A and D.*

*III. Axiom of Parallels (one axiom):*

1. *In a plane there can be drawn through a point outside a given line one and only one line which does not intersect the given line.*

*IV. Axioms of Congruence (six axioms):*

1. *If A and B are points on a straight line, and A' a point on any straight line, a', then on a given side of A' on a' there is a unique point, B', such that the line segment A'B' is congruent to the line segment AB.*

*V. Axiom of Continuity, or Archimedes' Axiom (one axiom):*

1. *Let A be any point on a straight line between designated points A and B. Choose points on the line $A_2$, $A_3$, $A_4$, etc., such that $A_1$ is between A and $A_2$, $A_2$ between $A_1$ and $A_3$, etc. Let the segments $AA_1$, $A_1A_2$, etc., be congruent. Then there is an $A_n$ such that B is between A and $A_n$.*

Some of the axioms are refinements of Euclid. Some, like Axiom V, are subtle additions necessary for the completeness of the system. Note the absence of "definitions." For Hilbert the basic terms (point, line, plane, between, is congruent to, etc.) are simply primitives of the theory. Definition is a process whereby new terms acquire meaning in terms of the primitives, not a means for fixing the meaning of the basic terms.

With his new axiom system, Hilbert can disentangle the Euclidean system, in the sense that he can now make clear just which of the fundamental theorems of Euclidean geometry require which axioms in their proof. By dropping one axiom, keeping the rest the same, adding to the reduced set the negation of the dropped axiom, and finding a model for the new axiom set, Hilbert can prove the independence of each axiom in turn. In finding such models for independence proofs, one is not restricted to models taken from geometry. Any set of known true propositions having the same logical form as the revised axiom set will demonstrate the consistency of this revised set, and hence the independence of the changed axiom.

Hilbert's formalization also makes much clearer the extent to which non-Euclidean geometries depend in their specification on the changes in the parallel postulate. To be sure, if we keep all the remaining axioms intact, then the Axiom of Parallels (Axiom III) uniquely characterizes Euclidean geometry, the No-Parallel Postulate Riemannian geometry, and the Many-Parallel Postulate Lobachevskian geometry. But he also shows, following the work of Dehn, the following somewhat surprising results: (1) It is true that the usual consequences of Euclidean geometry about the sums of interior angles of triangles follow from the Euclidean Parallel Postulate, even if we drop the Axiom of Archimedes. (2) If we drop Archimedes' Axiom, the No-Parallel Postulate still implies, with the remaining axioms, the usual Riemannian result that the sum of the interior angles of a triangle is greater than a straight angle. But, (3) if we drop the Archimedian Axiom, there are geometries, all compatible with the Many-Parallel Postulate, such that (*a*) the sum of the interior angles of a triangle is less than a straight angle (Lobachevskian geometry), or such that (*b*) the sum of the interior angles of a triangle is greater than a straight angle (non-Legendrian geometry), or such that (*c*) the sum of the interior angles of a triangle is equal to a straight angle (semi-Euclidean geometry). In semi-Euclidean geometry, in fact, all the theorems of Euclidean geometry except the uniqueness of parallels hold! What this shows is that the interdependence of the various axioms in proving standard important theorems is subtler than had been thought. It is, in fact, misleading to describe the alternative geometries as being characterized solely by the various parallel postulates, for it is the parallel postulate combined with all the remaining postulates which really determines the structure of the geometry deducible from the basic propositions.

## 4. Gauss's Theory of Surfaces

By combining resources drawn from the development of Descartes's "analytic" geometry and the infinitesimal calculus of Leibniz, Newton, and their successors, the mathematician K. F. Gauss was able to formulate a geometric theory of general curved surfaces of a power and richness far outrunning the two-dimensional axiomatic geometries, Euclidean or non-Euclidean. I am going to assume a knowledge of elementary calculus on the part of the reader, and if he knows this, he will probably have a fair familiarity with Cartesian geometry as well. But I do want to rehearse some of the basic concepts of analytic geometry before proceeding, expressing these notions in a language that would seem, perhaps, strange to Descartes himself.

The fundamental method of analytic geometry is repeated over and over again in contemporary geometry. Consider a space as a class of fundamental entities: points. The class of points has "structure" imposed upon it, constituting it a geometry—say the full structure of space as described by Euclidean geometry. We associate another class of entities with the class of points, for example a class of ordered $n$-tuples of real numbers, and by means of this "mapping" associate structural features of the space described by the geometry with structural features generated by the relations that may hold among the new class of entities—say functional relations among the reals. We can then study the geometry by studying, instead, the structure of the new associated system.

To make this concrete in the Cartesian case, let us first restrict ourselves to the flat, Euclidean plane described by Euclidean two-geometry. We associate with each point of the plane a pair of real numbers, by means of the usual Cartesian method of drawing two perpendicular straight lines on the plane, measuring the distance of a given point from each of these lines, and associating with each point the ordered pair of real numbers obtained by measurement.

Now consider a particular geometric figure, say an arbitrary straight line in the original Euclidean plane. Instead of considering the line, which is a set of points, consider instead the set of ordered pairs of real numbers whose members are the pairs of reals associated by our mapping with each point of the line. This set of pairs can be characterized by the fact that for each of its members, the second member of the ordered pair is a (common) linear function of the first member. For every ordered pair $\langle x,y \rangle$ in the set, in other words, $y = ax + b$, where $a$ and $b$ are constant reals, characteristic of the line in question and the original "coordinate axes" chosen. Knowing that every straight line in the plane is associated by our mapping with a unique linear equation, we can bring to bear on the study of geometry all that is known of the theory of equations. By similar methods, we can apply real analysis (the study of functions from $n$-tuples of reals into the reals) in all its power, including the differential and integral calculus, to resolve geometric problems. We can, for example, compute areas bounded by complex curves by the usual methods of integral calculus, once we know the appropriate "associated function" for the curve.

The method of Cartesian coordinates gives an elegant method for dealing with the flat, Euclidean plane. But how can we bring the methods of analytic geometry to bear when the surface is not flat, but curved? This was the problem Gauss set himself and ingeniously solved.

Consider an arbitrary curved surface embedded in the ordinary Euclidean three-dimensional space, i.e., in a three-dimensional space whose geometry is correctly described by the familiar Euclidean solid geometry. How can we usefully characterize this surface? Well, we can describe it by the functional relations holding between the three Cartesian coordinates of its points, coordinates relative to some selected set of three Cartesian axes in the embedding space. But, it would seem, we should be able to make do with *two* numbers for "coordinatizing" the points on the surface; for it is, after all, only a two-dimensional surface. If we could implant Cartesian axes on the surface, we could use these to coordinatize it. But Cartesian coordinates require perpendicular straight lines, and on a general curved surface there will, in general, be no straight lines. How should we proceed?

Gauss suggests the following procedure: (1) Choose two families of curves which "cover" the surface, i.e., such that each point of the surface belongs to one and only one curve of each family. (2) Number these curves, by means of the real numbers, in any way, such that the numbering is "monotonic," i.e., such that if curve $A$ is between curve $B$ and curve $C$, the number assigned to curve $A$ is between that assigned to curve $B$ and that assigned to curve $C$. (3) Assign to each point of the surface the ordered pair of numbers whose first and second members are the numbers assigned to the curve of one designated family which goes through the point and to the curve of the other family which goes through the point (Figs. 6, 7).

Given this coordinatization, figures on the surface are once more associated with sets of pairs of real numbers by means of the coordinatization. But the coordinatization is so arbitrary it is hard to see how to proceed further. The kind of clear association of curves with equations familiar from Cartesian analytic geometry (straight lines with linear equations, conic sections with quadratic equations, etc.) certainly won't go through here, given the arbitrariness of the basic coordinatization.

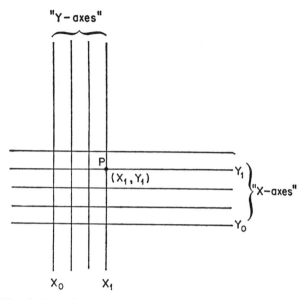

Fig. 6. Cartesian coordinates on the Euclidean plane. The plane is coordinatized by two families of parallel lines, the lines of one family being perpendicular to those of the other.

The "$X$-axes" are lines all of whose points have an identical $Y$ coordinate value, and the "$Y$-axes" are lines all of whose points have an identical $X$ coordinate value.

The coordinates of a point are found by finding the $X$ and $Y$ coordinate value of the $Y$ and $X$ axes upon whose intersection the point lies. For example, the coordinates of point $P$ are $X_1$ and $Y_1$

As an initial step, let us retreat from the problem of characterizing a curved surface embedded in three-space to two simpler problems: (1) characterizing a general curved line embedded in the plane, and (2) characterizing a general curved line embedded in three-space, but not confined to a single plane.

First, problem (1). In one sense, we already have a way of characterizing the general "plane curve." Pick a Cartesian coordinate system, consider the coordinates thereby assigned to each of the points of the curve, and find the functional relation between the second and first coordinate which characterizes this set of ordered pairs of real numbers.

But there is another approach. Its great virtue is that this new char-

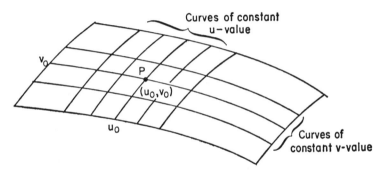

Fig. 7. Gaussian coordinates on an arbitrary curved two-dimensional surface. The families of curves are not, in general, families of straight lines. Nor are the $u$-curves and $v$-curves, in general, perpendicular to one another. But no two $u$-curves intersect, and no two $v$-curves intersect.

Every point is located at the intersection of a unique $u$-curve and a unique $v$-curve, which curves determine the point's coordinates.

acterization of the curve is completely independent of the selection of a particular coordinate system. First, pick a point on a curve. Then, characterize it by the ordered pair of new numbers $\langle s,k \rangle$ where $s$ is the distance of the given point from a designated endpoint of the curve and $k$ is the *curvature* of the curve defined as follows: At the given point of the curve there will be a unique circle which best approximates the curve at the point; this circle will have a specific radius of length $r$; then, $k = 1/r$ (Fig. 8). I should note that (1) we are making an assumption about the "smoothness" of the curve at every point, an assumption that can be given a rigorous definition by means of differential calculus, and that (2) our geometric definition of $k$ can be replaced by an equally adequate "analytic" definition, once again by means of differential calculus. The set of the ordered pairs of numbers which characterize the curve in the manner I have just described, gives, like the Cartesian coordinatization, a mapping from curves to sets of ordered pairs of numbers.

What is the relation between curves and their associated characterizations? It is this: A given set of ordered pairs of the form $\langle s,k \rangle$ uniquely specifies a curve, except that to any such set correspond all the curves that can be obtained from one another by rigid motions of a given curve —that is, by translations and rotations that keep the *shape* of the curve the same. Furthermore, it should be obvious that the set of ordered pairs associated with a given curve depends only upon the curve and upon the unit of length chosen. It is entirely independent of any choice of coordinate axes. I will call such a characterization of a geometric figure an *invariant* characterization.

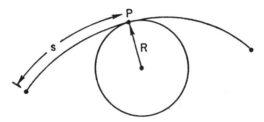

Fig. 8. The invariant characterization of a plane curve. The curvature of the curve at a point $k$ is $1/r$, where $r$ is the radius of the circle which best approximates the curve at $P$. The distance of $P$ from an endpoint of the curve is $s$. Then the function $k(s)$ characterizes the curve, except for the fact that any other curve obtained from this one by rigid motions has the same function.

A similar invariant characterization of curves not fixed to a given plane is possible, in terms of the curvature and torsion of the curve as a function of distance along the curve.

Now, problem (2). For an arbitrary curve in three-space, i.e., one not necessarily confined to a specific plane, an invariant characterization is only slightly more complex. Now instead of a set of ordered pairs of numbers, we need a set of ordered triples, $\langle s,k,t \rangle$. For each point on the curve, $s$ represents its distance from a designated endpoint and $k$ the curvature of the curve at that point. An additional number is $t$, the torsion of the curve at the point, whose geometric and analytic definitions may be found in any elementary text on differential geometry.

Now let us return to arbitrary curved surfaces, for which we have, at least, a quite arbitrary coordinatization in terms of the (generally non-Cartesian) coordinate curves on the surface. Gauss proceeds to show that: (1) Relative to a given coordinatization, we can find a number of functions, functions that assign to each point a number that is a function of its coordinates. These functions, relative as they are to the selected coordinatization, describe all the important geometric features of the curved surface. (2) From these functions we can form invariant scalar quantities; that is, numbers that can be assigned to the points on the surface, which characterize important geometric features of the surface and which do not depend for their values on the particular coordinatization chosen. (3) By studying these numbers we can distinguish importantly *different ways* a curved surface can be curved.

For the value they assign to each point the functions do depend on the coordinatization chosen, so let us assume that some such coordinatization has been performed. Once again, we are going to assume a certain smoothness for the surface. This can be well defined analytically, but we can be satisfied with the geometrical characterization that the surface is such that within a small enough region of a given point it can be "approximated" by a flat plane surface, the tangent plane at that point.

Gauss sets himself the following problem: We are given a curved surface, $S$, two points on the surface, $P$ and $Q$, and a curve connecting these points and wholly contained in the surface in question, $C$. How can we characterize the distance from $P$ to $Q$ along $C$ in terms of some arbitrary but given "internal" coordinatization of the surface $(u_1, u_2,)$? Solving the problem is fairly easy. The curve on the surface is at the same time a curve in the embedding three-space. Impose on this three-space an arbitrary *Cartesian* coordinatization. The length of a curve in three-space can be characterized in terms of the Cartesian coordinates of all its points by a simple application of the "Infinitesimal Pythogorean Theorem"—essentially, by elementary calculus and the Euclidean structure of the space. So we can characterize the length of the curve in terms of the Cartesian coordinates of all of its points. But each point on the curve has both a unique Cartesian coordinate and a unique internal coordinate $(u_1, u_2)$. So we can write the Cartesian characterization of the curve as a function of its internal characterization. A little elementary calculus applied to functions of functions, and we end up with a characterization of the length of the curve in terms of the internal coordinates of all of its points (Fig. 9).

Gauss's most important result is that, relative to a specified internal coordinatization, there is a single function $g_{ik}(u_1, u_2)$ whose values at each point are sufficient to allow us to calculate the length of any finite curve in $S$ in terms of the $(u_1, u_2)$ coordinates of all of its points. $g_{ik}$ is actually not a single function, but four related functions. That is, each point of $S$ must be assigned four numbers: $g_{11}, g_{22}, g_{12}, g_{21}$. Actually only three, since $g_{12} = g_{21}$. The g-function, called the *first fundamental form* for $S$, char-

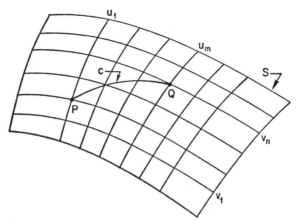

Fig. 9. Gauss's first fundamental form or $g$-function. On
the surface $S$ a curve, $C$, is drawn connecting two points,
$P$ and $Q$. The curve $C$ has a description in terms of the
$u$ and $v$ coordinates of each of the points on it.

The $g$-function, or first fundamental form, defined relative to the particular coordinatization, allows us to calculate
the distance along $C$ from $P$ to $Q$ in terms of the $(u, v)$
description of the curve $C$.

acterizes distances along curves in $S$ by means of the following two for-
mulas:

1. dist $(P,Q)$ along curve $C = \int_C ds$

2. $ds^2 = g_{ik}\, du^i du^k$

where in (2) we are adopting the standard convention that an index
repeated once up and once down is to be summed over, i.e.,

$$ds^2 = g_{11}(u_1,u_2)\, du^1 du^1 + g_{12}(u_1,u_2)\, du^1 du^2 +$$
$$g_{21}(u_1,u_2)\, du^2 du^1 + g_{22}(u_1,u_2)\, du^2 du^2$$

The first fundamental form is a function of the internal coordinates
that assign three numbers to each point $P$ of $S$ in terms of the internal
coordinates of $P$. With all its values known relative to a given internal
coordinatization, and with all the internal coordinates of a curve between
$P$ and $Q$ along $C$ in $S$ known, we can compute the distance along $C$ in $S$
from $P$ to $Q$. Note that while the $g$-function in a particular representation
is hardly an invariant characterization of $S$, at least its form depends
*only* upon the particular internal coordinatization chosen and not upon
the particular Cartesian coordinatization of the embedding three-space
chosen in the proof of $g$'s existence.

The *second fundamental form* is another multicomponent function of the internal coordinates of points in $S$. It is defined like this: (1) Consider a curve, $C$, drawn on $S$, and look at it at a particular point along it, $P$. (2) There is a vector, $\mathbf{k}$, at $P$, the *curvature vector* for $C$ at $P$, which (a) in general points out into three-space and is *not* contained in $S$, (b) which has its length equal to the curvature of $C$ at $P$, and (c) which points in the direction of the center of the circle that best approximates $C$ at $P$. (3) There will be a plane surface that best approximates the surface $S$ at $P$, called the *tangent plane* to $S$ at $P$. (4) There will be a line perpendicular to the tangent plane at $P$, called the *normal* to $S$ at $P$. The component of the vector $\mathbf{k}$ along the normal to $S$ I will call the *normal curvature* of $C$ at $P$. The size of the normal component of the curvature vector will depend upon how great the curvature vector is, which in general will depend upon the curve $C$ chosen and upon the way in which the confinement of the curve to $S$ "forces" the curvature vector to have a component along the normal to $S$.

This latter feature, the way $\mathbf{k}$ "decomposes" into pieces normal and tangential to $S$, is a feature of $S$ itself. It is described by the second fundamental form of $S$. This is, once again, a four-component function of the internal coordinatization of $S$ and is designated $L_{ik}(u_1,u_2)$. The form this function takes depends, again, upon the particular internal coordinatization chosen, i.e., the form of $L_{ik}$ is not an invariant feature of the surface. Shortly we shall find that $L_{ik}$ describes far more than the way curvature vectors "split up" into normal and tangential components.

One more important function on the internal coordinates and we are done. Return to considering a point $P$ of a curve $C$ on $S$ and to reflecting upon its curvature vector (Fig. 10). We have split the curvature vector into two components. One is along the direction of the normal to $S$. The other is in the tangent plane to $S$ at $P$. Split this latter, tangential component of $\mathbf{k}$ again into two components, along the directions of the vectors in the tangent plane which are tangent to the $u_1 =$ constant and $u_2 =$ constant curves of $S$ at $P$. Once again, the way this decomposition takes place will be determined by the nature of $S$ at $P$. The function describing this decomposition is called the *connection* or *affinity* function, and it is designated $\Gamma_{ik}{}^l$. It will be a function assigning eight numbers to each point of $S$ in terms of the point's internal coordinates. Like $g$ and $L$, as I have defined them, it will not provide an invariant characterization of $S$. It will, however, describe deep and important characteristic features of $S$ as we shall see below.

To show that there are at least some invariant scalar characterizations

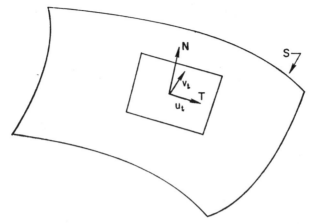

Fig. 10. The second fundamental form and the connection —or the $L$ and $\Gamma$ functions. At a point $P$ on the curved surface $S$, find the plane surface tangent to $S$ at the point. The vector perpendicular to this plane is the normal to the surface, **N**. The vectors in the tangent plane, $T$, tangent respectively to the $u$ and $v$ curves at $P$, are $\mathbf{u}_t$ and $\mathbf{v}_t$.

The second fundamental form, the $L$-function, allows us to determine the component of the curvature of a curve on $S$ through $P$ which is along **N**. The $\Gamma$ function, or connection, allows us to tell the components of the curvature vector along $\mathbf{u}_t$ and $\mathbf{v}_t$.

of a space is simple. Take a point $P$ on $S$. $S$ at $P$ will be best approximated by a set of circles tangent to $S$ at $P$. Let the radius of a circle tangent to $S$ at $P$ be $r$. There will always be an $r$ that is as great as any other $r$ and one that is as small as any other $r$, for all circles tangent to $S$ at $P$. Call these $r_{max}$ and $r_{min}$. Call $k_1 = 1/r_{max}$ and $k_2 = 1/r_{min}$ the *principal curvatures* of $S$ at $P$. They do not depend either upon the Cartesian coordinates that I have chosen for the embedding three-space or upon the internal $u$-coordinatization. They are invariant characterizations of $S$ at $P$.

Next take the product $k = k_1 k_2$. Count this as positive if the biggest and smallest tangent circles have their centers on the same side of $S$ and as negative if they have their centers on opposite sides. We call $k$ the *Gaussian curvature* of $S$ at $P$, and it is obviously an invariant quantity (Figs. 11, 12).

Now we want to distinguish *intrinsic* and *extrinsic* features of $S$. The intuition is this: The intrinsic features of $S$ are those that could be determined by a geometer completely confined to $S$, and who had no knowl-

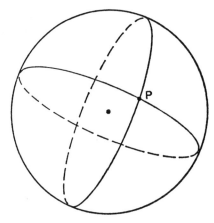

Fig. 11. Positive Gaussian curvature. Consider the surface of a sphere whose radius is $r$. At point $P$, or indeed, at any point on the surface, all of the circles which best approximate the surface have radius $r$, hence curvature $k = 1/r$. Since the centers of all these circles are on the same side of the surface (the inside of the sphere), the Gaussian curvature is positive, is constant everywhere, and has value $1/r^2$.

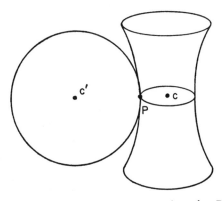

Fig. 12. Negative Gaussian curvatures. At point $P$ some of the circles that best approximate the surface and lie in planes normal to it have their centers on one side, and some on the other. For example, $c$ and $c'$ are the centers of two such circles. Counting one side as positive and the other as negative, then, the extremal circles lie on opposite sides of the surface. Their curvatures are then of opposite sign, and so the Gaussian curvature at $P$ is negative.

edge whatever of the manner in which $S$ was embedded in a three-space. The extrinsic features characterize the nature of this embedding. Now what could a geometer confined to $S$ do? He could measure all the distances between pairs of points on $S$ along chosen curves $C$ that he wished. But all such distances are determined solely by the $g$-function. So I will call a characteristic of the space intrinsic if it can be defined as a function of the $g$-values for points of $S$ alone.

The principal curvatures at a point are *not* intrinsic. Yet—and this was Gauss's most important *theorem* of his analytic geometry of surfaces —their product, the Gaussian curvature, is an intrinsic quantity. Next consider any two points on the surface. Consider all the curves on $S$ between these points. There will be curves that have the property that they are shorter curves between the point than any that can be obtained from them by "small" deformations, what the mathematicians will call "locally shortest curves." I will call them *geodesics*. Which curves are geodesics of the surface is an intrinsic matter only and does not depend upon how the surface is embedded in the three-space.

The geodesics are not only shortest-distance curves. Consider an ordinary vector on an ordinary flat plane. We can transport this vector to some other point keeping its length constant and keeping it parallel to itself. On a curved surface, the notion of parallel transport in general does not make sense. What you can do is this, however: (1) At a point $P$ consider a vector in the tangent plane to $S$ at $P$. (2) Consider a curve from $P$ to another point of $S$, $Q$. (3) At each point along the curve find the tangent plane to $S$ at that point. (4) As you move along the curve, find the vector in the associated tangent space which is "most parallel" to the vector in the tangent space of the point from which you've come. All this is quite crude, but it can be made technically precise. "Most parallel" transport amounts to moving from vector in tangent space to vector in tangent space in such a way that all the changes in the vector (viewed as a vector in the embedding three-space) are those "forced" upon it by the curvature of $S$ itself. The accompanying diagram gives an example that should make the intuitive ideas here fairly clear (Figs. 13, 14).

But what about geodesics? Take a point on a geodesic and attach to it a tangent space. Pick a vector in the tangent space tangent to the geodesic curve. Move along the geodesic and parallel-transport the vector as you move. The resulting set of vectors consists entirely of vectors that are tangent to the geodesic. This is true only of geodesic curves. For other curves, if you start off with a vector tangent to the curve and parallel-transport it as you move along the curve, you will end up with a vector *not* tangent to the curve. In this sense geodesics, besides being shortest-distance curves, are "maximally-straight curves." In a sense, they curve only insofar as their confinement to *S* forces them to curve.

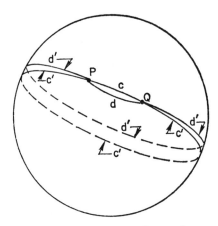

Fig. 13. Geodesics as "shortest distance" curves. On a sphere, the arc of the great circle (great circles being geodesics on the sphere), *C*, is the shortest curve connecting *P* and *Q*. It is shorter than *d*, for example.

The other part of the great circle, *C'*, is also an arc of a geodesic connecting *P* and *Q*, although certainly not the shortest distance curve connecting these points. It is, however, an extremal curve relative to curves that can be obtained from it by small deformations, for example, *d'*.

Parallel transport can be defined entirely in terms of the Γ's. Given what we have just seen about the connection of parallel transport with geodesics, and given the intrinsicalness of the notion of geodesic, it is not surprising that the Γ's are intrinsic features of *S*. They can, in fact, be defined as functions of the *g*-function (Christoffel symbols).

What is the connection between the material we have just examined and the earlier discussion of axiomatic non-Euclidean geometries? The connection is simple. For Riemannian and Lobachevskian two-dimensional geometry simply read the terms as follows: take 'is a plane' to mean 'is surface *S*' and 'is a straight line in the plane' to mean 'is a geodesic on *S*.' All the basic propositions of Riemannian geometry will then come out as true propositions so long as *S* is a surface of constant *positive* Gaussian curvature. And all the basic propositions of Lobachevskian geometry will come out true so long as *S* is a surface whose

Gaussian curvature is everywhere the same and is *negative*. Riemannian and Lobachevskian geometry, then, are the geometries of geodesic figures on Gaussian curved surfaces whose Gaussian curvatures are constant and positive or negative respectively; Euclidean geometry is the geometry of the Gaussian surface whose Gaussian curvature is constant and zero everywhere.

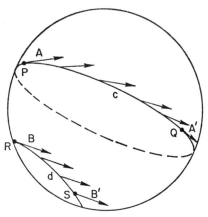

Fig. 14. Geodesics as "smoothest curves" on a surface. If we pick a vector, like **A**, tangent to a geodesic at a point, and then transport it parallel to itself along the geodesic, say along $C$ from $P$ to $Q$, we end up with a vector still tangent to the geodesic curve, like **A'**.

If we do the same thing along a nongeodesic however, the vector we end up with, say **B'**, is no longer tangent to the curve, $d$, despite the fact that it results from the vector **B**, which is tangent to $d$, by parallel displacement from $R$ to $S$.

I should end with a few more remarks about intrinsic versus extrinsic features, and a few remarks about the difference between local and global characterizations of surfaces.

Suppose we have two surfaces whose intrinsic features are everywhere the same. Will they be "equivalent" surfaces? Not if equivalent is taken to include the way the surfaces are embedded in the three-space. Despite the fact that both surfaces can be given internal coordinatizations such that their $g$-functions as functions of $(u_1, u_2)$ are exactly alike, their $L$-functions may differ radically.

The surface of a sphere differs *intrinsically* from the plane. We can coordinatize the plane by Cartesian coordinates. The $g$-function will be of the form $g_{11} = g_{22} = 1$ and $g_{12} = g_{21} = 0$ for all points on the plane. There is *no* internal coordinatization of the sphere which will give the $g$-function the same form. But we can find such a coordinatization for the cylinder, taking as coordinate axes the straight lines running up and down the cylinder and the circles running around it perpendicular to these lines. So the cylinder is intrinsically like the plane. But, of course, it is extrinsically different. Its distinct $L$-function for this coordinatization indicates that the cylinder is "bent around" in its embedding space in a way the plane is not (Fig. 15).

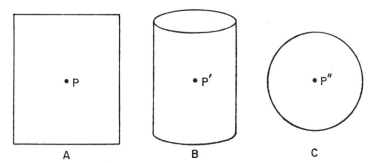

Fig. 15. Intrinsic vs. extrinsic curvature. The Gaussian curvature of the plane, *A*, at *P* is the same as that of the cylinder, *B*, at *P'*, and is zero. This is true for every point on the plane and the cylinder. So they do not differ in local intrinsic features. They do of course differ in (1) extrinsic curvature, which depends upon the way that they are embedded in three-space, and (2) global topological features.

The sphere, *C*, has a curvature at *P''* which is positive and nonzero. It differs intrinsically at every point both from the plane and the cylinder.

The examples of a cylinder and a plane will serve also to illuminate the important "global-local" distinction. How alike are a cylinder and a plane to someone who, confined to the surface, has no access whatever to the way the surface is embedded in the three-space? If he constricts himself to a small enough region about a given point, no measurements he makes will allow him to decide whether the surface on which he lives is a cylinder or a plane. This is because their intrinsic features are identical.

But suppose he allows himself free mobility throughout the surface. Clearly, then, his experiences will differ. For example, if we start from a point on a plane and travel along any geodesic (straight line) through that point without ever reversing direction, then we will never return to our initial point. On a cylinder, however, through each point there is a geodesic (the circle around the cylinder through that point) such that if we travel along that geodesic, never reversing direction, we will sooner or later return to our starting point. This shows that the intrinsic identity of a cylinder and a plane is a local matter. The two surfaces differ, even neglecting their embedding in three-space, in the global properties of *connectivity* determinable by a geometer confined to the surface and ignorant of the embedding. This is a feature of what I shall call below (II,B,6) the *topological* structure of the surface.

## 5. Riemann's General Theory of Curved Spaces

Building upon Gauss's analytic techniques, and upon his important notions of intrinsic-versus-extrinsic and invariant features, Riemann was able to construct a general analytic geometry of curved spaces of any dimension. This is traditionally called *Riemannian* or *differential* geometry.

The procedure is easy to comprehend. Let us start off with a space of some specified number of dimensions. I will have something to say about what this means in (II,B,6). Suppose that the space is sufficiently well-behaved that at each point one can approximate the space by means of a Euclidean $n$-dimensional space. This is similar to the assumption for the theory of surfaces that one could attach to each point of the curved surface a tangent plane, but we now have to view the Euclidean-in-the-small requirement less picturesquely; for we are *not* assuming that the $n$-dimensional "curved" space is embedded in any higher-dimensional flat space, and there is no actual "flat tangent space" to be attached. All that the abstract requirement amounts to is that the geometry about a point in the $n$-dimensional space can be approximated, if the region is sufficiently small, by the $n$-dimensional analogue of Euclidean two- or three-geometry.

Since we are not assuming any embedding of the space that we are dealing with in a higher-dimensional space, I shall have nothing to say about any extrinsic features of it. In particular, the second fundamental form, the $L$-function, will play no role in the theory. The space may, however, possess global connectivities more like that of a two-dimensional space bent in three-space than like that of the two-dimensional plane. I am not excluding, say, the possibility that following a geodesic through a point will eventually lead one back to one's origin, as happens with some geodesic trips on the cylinder but never on the plane.

Extrinsic features of the space have gone by the board, but we may still use the higher-dimensional surrogates of Gauss's devices for studying intrinsic geometric features to study the intrinsic geometry of the $n$-dimensional space. Let me rehearse once again some aspects of the intrinsic geometry of two-dimensional surfaces: (1) We labeled the point of the surface by means of two families of curves and numbered the curves with real numbers such that ($a$) each point of the surface was on exactly one curve from each family, which implies among other things that two curves from the same family never intersect, and such that ($b$) if curve $A$ was between curve $B$ and curve $C$, the number assigned to curve $A$ was between the numbers assigned to curve $B$ and curve $C$. (2) We found a function $g_{ik}(u_1, u_2)$ that assigned $4 = 2^2$ numbers (two of them identical) to each point, such that the distance between any two points of the surface along a curve could be calculated by means of a description of all the points of the curve in the new coordinatization and the value of the $g$-function at each of these points. (3) We found an additional function $\Gamma_{ik}{}^l$ that described the closest thing to parallel transport of a vector which we could define on the surface, and we discovered that this new function could be defined by means of the $g$-function and the coordinatization. (4) We discovered the existence of numbers that were independent of the coordinatization and which, at each point, described intrinsic features of the surface at that point (the Gaussian curvature).

Riemann suggests that we proceed in exactly the same way for the arbitrary "curved" $n$-dimensional space: (1) We coordinatize it by $n$ families of coordinate curves. (2) We find a $g$-function for the space. It is now a function of $n$ coordinates, but as before (1) dist$(P,Q)$ along $C = \int_C ds$ and (2) $ds^2 = g_{ik}du^i du^k$. Note that $g_{ik}$ will now have $n^2$ distinct components. (3) We invent a $\Gamma_{ik}{}^l$-function (now a $n^3$ component function instead of an $8 = 2^3$ component function) that describes the features of most-parallel transport on the space, and hence, the nature of straightest lines or geodesics. As before, the $\Gamma$-function can be proved to be definable in terms of the $g$-function and the coordinatization, and the straightest lines turn out to be the shortest as well. (Remember, they are only the shortest of all lines between two points obtainable from one another by small deformations. That is, if $C$ is a geodesic between $P$ and $Q$ it will be shorter than any "nearby" curve from $P$ to $Q$, but there may be some quite "remote" curve between $P$ and $Q$ shorter than $C$.) (4) We discover a set of numbers that characterize features of the intrinsic structure of space and are invariant.

The problem of scalar invariants for an $n$-dimensional space is nontrivial. A single number, like the Gaussian curvature for a two-surface, will no longer do to fully characterize the intrinsic features of the space. The resolution of the problem goes something like this. There is a multi-component function ($n^4$ components, counting identical components), $R_{ijk}{}^l$, called the Riemann curvature tensor. This function is usually represented in a noninvariant way, since its components at a point depend upon the coordinatization. A representation of it contains much of the information that can be had about the intrinsic structure of a space (relative to a particular coordinatization, of course). We can get some insight into it if we parallel-transport a vector around a closed loop of curves in a curved two-space: we may not end up with the vector we started with when we get back to the origin. To see this, examine the illustration below (Fig. 16).

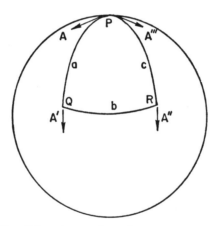

Fig. 16. Parallel transportation of a vector around a closed loop in a curved space. If we parallel transport vector **A** from $P$ to $Q$ along curve $a$, then from $Q$ to $R$ along curve $b$, then from $R$ to $P$ along curve $c$, we end up with vector **A'''**, which is *not* identical to vector **A**.

Pick a point in a general curved space. Define a vector at that point. Parallel-transport that vector about a closed four-sided circuit whose opposite sides are "parallel." One will arrive back at the origin, in general, with a vector that differs from the original. Let the area of the quadrilateral around which one is transporting the vector shrink to zero. The Riemann tensor at the point will describe the relation between the original vector and the limiting result of the transported vector as the area approaches zero.

In one special case the form of the Riemann tensor is particularly simple. In a "flat" $n$-dimensional space, all the components of the Riemann tensor are identically zero at all points, no matter what coordinatization of the points one adopts. This identical vanishing of the Riemann tensor at all points is also a sufficient condition for the space to be the $n$-dimensional Euclidean space.

But what about numerical or scalar invariants? From the Riemann tensor, by means of an algebraic operation known as contraction, one can ultimately form a single number $R$. $R$ is called the scalar curvature of a Riemannian space at a point. It has a geometric interpretation: $R$ divided by $n(n-1)$ is the arithmetic mean of the Gaussian curvatures of a set of two-surfaces in space which (1) intersect at the point in question and which (2) are generated by pairs of a set of mutually perpendicular vectors in the space at the point. $R$ is an invariant and it characterizes an intrinsic feature of the space. But an assignment of $R$ values to every point in a space does not fully determine its intrinsic form, for two intrinsically distinct curved spaces can have identical $R$ assignments.

To fully characterize a Riemannian curved space, we must pick a coordinatization and then assign the value of $g_{ik}$ for every $n$-tuple of coordinate values. If two Riemannian spaces are intrinsically alike then, although their $g_{ik}$- and $R_{ijk}{}^l$-functions might look quite different since we may have coordinatized them quite differently, there will always be a coordinatization of the second space such that relative to this new coordinatization the $g_{ik}$- and $R_{ijk}{}^l$-functions for the second space are identical to those for the first. Conversely, if the two spaces differ in their intrinsic features and if, relative to some coordinatization of the first, the two functions have a specified form, there will simply be *no way whatever* to coordinatize the second space to bring its $g_{ik}$-function into the specified form.

We should note in passing at this point that it is not necessary to restort to a particular coordinatization, and a specification of the $g_{ik}$-function relative to it, to characterize the intrinsic structure of a Riemannian space. There are coordinate-free methods that will do the same job, and much modern differential geometry is devoted to the problem of characterizing the intrinsic structure of general curved spaces without reference to an imposed coordinatization of the space. From this coordinate-free point of view, the various functions we have used to characterize the space in a coordinate-dependent way, such as the first fundamental form, the connection, and the Riemann tensor, can be defined without the necessity of first choosing a particular coordinatization for the space.

I haven't said anything yet about the physical possibility of the real world's having any sort of "curved" structure, nor will I until later (II,D). Nor have I yet suggested any method by which we might find out that the world was curved even if it were. Again, this is a discussion I will hold in reserve. What we have seen, however, is that one can give an elegant and coherent characterization of curved two-dimensional surfaces by means of Gauss's analytic methods; that the theory so constructed differentiates importantly between those features of a space independent of its embedding in a higher-dimensional space (intrinsic features) and those that depend upon the nature of such an embedding (extrinsic features); and that a consistent extension of Gauss's methods for dealing with the intrinsic features of a space can be constructed from the two- to the *n*-dimensional case, even if one forgoes the resource of embedding the space in question in a higher-dimensional flat space that was originally used by Gauss to construct his theory of intrinsic features.


## 6. *Variations on, Generalizations of, and Abstractions from Riemannian Space*

If the survey of geometry up to this point can be described as rapid, the pace in this section can only be called frantic. I would like to survey in the space of a few pages some of the more significant concepts of contemporary mathematical geometry. We will not be able to deal at length with any topic, but the almost intolerable brevity in this section will be compensated for throughout the later parts of the book; for as we utilize the concepts introduced in this section to characterize various physical theories and philosophical analyses, I will expand upon the explanation of the concepts and illuminate them by some concrete instances. Let me first warn the reader, that despite the number of topics in geometry touched on in this section, whole areas of contemporary geometry are totally ignored. Some of these, such as projective geometry, will never be discussed in this book. Others, such as the use of group theory to characterize the symmetry of Riemannian spaces, may be introduced as needed later on.

Let us first consider a number of "variant metric geometries" where, as in Riemannian geometry, a distance between two points in the space along a given curve is always well defined, but where this distance cannot be calculated in terms of an internal coordinatization of the points by means of a $g$-function with $n^2$ components ($n$ the dimensionality of the space) as in Riemannian geometry. The failure for there to exist such a quadratic first fundamental form amounts to rejecting the underlying Riemannian assumption that the space, curved as it may be, can be approximated in a sufficiently small region by a Euclidean tangent space.

*Spaces with Nonquadratic Differential Metric Forms:* The two fundamental equations relating distance along curves to internal coordinatization for Riemannian geometries were (1) dist($P,Q$) along $C = \int_C ds$ and (2) $ds^2 = g_{ik}du^i du^k$. Notice that no matter how many dimensions the space has, the form for $ds^2$ is always a sum of terms each of which contains *two* differentials only. This is the Riemannian generalization of the Pythagorean Theorem for curved spaces.

Riemann himself realized that one could define a "metric space" by keeping equation (1) but allowing more flexible definitions for $ds$, i.e., by allowing equation (2) to be of a form not like that for the most general Riemannian case. For example, mathematicians study so-called *Finsler spaces.* In a Finsler space of dimension $n$ the following equation replaces equation (2) above: $ds = F(x^i, dx^i)$, where $F$ is any function satisfying certain constraints. Under these appropriate constraints, such "differential forms" will allow one to define distance functions such that the distance between two points along a curve is always a positive number, but the distances defined in the space will not behave like those of Riemannian geometry. Since Finsler spaces haven't found much application in physics, I shall say no more about them.

*Spaces with Nonintegrable Metrics:* We have already noticed that if you move a vector about a closed loop, keeping it as parallel to itself as possible, you may very well end up at your origin point with a vector that differs in its direction from your initial vector assigned to the origin point. The curvature of a Riemannian space, in fact, can be interestingly characterized by the degree to which a vector can be "reoriented" by such closed-loop transport. This study is known as the study of the *holonomy group* of the space. For example, the holonomy group of the two-sphere is the full two-dimensional rotation group. This means that you can transform a vector at a point to any other vector, obtainable by rotating the given vector any number of degrees in the tangent plane of the sphere in which it is drawn, by moving the vector parallel to itself around a closed loop on the sphere.

But notice that this fact, for which I offer no rigorous definition or proof: Although a vector can have its direction changed by transportation parallel-fashion about a closed loop, the vector you end up with has the same *length* as the original vector. Hermann Weyl considered metric spaces for which this was not true. Physicists would call them "nonintegrable" metric spaces, borrowing a notion from potential theory in physics. In such a space the metric is characterized by two forms: the ordinary Riemannian quadratic $g$-function and additional linear form $\phi$.

What distinguishes such a space from a Riemannian space? Since such spaces, despite Weyl's hopes, have proven not very useful for physical world-pictures, let me give their "empirical" interpretation here and then forget about them. In such a Weylian world, if you took a small rigid rod congruent to a given rod and moved the first about a closed loop, leaving the second behind, when you returned to the origin the rods would not, in general, any longer be congruent to each other. Again, if you took at a given point two clocks that ticked at the same rate, left one behind and moved the other through a closed loop away from and then back to the origin, at the finish the clocks would no longer be ticking at the same rate—and not because of any special physical distortion of the moved clock while it was in transit.

The geometries we have just looked at differ from Riemannian geometry. They have defined on them a distance function, so that the distance along a curve between any two points is always well defined, but the distance depends on the coordinatization in quite a different way from the way the dependence can be formulated in a Riemannian space. I now want to discuss abstractions from metric geometry, geometries that differ from the Riemannian ones in a different way—distances are simply not defined on them at all.

Mathematics progresses to a great extent by ignoring things. We start with a given structure, for example Riemannian space. It has a lot of features: the distances between points along curves, the geodesics as "straightest lines," etc. Suppose we drop out one of these features entirely. Let us take a Riemannian space and simply pretend that a $g$-function, or any other alternative metric function, simply doesn't exist. What features does the remaining structure still have? If we describe them, we will have described a structure of which, in general, the original "full" structure will be only one particular example. For example, Riemannian and non-Riemannian Finsler spaces differ from one another, their metrics depend upon internal coordinatization in quite different ways. But they do have features in common, features that simply do not depend upon their metric structure. What are these features?

We will look at several "levels of abstraction" from Riemannian spaces. If we keep on abstracting, we eventually end up with a very primitive structure, indeed. It is is just a *set* of points, points being the fundamental "individuals" out of which all spaces are built. All we can say about this set is that a point is either a member of it or not a member of it, and a few other things that are true of this set because they are true of all sets. But before we reach this point there are several intermediate stages, sets of points that don't have all the structural features of a full Riemannian geometry but which have quite a rich structure nonetheless. Let us consider some of these stages obtained by systematically "undressing" Riemannian space.

*Affine Space:*   In a Riemannian space the two most important functions of the internal coordinates of the points of the space are the $g$-function, which "fixes" the metric or distance relationships between points and along curves for the space, and the $\Gamma$-function, which characterizes as-parallel-as-possible transport of vectors in the tangent spaces to the space. Using these two functions we could define geodesics as either (1) locally shortest curves between points or (2) straightest lines between points, and we could show that these definitions specified the same curves for a given space. We could, in fact, show that the $\Gamma$-function was *definable* in terms of the $g$-function.

Weyl suggested: Forget about the $g$-function. Could one then still define a $\Gamma$-function in a natural way? The answer is yes. Even without any distance function on the space being defined, we can still make sense of the notions of tangent spaces to a space, vectors in the tangent spaces, and most-parallel transport of tangent-space vectors from point to point. The defining characteristics of as-parallel-as-possible transport are quite intuitive, and the interested reader will have no difficulty comprehending them if he pursues the references cited below. The definition of "most parallel" requires no mention of distance in the space and is quite well characterized for spaces in which no metric function is defined at all. For such an *affine space,* we simply cannot ask, in general, "How far is it along C from P to Q?" We can however ask, "Is C the straightest curve between P and Q?" and expect to get an answer.

In an affine space the notion of a curve being a geodesic is perfectly well defined. So is a limited notion of distance in the sense that the metric ratios of intervals along geodesics are all well defined in the space. But the general notion of distance along a curve between two points is simply not defined in such a space.

*Differential Manifolds:*  Drop both the metric structure *and* the affine structure of a Riemannian space and what is left? Well, in a Riemannian space (1) we can assign internal coordinates to the points and (2) we can define functions on these coordinates and (3) we can define such notions as the derivatives of a function of the coordinates relative to the coordinatization. None of these notions presupposes the notion of a metric or an affine structure. If we attribute to our set of points just enough structure to make sure that these notions are well defined, we have made the set of points into a differential manifold.

One way of seeing how a differential manifold is weaker in its structure than a metric or an affine space is to look, briefly, at the notion of a feature of a space being *invariant under a transformation.* Suppose we have a metric space, and we consider it a rigid object that we can move about, say in some (possibly imaginary) embedding space. What stays the same about the space as we rigidly move it? Among other things, the distances between all the points of the space along specified curves. So we can, roughly, define the metric features of the space as those features that remain invariant (unchanged) by rigid transformations. The affine properties of the space are all those that remain invariant under affine or linear transformations, i.e., transformations that take straightest lines into straightest lines, even if they modify the distances among points, the angles between lines, etc. The differential features of the space are all those that remain invariant under any transformation of points into points so that the transforming function is differentiable relative to a specified coordinatization.

All rigid motions are affine transformations and all affine transformations are so-called diffeomorphisms. But there are affine transformations that change distances, and diffeomorphisms that change geodesics into nongeodesics. The theory of differential manifolds studies spaces insofar as they can be coordinatized and insofar as differentiation of functions defined on the coordinates is well defined. It ignores distances and parallelism, and is hence a further abstraction from Riemannian space.

*Topology:* Consider a space, and forget about the fact that it can be coordinatized. What is left? Well, it is still a set of points. So we can ask set-theoretic questions about it, such as, "How many points are there in the space?" There may be a finite number, an infinite number but such that the points can be enumerated by the integers (denumerable set), or an infinite number such that no enumeration is even possible (nondenumerable set). The existence of infinite, nondenumerable sets is the result of Cantor's Theorem in set theory. But is there any level of structure richer than mere set-theoretic structure, yet which does not get us the full structure of a differential manifold? Yes, there is *topological structure.*

Consider, for example, a two-dimensional space. What features might it have which we can describe without ever considering assigning the points coordinates? To begin with, there is its two-dimensionality. Then there is the fact that it either "goes on forever" or "has a boundary." It either has "holes" in it or it doesn't. And so on. These are the concepts topology makes rigorous, and the features of a space it describes.

Metric geometry studies invariants under rigid motions; affine geometry, invariants under geodesic-preserving transformations; differential manifold theory, invariants under diffeomorphisms; and topology, invariants under *bicontinuous transformations.* The notion of continuity is *the* primitive notion of topology. Usually, however, the notion of *open set* is taken as primitive instead. Since 'open set' is a primitive term we don't define it, but to get an intuition about it, think of the region interior to a disk on a plane, that is, all the points of the disk except its boundary.

A *space* is a set of points; an open set is a subset of the space. We do axiomatically characterize the set of open sets by insisting that (*a*) the empty set and the entire space are open and (*b*) the unions and intersections of open sets are open. A continuous mapping is a mapping from sets of points to sets of points such that open sets come only from open sets. *Why* this is a natural definition of continuity, intuitively, should be given some thought by the reader. The set of points left behind in a given set when an open set is removed is a *closed set.* A set can be both open and closed!

A *neighborhood* of a point is an open set containing the point. A point, $p$, is the *limit* of a denumerably infinite sequence of points, $p_i$, if for every neighborhood of $p$ there is an $N$ such that every $p_i$ with $i > N$ is in the specified neighborhood. The *closure* of a set is the smallest set containing the original set which is closed. The *boundary* of a set is the set of points in both the closure of the original set and the closure of the complement of the original set (the set complement is the set of all points *not* in the original set).

With such basic definitions as these we can begin to give rigorous characterizations of the topological features of a space. We shall only define a few crucial notions that will be useful to us later on, and maybe provide enough illustration of the definitions to give the reader a feel for their intuitive content. Topology is now a "heavy industry" of mathematics, and I am afraid that the reader previously unfamiliar with it will end up with no more understanding from this section than the casual tourist has of the assembly line he views from the visitor's gallery.

*Dimensionality:* Topology gives us the first rigorous definition of the dimensionality of a space. The definition I give was developed independently by Brouwer, Menger, and Urysohn, and is a so-called recursive definition:

    *a.* The empty set, i.e., the set with no points in it, has dimension $-1$.
    *b.* The dimension of a space is the smallest integer $n$ such that every point of the space has arbitrarily small neighborhoods whose boundaries have dimension less than $n$.

The set of a single point has dimension 0, the line dimension 1, the plane dimension 2, etc., just as we might wish. Let $X$ and $Y$ be two spaces. A *mapping* from $X$ to $Y$ is an assignment of a set of points in $Y$ to each point of $X$. A mapping is 1–1 if each point of $X$ has one and only one point of $Y$ assigned to it by the mapping. A *homeomorphism* is a continuous 1–1 mapping. It follows from this definition that if the dimension of $X$ is $m$, and $Y$ is mapped from $X$ by some homeomorphism, then the dimension of $Y$ is $m$; that is, dimensionality is an invariant under homeomorphic transformations. In addition, as Brouwer proved, a Euclidean $n$-space has dimension $n$ according to this definition, thus tying up the formally defined notion with the intuitive concept of the dimension of a space.

It should be noted that other possible definitions of dimension can be given as well. For separable metric spaces, including all those spaces proposed as possibilities in current physical theory, all the definitions are equivalent, although they may disagree with one another about the dimensionality of a space for more general topological spaces.

*Connectedness:*   A space is *connected* if it cannot be decomposed into two open subsets such that (*a*) each is nonempty and (*b*) they have no point in common. A space is *simply-connected* if every simple closed curve embedded in it, that is, every closed curve in it which does not intersect itself, can be shrunk continuously to a point—where the notion of a continuous shrinking can be rigorously defined as one might expect. If a space is connected but not simply-connected, it is *multiply-connected*.

*Compactness:*   If a space is a metric space, we can define its *diameter* as the maximum distance between any two of its points. If this diameter has a finite value, the space is said to be *bounded*.

A collection of open sets, $A_i$, is said to cover a space, $S$, if every point of $S$ is a point of some $A_i$. A space is *compact* if and only if for every covering of it there is a finite subcollection of the sets of the covering such that this finite subcollection of sets also covers $S$. An important theorem, the Heine-Borel Theorem, tells us that a set is compact if and only if it is closed and bounded. The closed disk is compact whereas the whole Euclidean plane is not. The plane, however, is *locally compact,* in the sense that each neighborhood of a point has a subneighborhood containing the point that is such that its closure is compact. The plane is locally compact since each point belongs to some closed, bounded disk.

*Orientability:*   A final topological notion that we shall need is that of the *orientability* or *nonorientability* of a space. The formal definition is too complex for us to present here, but for our purposes we can take the distinction to be this: In a nonorientable space the situation can occur that within a region there are two figures that cannot be made congruent by any continuous transformation of one of them which changes it, continuously, into another figure within the region. However, there will be a continuous transformation of one figure through figures located outside of the specified region such that, ultimately, it is continuously transformed into the other figure when the transformation brings it back into the specified region. That is, although no "local" continuous transformation can take figure *A* into figure *B*, by "moving" figure *A* continuously about a path "around" the space and continuously transforming it in the process, eventually figure *A* can be brought into congruence with figure *B*.

In an orientable space this can't happen. Any two figures that can be brought into congruence by a "global" continuous transformation, can be brought into coincidence locally as well. As the illustration below indicates, the Euclidean two-surface is an orientable surface, while the Möbius band is a nonorientable two-surface (Fig. 17).

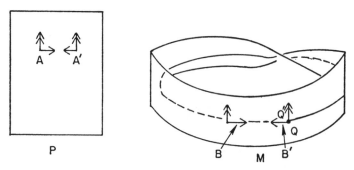

Fig. 17. Orientable versus non-orientable surfaces. On the orientable plane, *P*, there is no way of continuously transforming *A* into *A'*. On the nonorientable, one-sided figure, *M*, called the Möbius band, however, *B'* can be superimposed continuously on *B* by being taken "around" the strip from *Q* to *Q'*. At *Q'*, *B'* will look just like *B*. But since the surface is one-sided, *Q* is *Q'*. So a continuous trip "reverses" the apparent orientation of *B'*, showing that, actually, there is no global differential orientation between *B* and *B'* on the Möbius band.

The problems of the connectivity and orientability of a space can be studied in a very elegant way by the means of algebraic topology. This is what is done: To each space a particular algebraic structure is associated, say a group. Then, an attempt is made to capture topological relations among the spaces by looking for associated algebraic relations among the algebraic structures assigned to each space. The study of these techniques and their limitations will, fortunately, not be needed by us.

There is one additional point I should note here, though. The specification of the intrinsic features of a metric space, in terms of Riemannian invariants, and the characterization of its topological structure are not completely independent matters. One can show, for example, that a compact two-surface with constant Gaussian curvature and in which all geodesics can be "completed" is necessarily a two-sphere, topologically. A sufficiently-detailed specification of metric features, then, may constrain the topological structure of the space.

# SPACE AND TIME VERSUS SPACETIME—
# A PRELIMINARY ACCOUNT

The ultimate goal of Chapter II is an epistemological critique of those physical theories that ask us to seriously consider the physical applicability of non-Euclidean geometries, in particular, the general Riemannian analytic geometries of curved $n$-dimensional spaces. But *the* theory of "curved space" which has viability as a possibly true physical theory of the world is general relativity, and general relativity is not a theory postulating curved space, it is a theory postulating curved spacetime. Thus, in order to make the philosophical critique plausibly applicable at all, we will have to have at least some idea of just what spacetime is. In particular, how does a theory that postulates the existence of spacetime differ from one that presupposes the notions of space and of time?

The nature of spacetime, and the believability of theories that postulate it, will be virtually the entire content of Chapter IV. So I will not at this point offer either a detailed analysis of the structure of spacetime or any reasons at all why one should believe that there is such a thing. But I have to say something. A good bit of the remainder of Chapter II, and some parts of Chapter III, presupposes a knowledge on the part of the reader of what spacetime amounts to, so in the next few pages I will offer an abbreviated account of the structure of spacetime. Later, in (II,D), I will give a short account of how, on the basis of the theory of general relativity, one finds out what the actual spacetime of the world is like and, once again on general relativity's account, just why it is the way it is. Much that I say will have to be taken on faith for the time being. But the reader can have some hopes that what he is forced to accept on authority here will be given argumentative support in Chapter IV.

## 1. *Minkowski Spacetime*

We can begin the construction of the Minkowski spacetime, the spacetime appropriate to the special theory of relativity, and the comparison of it with the prerelativistic notions by making some remarks about the "basic individuals" of which "space" (in the abstract sense) is constructed.

I have been talking previously of a space as being a set of *points*. But for spaces that are supposed to be more than mathematical abstractions, which are actually, somehow or other, to represent the world, what should we choose as the *points?* There are importantly-controversial philosophical issues involved here; I shall be discussing them at length toward the end of Chapter II and throughout Chapter III. For the moment, though, I shall try to ignore these issues. I will frame what I have to say in a language that, although it *suggests* a definite philosophical viewpoint about the reality and nature of the points, is not meant here to have any deep philosophical import or to beg any important questions.

It is convenient, even in characterizing prerelativistic theories, to start off with a relativistic technique. Let us imagine some idealized *events* in the world, for example, the collision of two perfectly unextended point masses or the intersection of two nonparallel perfectly breadthless light rays. We will take an idealized event as "marking" a definite location in spacetime. It will be convenient, however, to have such locations where events do not ever occur; so we will use a trick, at least as old as Leibniz, and speak not only about actual idealized events but about possible ones as well. The points of spacetime, then, will be all the locations of possible ideal events. Since the events are extensionless, so are their locations. We will adopt a traditional, but misleading, language and speak of the locations of possible idealized events as themselves being *events*. Spacetime is, then, the set of all events. But this is not meant to imply that real events, or even possible events, are the individuals of the set which is space. Rather, these individuals are *locations* of possible events which we are now calling 'events.'

The fundamental assumption of prerelativistic spacetime theory is that these fundamental individuals, the events, are not truly fundamental. Every event location can, in fact, be "analyzed" as an ordered pair of individuals, a spatial location and a temporal moment. Prerelativistically, what is the location of an idealized event? It is a specification of two "independent" features: (*a*) where the event took place, i.e., the spatial location of the event, and (*b*) when the event took place, i.e., at what moment of time it occurred. So, if *e* is an event location, we can look upon *e* as $e = \langle p,t \rangle$ where *p* is a place and *t* is a moment of time. Given the fundamental notions of places and times, instead of the set of events we can consider the set of places and the set of temporal moments. We can describe the structure of these sets and we can then *reconstruct* the structure of the spacetime set by examining its "composition" out of space and time.

According to the Newtonian theory, the structure of *P*, the set of spatial locations, is simple. It is $E^3$, i.e., Euclidean three-space. The structure of the set of temporal moments, *T*, is even simpler; it is $E^1$, the structure of the one-dimensional real line. The individuals of spacetime, events, are ordered pairs of places and times. So the structure of the spacetime is simply $E^3 \times E^1$, the Cartesian product of space and time, or the set of ordered pairs, the first element of which is selected from $E^3$ and the second from $E^1$.

We can extract the spatial structure from the spacetime structure as follows: Take spacetime as a class of events and as having the structure $E^3 \times E^1$. Define a relation among events, 'sim,' where $e_1$ sim $e_2$ if and only if the temporal location of $e_1$ equals the temporal location of $e_2$. 'Sim' will be an *equivalence* relation, that is, a transitive, reflective, symmetric relation on the set of events. We can use it to split up the set of events into "equivalence classes," where two events are in the same class if and only if they occur at the same time. The equivalence classes generated out of the spacetime by 'sim' will all have the structure $E^3$. Intuitively, 'sim' is the relation that holds between two events when they are *simultaneous*. An exactly parallel procedure, using the equivalence relation 'occur at the same place' instead of 'occur at the same time,' will split up the spacetime into a collection of sets of events at a single location, all of which have the structure $E^1$.

Given a pair of events, $e_1$ and $e_2$, what significant questions can we ask about them in this picture of spacetime? (*A*) We can ask if they occurred at the same time. We can ask, given that they did not occur at the same time, just how large the temporal interval between them was. (If they did not occur at the same time we can also ask which of them was *later* than the other. I will forgo probing into this question until Chapter V.) (*B*) We can ask whether or not the events occurred at the same place. If they did not, we can inquire about the spatial separation between them. Reflection on the spacetime structure we have been examining will show that the question, "How far apart in space were $e_1$ and $e_2$?" is a meaningful question whether or not $e_1$ and $e_2$ were *simultaneous* events. We shall later see, throughout Chapter III, how this feature of what I will call from now on 'Newtonian spacetime' led to controversy during the prerelativistic period of science, and how it still has some puzzlement attached to it.

We can now construct *Minkowski spacetime* by contrasting it with the Newtonian spacetime I have just described. Once again, the spacetime is a set of events. But the structure imposed upon this set is radically different from the one I have just described. First, some remarks about what Minkowski spacetime is *not:* (1) It does not have the structure $E^3 \times E^1$. (2) It is not meaningful to ask for the temporal separation between two events, or even to ask whether or not the events are simultaneous. (3) It is not meaningful to ask for the spatial separation between two events, or even if the two events occurred at the same place.

What, then, is the structure of Minkowski spacetime and what questions have meaning relative to this structure?

Minkowski spacetime is a space (*a*) that has four dimensions; (*b*) that is a differential manifold and an affine space; (*c*) that has the *topological* structure of $E^4$, the Euclidean four-dimensional space; but (*d*) that has a radically different metric structure from $E^4$. In Minkowski spacetime we do not discuss distances between events, but rather the *interval* between them. The interval between two events along a curve, *A*, is defined like this:

1. interval $(P,Q)$ along $A = \int_A ds$

2. $ds^2 = (dx^1)^2 + (dx^2)^2 + (dx^3)^2 - c^2 dt^2$

where $c$ is the velocity of light in a vacuum and is a constant. If we take $c = 1$, which is simply a matter of adopting new units for space (or time) measurement, equation (2) simplifies to $ds^2 = (dx^1)^2 + (dx^2)^2 + (dx^3)^2 - dt^2$.

Let me expand upon the meaning of equation (2). Once again, let me warn the reader. At this point I am simply trying to briefly describe the spacetimes appropriate for relativistic theories. He shall have to forgo a rationalization for adopting such a spacetime model until Chapter IV.

Minkowski spacetime is a set of events. It is four-dimensional. Between any two events of the spacetime, and along any curve, there is a quantity, the interval between the events along the curve, which is a number and which is an *invariant* property of the spacetime. The spacetime is such that between any two events there is a curve such that the interval between these events along this curve has an *extreme* value, relative to the interval between the events along any neighboring curve. That is, $int^2(P,Q)$ along this curve is either (*a*) greater than $int^2(P,Q)$ along neighboring curves, or (*b*) less than $int^2(P,Q)$ along neighboring curves, or (*c*) $int^2(P,Q)$ along the curve is zero while it is positive or negative along all neighboring curves. These "extremal curves" are the geodesics of Minkowski spacetime.

The interval between events along a curve is not a *distance* between them, for distances are always nonnegative whereas the square of the interval can be positive, zero, or negative. If the square of the extremal interval between $P$ and $Q$ is positive, we say that $P$ and $Q$ have *spacelike separation,* if it is negative we say that they have *timelike separation,* and if it is zero we say that $P$ and $Q$ have *lightlike separation* (Fig. 18). The set of events that have lightlike separation from a given event, $e$, is called the *light-cone for* e. Any set of events which is a submanifold of the space and all of whose point events have spacelike separation from $e$ is called a *spacelike hypersurface for* e.

We can characterize some of the physical meaning of these relations. If $P$ and $Q$ have lightlike separation, then a light ray emitted from one of the two events can reach the other, or else $P$ and $Q$ are the same event (remember that $P$ and $Q$ are events, *not spatial locations*). If $P$ and $Q$ have timelike separation, then a causal signal whose velocity of propagation is less than that of light can get from one of these events to the other. If the events have spacelike separation, then no causal signal can get from one event to the other unless its velocity of propagation exceeds that of light in a vacuum. Since, as we shall see in detail in Chapter IV, a fundamental assumption of relativistic theories is that there are no such faster-than-light signals, spacelike-separated events have no possible causal signal connecting them at all.

The extremal interval between events is an invariant. The spatial

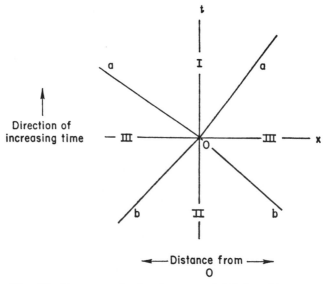

Fig. 18. Some very basic elements of Minkowski space-time. In this diagram, two spatial dimensions have been suppressed, so properly speaking it is a diagram of a Minkowski spacetime with only one spatial dimension.

The lines *a* represent light signals sent from *O*, and the events along these lines are those with lightlike separation from *O* in the future time direction. The events along the lines *b* are those with lightlike separation from *O* in the past time direction.

The events in region I are those with timelike separation from *O* in *O*'s "absolute future," and those in region II are the events with timelike separation from *O* in *O*'s "absolute past."

The events in regions III are those with spacelike separation from *O*, i.e., those not connectible to event *O* by any causal signal whatever.

separation and the temporal separation are *not*. We can get a notion of spatial and temporal separation, however, by introducing a co-ordinate system. Two basic features of Minkowski spacetime relevant to coordinatizing are these: First, since the spacetime is four-dimensional, we can supply an internal coordinatization that labels each point, i.e., each event, by four numbers. Call these $x^1, x^2, x^3$ and $t = x^4$. Second, although quite arbitrary coordinatizations are possible, the spacetime is such that certain particularly simple coordinatizations can be found. These "special" coordinatizations are closely analogous to the Cartesian coordinates imposable on Euclidean $n$-space by the means of $n$ mutually perpendicular straight lines. The fact that such a co-ordinatization can be found for Minkowski spacetime indicates something important about it, its "flatness," just as the fact that Cartesian coordinates can be imposed on a Riemannian $n$-dimensional space indicates that the space is a Euclidean flat $n$-dimensional space.

As we shall see in Chapter IV in detail, relativistic spacetimes are characterized by the fact that their splitting up into space and time is dependent upon the state of motion of the "observer." That is, two observers in motion with respect to one another will, in general, disagree about the spatial and temporal separation of a pair of events but will agree about the interval between the events. We shall also see that for special relativity there are observers whose states of motion are distinguished from those of observers in general. These are the observers who are in uniform or "inertial" motion. It is these special observers who will be able to naturally impose upon the events of the spacetime their "preferential" coordinatization. This is the coordinatization in which the "differential of the interval" has the form given it in equation (2) above. Relative to it, the equation of the geodesic connecting spacelike-separated events is a linear equation, just as the equation for a straight line is a linear equation in a Euclidean $n$-space when we impose upon the space a set of arbitrary Cartesian coordinates. For an inertial observer, the spacetime does split up into $E^3 \times E^1$, but this splitting is by no means invariant—which events go into which spaces with which other events varies from observer to observer. But, relative to a given observer, the splitting up can be done. The result is a relativized spacetime that can, like the nonrelativized spacetime of Newtonian theory, be viewed as "Euclidean three-space persisting through time." The spatial three-spaces can be Cartesian coordinatized, as can the temporal one-space, and the result is a coordinate system in which $ds^2 = (dx^1)^2 + (dx^2)^2 + (dx^3)^2 - (dx^4)^2$.

The close analogy between Euclidean four-space and Minkowski spacetime leads us to call Minkowski spacetime a *pseudo-Euclidean space*. *Euclidean* because Cartesian coordinates can be imposed upon it once an observer in a particular inertial state of motion is selected; *pseudo* because the invariant quantity relating pairs of events is not the always nonnegative distance invariant of Euclidean $n$-space, but instead, the invariant interval of Minkowski spacetime whose square can be positive, zero, or negative, depending upon whether the events are spacelike-, lightlike-, or timelike-separated from one another.

## 2. Riemannian Relativistic Spacetime

The preceding section presented an extremely condensed treatment of the distinction between the Minkowski spacetime of special relativity and the space-through-time of Newtonian theory. Missing details will be offered and arguments for considering such a theory in the first place will be considered in depth at a later point (IV). The primary purpose of introducing spacetime notions at this stage is this: We are discussing those features of geometry about which philosophical controversy has raged over the epistemic basis for rational belief in a particular geometric theory. Much of the recent discussion of this issue has come about because, for the first time, a serious contender for "a true physical theory of the world" has come forth which postulates a non-Euclidean world. But the general theory of relativity offers an account of the world in terms of curved spacetime, not curved space. So we have to have some idea of what spacetime is. In this section we will examine the transition from the flat spacetime of special relativity to the curved spacetime of general relativity. Once again, I shall only *describe* this spacetime here. I will offer very little in the way of explanation of just how the spacetime picture functions in the theory, reserving that for later (II,D,2,3), and we shall simply assume, as we did above, that the move from space-through-time to spacetime has already received its rationale.

Understanding the move from Minkowski to Riemannian spacetime is quite easy. We can bring to bear on this question all the apparatus of the earlier sections (II,B,4,5), where the move from flat to curved spaces in general was discussed. The spacetime of general relativity is, again, a four-dimensional, differential manifold equipped with an affinity (a geodesic structure) and a metric. But, as in Minkowski spacetime, the metric is the *interval* metric, not the ordinary Gaussian-Riemannian distance metric. Hence, it is not an assignment of an invariant number to a pair of points and a curve which always assigns a nonnegative number, but is instead an assignment of invariant numbers to pairs of points and curves between them whose squares can be positive, zero, or negative, and which can be zero between two points even if the points are distinct from one another.

What is the fundamental assumption of the theory of Riemannian spaces? It is this: Although the *n*-dimensional space may itself be non-Euclidean, it can be approximated, within a sufficiently small region about a point, by an $n$-dimensional Euclidean space containing the point, i.e., the tangent space at the point. The fundamental assumption of curved spacetime is exactly analogous: Although four-dimensional spacetime may itself be a non-Euclidean, or better, a non-pseudo-Euclidean spacetime, it can be approximated, within a sufficiently small region about a point, by a four-dimensional pseudo-Euclidean, i.e., Minkowski, spacetime containing the point.

Once this fundamental assumption is made, the description of the structure of relativistic Riemannian spacetime is clear. For each such spacetime there are an infinite number of arbitrary internal coordinatizations of the events that are the spacetime's points. Relative to such a coordinatization there exists: ($a$) A $g$-function that characterizes the interval between two points in the spacetime along a curve between them; just as the $g$-function in Riemannian geometry fully specifies, relative to a coordinatization, the distance between two points along a curve in terms of the coordinate description of points and curve. ($b$) A $\Gamma$-function that characterizes the notion of "most-parallel transport" for vectors assigned to points in the spacetime and the notion of "straightest lines" or geodesics. ($c$) An $R$-function whose expression, relative to a coordinatization, characterizes important features of the intrinsic geometry of the space.

The spacetime is flat if and only if there is a coordinatization of it such that all the components of the $R$-function are zero everywhere in that coordinatization. In that case, the components of the $R$-function are all zero everywhere relative to any coordinatization. As usual, the $\Gamma$- and $R$-functions can be defined in terms of the $g$-function. For a flat (Minkowski) spacetime, one can view the $g$-function as having the form $g_{11} = g_{22} = g_{33} = -g_{44} = 1$ and $g_{ij} = 0$, $i \neq j$ at each point, relative to the "pseudo-Cartesian" coordinatization. Again, as usual, from the $R$-function one can construct some invariant characterizations of the spacetime, in particular the scalar curvature, $R$, by means of the algebraic operation of contraction.

So the metric structure of relativistic Riemannian spacetime is clear. Of course, I haven't yet said anything about what *physical meaning* there is in the assertion that the spacetime of the world is a "pseudo-Riemannian" or "curved" spacetime. I will proceed to this below (II,D, 2,3).

One additional note I can make at this point. Although Minkowski spacetime is metrically quite different from a four-dimensional Euclidean space—since its metric is that of interval and not that of distance, and since its metric is no longer real, and nonnegative between any two points or along any curve—it still is *topologically* just like $E^4$, Euclidean four-space. Meaning, for example, that it is noncompact and that it is simply-connected and that following any geodesic in a fixed direction always takes one on an infinite trip. Once one has opened oneself to the possibility of a *curved* spacetime, however, the possibility of a spacetime with a different topological structure, both in the small and in the large, suggests itself. We will examine some of these possibilities and their physical and philosophical consequences in detail at a later point (IV,D).

Notice, before I go on, that in the characterization of curved spacetime, no embedding of the four-dimensional spacetime in some higher-dimensional flat space or spacetime was ever mentioned. Insofar as general relativity asks us to contemplate a world in which spacetime is curved, it is *intrinsic* curvature only that is being discussed.

# THE PHYSICAL POSSIBILITY OF
# A NON-EUCLIDEAN WORLD

## 1. *Prerelativistic Speculations*

As soon as Riemann's construction of curved three-dimensional geometry as an extension of the Gaussian theory of curved surfaces was developed, speculation naturally ensued as to whether the actual three-dimensional space of the world might, in fact, be non-Euclidean. Once we have made the distinction between intrinsic and extrinsic features of a space, the additional possibility of our actually being able to *detect* a deviation of the space of the world from perfect flatness also suggests itself. For, if the space of the world has intrinsic curvature, we could perhaps determine this fact without ever gaining access to a higher-dimensional space in which the space of the world was embedded, indeed, without ever contemplating the possibility of there being such an embedding space.

Gauss himself used his position as head of a government land-surveying department to make some empirical tests of the flatness of space. He chose three mountain peaks and measured the sum of the interior angles of a triangle constructed of light rays traveling from peak to peak. As far as he could tell, the sum was a straight angle and he concluded that, to within the limits of accuracy of his measurements, space was indeed Euclidean.

In 1870, the English mathematician-philosopher William Kingdon Clifford published some speculations about curved space. He assumed that on the whole and in the large space certainly was a Euclidean three-dimensional manifold. But, he speculated, there might be small regions of curved space which traveled through the overall flat manifold, much as bumps in a thin rubber sheet might move if they were caused by beetles under the sheet. His speculations went further, for he considered the possibility that what we call "matter in space" might actually be small regions of relatively mobile "deviation from flatness." I will have some more to say about such theories below (III,E).

Prior to the relativistic introduction of spacetime as a replacement for Newtonian space-through-time, all the speculations about non-Euclidean geometries actually describing the world were, of course, speculations about the possibility that the world consisted of curved-space-through-time. But the first serious scientific theory making intensive use of Riemannian non-Euclidean geometries to describe actual features of the world came with the attempt by Einstein, in the theory of general relativity, to account for important features of the world in terms of relativistic Riemannian curved spacetime, and it is to this that we now turn.


## 2. Some Fundamental Features of General Relativity

The theory of general relativity arises out of Einstein's search for a relativistically acceptable theory of gravitation. That the older, Newtonian theory is unacceptable, given the results of the special theory of relativity, is clear. I have already noted above (II,C,1) that (a) the special theory of relativity assumes that there is a maximum velocity of propagation of causal signals and (b) that it rejects an invariant notion of simultaneity as a relation among events. But since the Newtonian theory of gravitation assumes (a) instantaneous action-at-a-distance for the gravitational interaction and (b) the existence of an underlying Newtonian spacetime with its invariant simultaneity relation, the Newtonian gravitational theory is plainly incompatible with the general features of a relativistically acceptable theory.

The Einstein general theory of relativity offers a novel theory of gravitation immune to these objections, and in doing so, it utilizes for the first time the concept of a curved spacetime. But Einstein's theory is not the only possible gravitational theory compatible with relativistic considerations. A theory of Nordström's, for example, antedates the general theory of relativity, and it makes no use of Riemannian geometry at all. Later theories, such as Dicke's "scalar-tensor" gravitational theory, are relativistically acceptable as well. In addition, whereas special relativity has been shown to be highly confirmed empirically in some of its most important innovative features, the experimental evidence for general relativity is still, more than half a century after the exposition of the theory by Einstein, not by any means overwhelming. The Einstein theory, nonetheless, does offer us a persuasive argument that realistic theories of the world postulating non-Euclidean spacetimes are possible. This will be sufficient to motivate our concern with the question, "To what extent can one empirically decide between a Euclidean and a non-Euclidean picture of the world?"

The study of general relativity is a major scientific undertaking. All I am able to do is outline some important motivations behind the construction of the theory and some of its most important consequences with regard to observable features of the physical world. Topics I only mention here in the briefest possible way, such as the role of global or overall topology in the theory, will receive further attention in later sections (IV,E). But many fascinating questions, such as the role of conservation principles in general relativity, the notion of symmetry in general relativity, etc., will receive only scant notice throughout the book.

Let me survey quickly some basic intuitions behind the theory, then outline the form the theory takes and a few of its more relevant features, and finally, in the next section (II,D,3), discuss the problem of empirically determining the actual structure of a given region of spacetime, assuming the theory to be correct.

We wish to study the laws of gravitation. We shall assume that the effects of gravitation can be fully specified by finding (a) the consequences of gravitation for spacetime structure, (b) its consequences for the dynamics of particles, and (c) its consequences for the laws governing the propagation of light. Newtonian theory dealt only with (b), of course. Einstein was the first to suggest that, given the results of special relativity, (a) and (c) should be taken into account as well.

Let us start with a famous "thought-experiment" of Einstein's. We are enclosed in a sealed box and equipped with such apparatus as some point masses, clocks, light-ray generators, etc. Performing experiments with these devices, we try to determine the nature of objects outside the box insofar as they exert gravitational forces.

Suppose we release a point mass and we observe it to accelerate in a given direction relative to the box. Suppose that this acceleration seems to be (1) independent of the mass of the point mass and (2) independent of its particular constitution (electric charge, etc.). We are likely to conclude that there is a massive object outside the box exerting a gravitational force on the object. But we could offer an alternative explanation of the behavior of the unsupported point mass: There is no massive object outside the box; rather, the box is being accelerated (in the direction opposite to the apparent fall of the object) by attached rocket engines, say. Since all point masses accelerate in a gravitational field equally, no matter what their mass of internal constitution, the uniform relative acceleration of mass to box, due to the *inertia* of the mass and the "real" acceleration of the box, will be indistinguishable from an acceleration of the mass relative to the box due to an outside, massive, gravity-exerting object.

We must be very cautious here. The equivalence of "gravitational fields" to "accelerated coordinate frames" in generating relative accelerations of unsupported masses to the surrounding coordinate frame is *not* a full equivalence. Here are two reasons why: (1) The acceleration of the box will be uniform. No matter how far the particle falls relative to the box (before it hits the floor) it will accelerate relative to the box with a constant acceleration (assuming, of course, that the rockets are applying constant force to the box). In a "real" gravitational field the acceleration of the particle would increase, if only slightly, as it falls, since in falling it is decreasing its distance to the center of mass of the attracting body. (2) Suppose we drop not one, but two objects. We start them at rest, a slight horizontal distance from one another. If the box is being accelerated, the two particles fall in straight parallel lines with respect to another. If the falling is being caused by an external real object, the lines of descent of the particles converge, if very slightly, since both objects are falling toward a common point, the center of mass of the outside object. Einstein describes the "equivalence" of a gravitational field to the acceleration of a coordinate frame as a "local" equivalence. If you confine yourself to a small enough region of spacetime, you can't, by means of dropping point masses, distinguish being in an accelerated box from being in the gravitational field of an outside massive object (Figs. 19, 20).

Note how this equivalence depends upon the proportionality for all objects of the inertial mass, $m_i$, of an object (the mass that appears in Newton's $F = ma$) and its "passive" gravitational mass, $m_p$ (the mass that appears in $F_{grav} = Gm_pM/r^2$, where $M$ is the mass—"active" gravi-

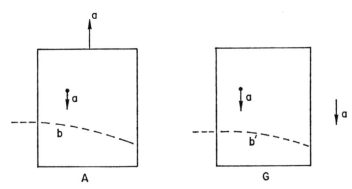

Fig. 19. The principle of equivalence. From the point of view of an observer confined to the box, the effect on a single unsupported test object of the box's being accelerated upward with an acceleration *a*, relative to the inertial frames, is the same as the effect of a uniform gravitational field on such an object would be. In both cases the object is accelerated downward relative to the box with an acceleration *a*, whatever the mass of the object or its particular constitution.

Einstein suggests that this equivalence of acceleration of the observer's frame with a gravitational field be extended to optical phenomena—as illustrated by the curved path of a light beam projected across the box and observed by an observer at rest in the box—and to metric features of time and space as well, these last because of the consequence of special relativity which relativizes distances and time intervals between events to the state of motion of the observer.

tational—of the attracting object). This proportionality has been established by Eötvös and Dicke to a very high degree of accuracy.

One of Einstein's boldest and most brilliant speculative leaps concerned this question: How does gravity affect the propagation of light? If we shine a flashlight across the box perpendicular to its motion, the path across the box will not be a straight line relative to the wall of the box, should the box be accelerated upward. We can calculate the deviation from straightness by means of special relativity, say. But, for dynamical phenomena, gravitational fields are locally equivalent to the acceleration of coordinate frames, as we have seen. So Einstein assumes that the same local equivalence holds for the propagation of light. Thus to calculate the behavior of either a point mass or a light ray in a gravitational field: Find the accelerated framework that would be locally

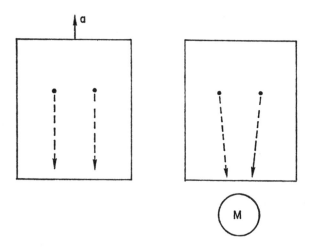

Fig. 20. Why the principle of equivalence is only local. The acceleration of the observer's coordinate frame is only locally equivalent to any real gravitational field. Suppose we drop two particles having horizontal separation. The observer in the accelerated box sees the particles as following parallel courses, whereas the observer in the box subject to a real gravitational field generated by a mass, $M$, sees the paths of the objects approach each other. If $M$ is very distant this effect may be small, but it still exists since both particles fall toward a common point, the center of $M$.

equivalent to the gravitational field at the point in question. Calculate the behavior of the masses and light rays relative to this accelerated frame, using the kinematical results of special relativity. Then "translate" these results back to the original reference frame with respect to which the new frame was accelerated. This will give the dynamic effect of the gravitational field on the local behavior of the masses and light rays, expressed in the original frame of reference.

I suggested that gravity will have an effect on spacetime features as well. A thought-experiment will reveal some of the intuitions behind this hypothesis. Suppose we have two clocks, one rigidly mounted on the floor of the box and one mounted on its roof. The two clocks are synchronous with each other when they are next to one another. When the floor clock ticks, it sends out a signal that is received at the roof clock. When the box is unaccelerated, the signals are received at the roof at the rate they are emitted at the floor. Now accelerate the box in the direction of the floor-to-roof vector. As an emitted signal gets from floor

to roof, the velocity of the roof clock increases relative to the initial velocity of the floor clock, because of the acceleration. As the accompanying illustration shows (Fig. 21), the net effect is a Doppler Red Shift in

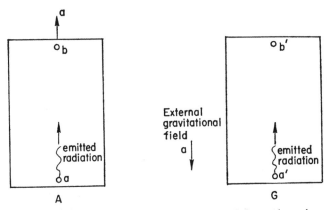

Fig. 21. The gravitational red shift derived from the principle of equivalence. First consider the accelerated box, *A*.
Light is emitted by atom *a* in the bottom of the box. But
by the time the light reaches atom *b* at the top of the box,
the box is moving with a velocity in the upward direction,
relative to the inertial frame in which *a* was at rest when
it emitted the light. This is due to the upward acceleration
of the box. The result is that the observer at *b* sees the
light "shifted to the red" in its wavelength, i.e., slowed in
its frequency. This is the ordinary Doppler red shift familiar
from prerelativistic physics.

Then consider *G*, the box in the uniform gravitational
field. By the Principle of Equivalence, the atom *a*, lower
down in the gravitational potential than atom *b*, should
have its light, observed at *b*, shifted to the red with respect
to the light emitted at *b* and observed at *b*. This is the
gravitational red shift; it is suggestive of the effect on the
metric of time and space of gravitating objects which is
predicted in general relativity.

which the rate of signal reception is less, measured on the roof clock,
than the rate of emission according to the floor clock. But the rate of
emission of the floor clock is synchronized with the clock's "ticks," so
the observer stationed at the roof observes the floor clock to be "running
slow." If we take these clocks to be ideal, accurately measuring the
"rate of time" at their locations, one can picturesquely say that time
is slower at the floor than it is at the roof. But accelerating the box

should be equivalent to imposing a gravitational field on an unaccelerated box. So a clock deeper in a gravitational potential than another clock should appear to run "slow" when the clock deeper in the potential is timed by the clock higher up. This is the famous gravitational red shift predicted by general relativity. Its existence has been experimentally confirmed to a very high degree of accuracy. One mustn't overrate this experiment as a test of general relativity, however, for such a gravitational red shift is compatible with (and predicted by) many distinct theories of gravitation.

So we see that gravity should have some effect on time, if by this we mean that being in a gravitational field will have an effect on the time interval between events as measured by devices not so immersed in the field. There are other thought-experiments, some suggested by Einstein himself, that motivate the hypothesis that gravity will affect measurements of spatial features of the world as well. I will forgo discussing these in order first to consider one more connection between spacetime and gravitation, and then to outline Einstein's theory that was a result of these intuitive considerations.

A gravitational field theory traditionally has two aspects: A side that tells how the field is generated by the distribution of masses in the world, and a side that tells how, given a field, the behavior of test objects is affected by the field. Let us look at the second side. Einstein ultimately takes as the gravitational field the structure of spacetime itself. There is an important feature about gravity, one I have already noted, that allows such an account. This is the fact that the path of a particle in a gravitational field is the result of (a) the structure of the field and (b) the initial position and velocity of the particle. Its motion is entirely independent of the mass of the particle or its internal constitution. In this way, gravity deviates remarkably from, say, electrostatics. Einstein uses this fact to make the following bold argument:

In the absence of a gravitational field, a point mass not acted upon by forces will travel a straight line, that is, a timelike geodesic of Minkowski spacetime.

Let us assume that in the presence of a gravitational field point masses also travel timelike geodesics, but now the geodesics of a non-Euclidean, Riemannian curved spacetime. If particles of differing mass or differing constitution traveled different trajectories from the same set of initial positions and velocities in a gravitational field, we could not do this. For there is only one geodesic with the right "starting point," whereas there would be many trajectories. And if we considered geodesics in space, not spacetime, the theory would fail as well; for particles with differing initial velocities do travel different trajectories, and there is only one spatial geodesic through a point in a given direction. But spacetime geodesics will do nicely. In a Riemannian spacetime there is only one geodesic through a point which has a given direction at that point. But a *spacetime direction* captures both the notions of spatial direction and of initial velocity in its definition, since if we specify a vector in spacetime we not only specify ratios of spatial intervals (initial direction of motion of the particle) but ratios of spatial to temporal intervals as well (initial magnitude of velocity of the particle). Just as a specification of initial position and velocity fully determines the trajectory of a particle in a gravitational field, so the specification of a point and a spacetime direction in a spacetime uniquely characterizes a geodesic in the spacetime. So why not take spacetime geodesics as the paths of otherwise freely-moving point masses in a gravitational field?

What about the paths of light rays? As we have seen, in special relativity any two points connectible by a light ray have a null-interval between them along the extremal (geodesic) path connecting them. We hold to this assumption in general relativity, simply letting the geodesics be the geodesics of Riemannian as opposed to Minkowski spacetime. In a gravitational field a light ray travels along a geodesic such that the interval between any two points on the geodesic is zero; that is, light travels the null-geodesics of the spacetime. The gravitational field, then, is the spacetime metric structure. To fully specify a gravitational field in a region of spacetime one must (*a*) coordinate the points (events in the region and then (*b*) find the *g*-function that describes the intrinsic geometry of the region relative to the given coordinatization. This tells us what the gravitational field is—it is the intrinsic geometry of the spacetime—and how the dynamics of particles and light rays are affected by the field. It doesn't yet tell us what the lawlike connection is between a gravitational field of a particular kind and a particular distribution of masses that is the field's "source."

Before we turn to that problem I might say something about the hypothesis of point masses traveling timelike geodesics in a gravitational field. In the original Einstein theory this is simply taken as a hypothesis of the theory. Later, Einstein and his collaborators tried to show that since every point mass was not only a particle being affected by a gravitational field but also a *source* of a gravitational field, the geodesic motion of such masses could be demonstrated from the field equations without an additional "dynamic" hypothesis. The trick is to consider the mass as a "singularity" in the field and to try to show that the motion of such singularities is governed by the equations that describe the relation of field to source-mass distribution without the need for an additional dynamic postulate.

Now we have seen the answers to the question, "Given a specified gravitational field, what effects does it have on the material objects present in the region of the field?" The answer is twofold. First, insofar as material objects are instruments to measure intervals in spacetime, the measurements obtained with them in a region in which a gravitational field is present will be such as to indicate a pseudo-Riemannian rather than a Minkowski spacetime in the region. Second, the field will have the dynamical effect of deflecting moving point masses and light rays; for the former follow timelike geodesics and the latter follow null-geodesics when no forces other than gravity are acting upon them, and a gravitational field is a distortion of these geodesics from Minkowskian straight lines.

But how is the field itself determined? In general relativity, just as in traditional gravitational theory, there is an equation connecting the nature of the gravitational field throughout a region with the distribution of mass throughout the region. In Newtonian gravitational theory this equation is Poisson's Equation. Einstein searches for a relativistically acceptable substitute for the older, plainly nonrelativistic equation.

To find his relativistic substitute, he looks for an equation that satisfies a number of antecedently-specified "criteria of adequacy." First, he wishes the equation to be written in a form adequate to represent the equation no matter what arbitrary internal coordinatization of the spacetime he chooses. This indicates what the mathematicians call a tensor equation. Second, the equation must state the connection between the gravitational field having a certain form and the mass of the region having a certain distribution; so the equation should have on one side a form built up out of the basic quantities that describe the gravitational field, and on the other side a form that represents the mass distribution. But the field *is* the intrinsic geometry of the spacetime, so one side of the equation should have a mathematical form built up out of the $g$'s describing the intrinsic geometry. On the other side we have not the ordinary "mass density" function of the prerelativistic theory, but a more complicated representative of the mass of the universe—the stress-energy tensor. This is due to the special-relativistic relation between mass and energy, and the necessity for taking energies, pressures, etc., into account in the determination of the field. Third, the general-relativistic equation should approximate the older nonrelativistic equation when masses and velocities are small and this places additional constraints on the relativistic equation. Finally, there is a sort of "pseudo-conservation principle" that Einstein suggests as an additional lead to finding the structure of the equation.

Utilizing these principles Einstein comes up with a field equation. In its most general form it looks like this:

$$R_{ik} - \tfrac{1}{2}g_{ik}R + \lambda g_{ik} = T_{ik}$$

$T_{ik}$ is the stress-energy tensor, $g_{ik}$ the g-function, $R$ the scalar curvature, $R_{ik}$ the so-called Ricci tensor derived from the Riemann curvature tensor by the algebraic operation of contraction, and $\lambda$ a constant. The equation Einstein prefers, and which has been studied most thoroughly, is one like that above but with the '$\lambda g_{ik}$' term deleted.

The field equation expresses a lawlike connection between the structure of the gravitational field, i.e., the intrinsic geometry of the spacetime, and the distribution of mass-energy throughout that spacetime. It gives us the remaining half of the geometric field theory of gravitation.

Before moving on, however, I might make two remarks about what the field equation does and does not do, remarks that will be expanded upon at some length later (III,E).

First, should we view the equation as telling us that the intrinsic geometric structure of spacetime is *caused* by the distribution of mass-energy throughout the world? A little reflection will show that this can be quite misleading. The form on the right-hand side of the equation, the stress-energy tensor, takes into account the distribution of mass-energy in the world utilizing the *metric* features of this distribution. It is not only how much mass there is, but also how it is distributed that counts. The possible distribution of mass-energy throughout a spacetime depends upon the intrinsic geometry of that spacetime. One could not, for example, have a uniform "flat" distribution of mass throughout a region that was itself not a flat spacetime.

It is more enlightening to look upon the equation as giving a lawlike "consistency" constraint upon the joint features of the world—spacetime structure and mass-energy distribution. The equation tells us that given *both* a certain intrinsic geometry for spacetime *and* a specification of the distribution of mass-energy throughout this spacetime, the joint description is the description of a general-relativistically possible world only if the two descriptions jointly obey the field equation.

Second, is there a one-to-one correspondence between mass-energy distributions and particular structures for spacetimes? No. The field equation, like the equation of classical prerelativistic gravitational theory, is a *differential equation*. Differential equations can be solved. But, in general, their solution requires the invocation of boundary conditions. For example, the field generated by a point electric charge in classical electrostatics will be quite different if the charge is in otherwise empty space or if it is one meter in front of an infinitely extended conducting sheet.

The same is true in general relativity. In general, to solve the field equation one must specify boundary conditions. With distinct specifications of boundary conditions, the same mass distribution can be associated with quite different spacetimes by the equation. It is possible, for example, to have spacetimes empty of ordinary matter with flat and with curved overall intrinsic structure. Which solution to the equation one gets for the overall geometry in the "empty" world depends upon which boundary conditions one specifies. We will see the great importance of this in (III,E) below. Incidentally, I should mention at this point that the stress-energy tensor in the field equation is supposed to take into account the distribution of only the mass-energy other than the mass-energy of the gravitational field itself.

### 3. *Mapping the Spacetime Field*

Suppose we accept general relativity as a correct description of the world. The question I will examine in this section is, how could we, using various test devices, empirically determine the geometric structure of a region of spacetime? Once again, we are assuming that the general relativity theory is correct. This question is of some importance, for the adequacy of empirical methods for answering geometric questions about the real nature of space and spacetime in the actual world will be our fundamental philosophical concern shortly.

How many numbers must we calculate to determine the intrinsic structure of spacetime at a point? First choose an arbitrary coordinatization of the events. Relative to this coordinatization, the spacetime is fully described by the values of all the components of the g-function at each point. There are $4^2 = 16$ such components, but since $g_{ij} = g_{ji}$ when $i \neq j$, we actually need determine only ten values.

What do we have to work with? Our experimental apparatus will consist of "ideal" physical entities: ideal rigid rods, whose lengths are unaffected by such things as temperature changes; ideal clocks, whose rates of ticking are uninfluenced by stray electric fields, etc.; ideal point masses traveling freely, except for the influence of gravitation; and ideal light rays traveling in perfect vacuums with perfect lack of width and breadth.

The problem can be easily understood by considering the spatial analogy. Suppose we were two-dimensional creatures living on a curved surface. How could we determine the intrinsic geometry of the surface without utilizing any reference to a Euclidean embedding space? The solution is fairly clear. Choose a large number of point locations on the surface and measure all the extremal distances between them. If one chooses a network with a sufficiently large number of points, of sufficient "fineness," one will be able, by examining the table of interpoint distances, to determine the intrinsic geometry of the surface as accurately as one likes. A few measurements of airline distances between cities on the globe, for example, would rapidly convince one that the globe was not a Euclidean plane, for there would be no way of distributing the points representing the cities on a plane surface so as to make the intercity distances consistent with one's table.

Since we are dealing with spacetime we measure intervals, not distances. But we can infer intervals from distance and time measurements relative to a particular observer. Consider, for example, an observer in a state of motion. Suppose he chooses two events simultaneous *relative to him*. He finds the minimal distance along curves between them by using his transportable rigid rods. He thereby determines the interval between the events. Some other observer will see the events as nonsimultaneous, but if he measures the time interval between the events with his ideal clock, and their spatial separation (relative to his state of motion) with the rigid rods, and then computes the interval, he will arrive at the same result as the first observer about the *interval* between the two events. It is intuitively clear, I think, that an observer equipped with ideal rigid rods and ideal clocks can determine the g-function for his region, relative to a particular coordinatization, by making sufficiently-many interval measurements between events in that region, and so map out the intrinsic geometry of the spacetime about him.

While rods and clocks provide a sufficient basis for mapping the spacetime, this is not the only sufficient basis. Assuming our theory correct, point masses not acted upon by forces other than gravity travel timelike-geodesics in the spacetime, and light rays travel null-geodesics. We can use these facts to construct other fully adequate experimental bases for mapping the spacetime. For example, one can determine the metric using only light rays and clocks, or using only light rays and freely-moving point masses. The last approach, in which one uses timelike- and null-geodesics to determine the structure of the space, is preferred by some physicists since it makes no reference to "rigid rods" and "ideal clocks," and since the idealized point masses and light rays seem to them a less extravagant and more "natural" basis on which to construct the mapping. I will not outline the mapping procedures necessary to determine the ten values of the $g$-components, but the interested reader can pursue the various approaches in the literature by following up the references cited.

# SOME OLDER EPISTEMOLOGICAL
# VIEWS ABOUT GEOMETRY

The last section concluded the exposition of mathematics and physics that we will need for Chapter II, and we can here at last turn to philosophical issues. In the remainder of this chapter I will consider the following issues in turn: First, I will briefly expound a few "traditional" views about the epistemology of geometry, views inspired both by general philosophical positions and by the actual state of the development of geometry at the time the philosophy was offered. Next, I will look at a famous claim of Henri Poincaré, to the effect that which geometry we hold to be the case "in the world" is merely a matter of *convention*. I will outline a number of important replies to Poincaré immediately thereafter, but will forgo examining them in detail until the last part of the chapter.

In (II,G), I will examine a number of ways philosophers have tried, wrongly I believe, to offer an account of what it means to claim that geometry is conventional. I believe that the approaches I will examine in this section are the pursuits of red herrings, but it will be necessary to make a few remarks about them since they form quite a large bulk of recent philosophical work in this area. Finally, in (II,H), I will return to the debate about conventionalism as it was formulated by Poincaré and his critics. I hope to show in this section that following up the issue of conventionalism with regard to geometry will lead into one of the deepest and most perplexing areas of philosophy. At the end I will outline some "ways out" of the quandaries which have been proposed, but only to indicate my own dissatisfaction with each escape from perplexity. The results, let me warn the reader here, will be inconclusive; for I propose no ultimate solution to these questions.

## 1. *Traditional Apriorism*

It is hard to say what, if any, epistemological views about geometry were held before the Greek philosophers. Men knew and used geometric truths, to be sure, but the absence of a systematic geometric theory was accompanied by an equal dearth of philosophical reflection on the origins and justifications of belief.

By the time the Greeks, culminating in Plato and Aristotle, began to speculate about the justification of belief, geometry was already sufficiently like its ultimate (for them) Euclidean form to present them with a theory of some aspect of the world which led them toward what later came to be called a rationalistic theory of knowledge. The propositions that constitute the body of geometric knowledge, for present purposes identifiable with the theorems of Euclidean geometry, seemed to the philosophers to have both an *exactness* of content and a *security of belief* not possessed by the propositions of either the rudimentary physical and biological sciences or those of everyday belief about the material world. It is not that the sum of the squares of the sides of a right triangle equals *more or less* the square of the length of its hypotenuse; rather, the equation is *fully exact*—exact in a way that the "more or less true" propositions of ordinary discourse can never hope to obtain. And it is not as though we have *some reason for belief* in the Pythagorean Theorem; we have, rather, *absolute secure certainty* in its truth.

In what does this certainty consist? For Plato, Aristotle, and the succeeding generations of "rationalist" philosophers, belief in the truth of geometric propositions is never to be genuinely secured by an "upward," merely probable inference from the data of everyday experience. Such inductive reasoning, as Aristotle freely admits, may have heuristic value in leading us to the contemplation of a "perhaps true" geometric proposition. A cleverly drawn diagram, for example, can attract our attention to significant geometric features previously overlooked. But we can be said to have genuine geometric knowledge only when the proposition whose truth is to be established has been shown to be derivable, by indubitable deductive inference alone, from the fundamental propositions underlying the axiomatic geometric system. And once we have seen such a deduction, and understood it, our belief in the truth of the proposition is ineluctable and unshakable. We are immediately possessed of a certainty of its truth which is totally unobtainable with regard to ordinary propositions.

For our purposes we can forgo most of the details of the rationalist accounts of geometric knowledge. Plato held the propositions of geometry to be, properly speaking, about the objects of a timeless, immaterial world of which the material world was a mere copy or shadow; he accompanied this doctrine with the astonishing view that our knowledge of such truths as the propositions of geometry was, ultimately, a remembrance of direct, almost sensory, acquaintance with the ideal objects in some previous, nonmaterial existence of the soul.

Aristotle's account is rather less extravagant, and several of his doctrines are of interest to us. He realized quite clearly the hierarchical nature of the axiomatic system: belief in less fundamental propositions is founded upon their deduction from more fundamental propositions. The argument from more to less fundamental is such that it is self-evidently truth-preserving. If we are once assured of the truth of the premises, the more fundamental propositions, and once in possession of the argument from the more to the less fundamental, assurance of the truth of the conclusions follows automatically. But what, in fact, does assure us of the truth of the premises? Aristotle held that at the foundation of all knowledge one must find a group of most-basic propositions whose truths are self-evident and self-justifying. No derivation of these fundamental propositions from more basic ones can be given, and none is needed. For these fundamental propositions constitute the *self-evident* basis of all knowledge. Aristotle doesn't tell us too much about these, although he does offer as candidates for most-fundamental propositions the laws of noncontradiction and of the excluded middle in logic; he also offers a curious and illuminating "pragmatic" rationale for belief in them, to the effect that anyone who denied them was totally incapable of communication or argument and, hence, not worth contending with. Most rationalistically inclined successors of the Greeks, however, have been perfectly willing to accept the basic Euclidean axioms and postulates as self-evident truths. They were, of course, puzzled by the parallel postulate, the Fifth Postulate, since its truth seemed to them far less intuitively obvious than that of the remaining axioms and postulates. But for just this reason, they never did abandon the hope that the Fifth Postulate itself could be deductively inferred from the remaining fundamental propositions.

We need not, I think, pursue the vagaries of the rationalist accounts of geometry in too much detail. Two features of geometry and the philosophers' attitudes toward it will be sufficient for further discussion. First, for a very long period of cultural history geometry provided the one scientific theory that seemed to have anything like a finished form. Its propositions seemed exactly, not just approximately, true and they formed a coherent, systematic structure unapproached by that of any of the other sciences. So great was the advantage of geometry over the other sciences, that for many years it provided the philosopher's ideal paradigm of just what a science ought to be. Witness, for example, Spinoza's famous attempt at an all-inclusive metaphysical theory of the world, demonstrated *more geometrico* or "in the manner of geometry."

Next, the foundation of geometric belief, as far as the philosophers were concerned, was the demonstration, by means of indubitably truth-preserving deductive argument, of the proposition to be justified, from a set of fundamental propositions whose truth (being self-evident and indisputable) was not in need of any epistemic support. The evidence of the senses, of experience and experiment, was at best a useful device for suggesting hypotheses to be proved. But only the proof supplied genuine epistemic support for belief in the proposition in question, and the support the proof supplied made the belief in question an infallible one.


## 2. Kant's Theory of Geometry

There is one philosophical theory of geometry of a rationalist sort which deserves particular attention. This is Immanuel Kant's theory, expounded in the "Transcendental Aesthetic" section of the *Critique of Pure Reason* and in his *Prolegomenon to any Future Metaphysic*. Kant's doctrine is interesting enough in its own right to deserve at least fleeting attention from us. In addition, his particular version of the rationalist account of geometry is of special importance because it is one of the few offered in reply to an antecedent empiricist criticism.

Kant is a metaphysician. He is interested in establishing the most general and all-pervasive metaphysical truths about the world—truths, he believes, like "Every event has a cause." But metaphysics in general has come in for some extremely destructive criticism, particularly from David Hume. We can briefly outline the position of Hume which most disturbs Kant in a few fundamental "empiricist" claims (1) All propositions are either a priori or a posteriori. That is, either they are immune to rejection or revision in the face of experience, or instead, they rest for their epistemic support upon generalization from experience, and they can be tested (and if necessary rejected or revised) by reference to further sensory experience. (2) The only a priori propositions are those that merely unpack what is contained in the meanings of the words. They are, at most, disguised logical truths, disguised by the packing of meanings into complex terms. For example, "All bachelors are unmarried" is a priori, but this is simply because 'bachelor' means 'never having been married adult male.' Such a priori truths, according to Hume, "merely express relations among ideas." Kant calls them analytic. They tell us nothing about the nature of the world, and are fundamentally empty of genuine content. (3) But the traditional propositions of metaphysics are supposed to be at the same time a priori *and* nonanalytic. They are supposed to be irrefutable by any possible empirical experience, but are still supposed to have informative content, telling us something nontrivial about the nature of the world. (4) Hence, metaphysics is nothing but "sophistry and illusion" and should be discarded as linguistic pretense, having no genuine cognitive value.

Kant agrees with Hume that metaphysical propositions are supposed to be both a priori and nonanalytic, or synthetic in Kant's terminology. But if Hume is correct in (2) above, then metaphysics is impossible. Kant sets himself the complex task of demonstrating the possibility of metaphysics, showing *how* metaphysics is possible, finding the limits that demarcate real metaphysics from linguistic pretense, and mapping out the fundamental propositions of the acceptable metaphysical corpus. We shall be interested only in the first two of these projects, and only insofar as they bring Kant's attention to geometry.

The existence of geometry (and arithmetic), according to Kant, gives the lie to Hume's doctrine. For geometry consists of propositions that all men must on reflection admit to be a priori, since they are propositions derivable by impeccable logic from self-evident fundamental propositions, and that all men must, on reflection, admit to be synthetic. The proposition that the sum of the interior angles of a triangle is a straight angle must, says Kant, surely be taken as a synthetic proposition, for 'having the sum of its interior angles equal to a straight angle' is not part of the meaning of 'is a triangle,' since 'is a three-angled straight-sided figure' exhausts the meaning of the latter phrase.

Now nobody is going to discard the splendid edifice of geometry simply because, according to Hume's doctrine, it can be nothing but sophistry and illusion. The only alternative is to reject Hume's account as unsatisfactory and to admit the possibility of synthetic a priori propositions, on the basis of geometry as an actual, indisputable collection of true, a priori, yet synthetic propositions. But if some such propositions, those of geometry, are possible, how can one reject out of hand the body of metaphysics simply because it too consists of propositions that are both synthetic and a priori?

I will not pursue Kant's investigations into the nature and limits of metaphysics, but I will instead look at his answer to the second fundamental question of his metaphysical task: Granted that synthetic a priori propositions are possible, *how* are they possible? How can there be propositions that describe all experience in a "nontrivial" way but are not refutable by any possible experience? In particular, how is an a priori, synthetic science of geometry possible?

Kant's solution to this problem is remarkable in its ingenuity. How can geometry apply to all our empirical experiences despite the fact that its basic propositions are totally immune to the possibility of revision in the face of those experiences? The answer is that "experience" is the experience of some sentient being. Insofar as the physical world is the empirical physical world open to our sensory examination, it is a world as experienced by us. According to Kant, two elements are brought to any act of experiencing the world.

On the one hand, the content of any sensory experience plainly has a factor that is "from the outside." The details of the content of a particular sensory experience depends upon the world's being the way it is independent of our perceptual apparatus. On the other hand, the world we experience is that organized by the antecedently-given structuring apparatus of our perception. In order to have experience at all, that is, the data of sense must be structured into a coherent whole, and this structuring is a function of our mental apparatus.

In particular, the spatiality of things, which we naïvely attribute to things-in-themselves as they exist unperceived, is according to Kant actually a structure brought to the perceptual experience by us. Space is a "manifold of intuition," a structuring organization of the mind which the mind uses to bring the otherwise incoherent data of sensation into a systematic order.

But geometric truths are truths about space. So it is no wonder that they are both a priori and synthetic. They are synthetic because the structuring of the perceptual world is no mere matter of pure logic and the meaning of words, but a genuine discoverable feature of the world, knowable by sensation. They are a priori since space is brought into experience by the antecedently-given and ever-unchanging organizing devices of the mind and since geometry is truths about experience. We need have no qualms about asserting the "in principle" irrefutability of any geometric proposition by any possible sensory experience. Insofar as an experience is a sensory experience of the physical world, it is an experience organized into a perception by the "spatializing" structure of the mind. And insofar as the laws of geometry are the true description of the workings of this organizing structure, we can have total assurance that the spatial features of the experience will conform to the principles of geometry.

At least one post-Kantian philosopher, Bertrand Russell, has attempted to reconcile a basically Kantian point of view with an acknowledgment of the existence of non-Euclidean geometries as possible descriptions of the actual physical world. Russell argues, that while Kant was wrong in believing the full metric structure of space to be synthetic a priori, we can be guaranteed at least, a priori, that the true geometry of the world is one of the Riemannian spaces of constant curvature. Since I don't think the argument very plausible, I won't pursue it.

## 3. *Traditional Empiricism*

As we have just seen, the beginnings of empiricist theories of geometry antedate the discovery of the non-Eucildean geometries. Hume, for example, reflected somewhat upon the epistemic justifications of geometric beliefs and offered an empiricist account, but this was prior to Kant's theory, which was itself antecedent in time to the first non-Euclidean geometries. Gauss, in fact, in an act which reflects rather badly on philosophers, hesitated to publish his original non-Euclidean thoughts for fear of the resentment it would arouse in the disciples of Kant. Not surprisingly, the discovery of the non-Euclidean geometries, axiomatic and analytic, dealt a heavy blow to rationalistic accounts of all types. The strongest motivation of the rationalist theories is the apparent aprioricity of Euclidean geometry. But if alternatives to a Euclidean world actually exist, then how could one be sure prior to observation and experiment that the fundamental assumptions of the Euclidean theory would, in fact, not be found by experience to be false?

We have already seen a few features of traditional empiricist accounts in the brief survey of the Humean doctrine, against which Kant proposed his theory of the synthetic a priori. Most of the remainder of Chapter II will be a pursuit in depth of a critical examination of some important features of empiricist justifications of belief in theories in general and in geometry in particular. At this point, however, I would like to rehearse a few of the most fundamental empiricist "dogmas," so that we will have a firmer idea of what the following critical sections take as a starting point. Nearly all of the interesting work on the epistemology of geometry which exercises contemporary philosophers starts out with a rather strong sympathy for empiricist positions but forthrightly faces up to the difficulties facing empiricism when its claims are examined honestly and in detail.

The most important empiricist doctrines for our purposes are the partitioning of all propositions into two classes—analytic a priori and synthetic a posteriori—and the doctrine of meaning. We will focus on the former in this section, for the latter will be explored in some detail below (II,F,H).

All meaningful propositions are of two kinds. They may be a priori, irrefutable by any possible sensory experience. In this case, however, they are merely analytic; in Hume's language they "express only relations of ideas." In more modern terminology, they are true "solely in virtue of the meanings of the terms involved in them" or sometimes, "they are reducible to propositions of pure logic by definition alone." The paradigm I have already used, "All bachelors are unmarried," will serve to illustrate this doctrine. Whatever the details of one's "unpacking" of the meaning of analytic, all are agreed that insofar as a proposition is a priori it is empty of descriptive content. The a priori propositions are irrefutable by any possible sensory experience. But this is because they are analytic—they say nothing about the world, have no genuine descriptive content, and tell us nothing that we could use to make useful predictions about the world or to control it. They are true in all "possible worlds" because they tell us nothing about the world; at best, they reflect conventions of linguistic usage.

Propositions may also be synthetic and a posteriori. Those propositions that do have genuine empirical content are all a posteriori; belief in their truth rests upon inference from experience. They may be "reports of basic observation," generalizations from experience arrived at by inductive reasoning, or theoretical hypotheses, "abduced" from observation by whatever principles of rational inference allow us to certify our belief in high level hypotheses in terms of their abilities to explain the data of observation and experiment. But in any case, they are ultimately grounded in sensory experience, in the sense that (a) to rationalize belief in them requires reference to the result of observation and experiment, and that (b) they remain, no matter how great our assurance in their truth, ultimately subject to empirical test, ultimately rejectable as false on the basis of some possible (even if never encountered) observational data.

Insofar as geometry is a theory whose description of the world is useful to us for prediction and control, as its propositions have genuine empirical content, geometrical propositions are all, even the most fundamental and "intuitively self-evident," subject to trial by experience. And all are subject to the *possibility* of finding themselves refuted by the test of experiment. That is, insofar as geometry is synthetic, it fails to be a priori.

As noted above, this doctrine about epistemic justification is usually supplemented by a doctrine about meaning. The fundamental assumption is that a term is meaningful only insofar as the term can be "associated" with some possible sensory experience. The variations on this theme—and the serious difficulties it faces in the case of terms that are according to most empiricists prima facie "empirically meaningful," but which obtain their use only through their appearance in theories whose acceptability is only indirectly related to the "data of experience" —provide a major theme of contemporary explorations of empiricism. It won't serve our purposes very efficiently to pursue this topic here, since I will approach it in some detail through a controversy about the particular case of geometry in the remaining parts of the chapter.

But the basic questions should be held constantly in the reader's mind: "What should an empiricist say about the *meanings* of terms and propositions of science? How should a doctrine of meaning be based upon a connection between the usage of terms and sentences and the possible features of sensation and experience?" We shall see that the question of the epistemic status of geometry and the questions concerning the meanings of geometric terms and propositions are inseparable; pushed deeply enough, these very interconnections force one to face squarely the ultimate difficulties of an empiricist account.

# THE CONVENTIONALIST THESIS
# AND ITS FIRST CRITICS

In this section I will expound a famous epistemological critique of geometry by Henri Poincaré. His position is usually labeled a doctrine of geometric conventionalism, but as we shall see, understanding just what a conventionalist claim amounts to is far from trivial. Part 1 will outline Poincaré's position, Part 2 will offer a reply to it originally formulated by Eddington and developed at great length by Reichenbach. (The conflict between the views of these two men and that of Poincaré forms the bulk of the investigations of [II,H].) Part 3 of this section will be a preliminary explanation of the meaning of conventionalist doctrine when it is extended beyond the conventionality of the metric features of spacetime into a doctrine of the conventionality of topological features. The treatment here will be quite brief, as I shall have to return to this question in some detail at a later point (IV, D,3).

## 1. *Poincaré's Thesis of Conventionalism*

Poincaré's epistemological critique of geometry is best expressed in three famous articles on geometry collected in *Science and Hypothesis*. In outline, his argument goes something like this: (1) It might be thought that non-Euclidean geometries can be rejected out of hand because they are infected with internal inconsistency. (2) This line is clearly refuted by the relative consistency proofs that demonstrate that the non-Euclidean geometries are at least as consistent as the older Euclidean theory. (3) It might then be contended that, consistent as they are, non-Euclidean geometries can have no applicability as empirically establishable theories about the actual world because the world of our sense experience obeys the Euclidean laws. (4) A proper understanding of how the geometric theory of the nature of the physical world is founded upon our sensory experience shows that this argument deeply misconstrues the relation between the immediate data of experience and the postulated geometric theory. The relationship shows conclusively that sensory data compatible with a non-Euclidean physical world are perfectly possible and, in fact, they can be easily described. (5) The above two arguments seem to show clearly that an empiricist epistemology for geometry is to be preferred. If the consistent geometries are all possibly compatible with some sensory experiences and incompatible with others, the means for establishing which of the geometries truly describes the actual world seems clear—perform a large range of experiments and on the basis of the observations made choose the geometry that is compatible with these actual results. In this process, of course, one may have to make "inductive leaps" from the data to the geometric hypothesis, but this is clearly a feature of the empirical establishment of theories in general and in no way shows an inadequacy in the empiricist account in this case. (6) Nevertheless, the empiricist account is wrong. For, given any collection of empirical observations, a multitude of geometries, all incompatible with one another, will all be *equally compatible with the experimental results.* (7) The inevitable consequence of such an "under-determination" of hypothesis by facts, no matter how rich a collection of observational facts we accumulate, suggests a *conventionalist* doctrine. One must simply *choose* the geometry one uses to describe the world, *by convention.* (8) Since Euclidean geometry is "simpler" than the non-Euclidean geometries we will always, in fact, conventionally choose to use Euclidean geometry to describe the actual physical world. Let us examine each stage of this somewhat complex argument in turn.

*Steps (1) and (2).* This requires simply the application of the well-known relative consistency proofs of the axiomatic non-Euclidean geometries to Euclidean geometry. If one is considering as possible candidates for the "real" geometry of the world those Riemannian geometries of

nonconstant curvature which are not representable by the earlier axiomatic geometric theories, one needs additional assurance of the consistency of the more general analytic geometries; but the models for
two-dimensional Riemannian geometry in terms of curved two-surfaces
embedded in a Euclidean three-space should convince us that they too
are consistent. Poincaré himself, incidentally, offers in these papers his
own relative consistency proof of Lobachevskian geometry in terms of
a nonisometric model in a fragment of Euclidean three-space.

*Steps (3) and (4).*   Can we reject non-Euclidean geometries as descriptive of the actual physical space of the world because our sensations
of the world are prima facie Euclidean in nature? This would be, I suppose, a sort of neo-Kantian apriorism. Poincaré answers with a resolute,
"No!"

It is grossly misleading, he argues, even if one believes the physical
space of the world to be the traditional Euclidean three-space, to take
Euclidean three-space as the "structural form" of the perceptions that
inform us about the nature of the space in which we live. For the time
being, think of our sensory access to space as being visual and tactile
sensations. Do these have "intrinsically" a three-dimensional Euclidean
structure? This seems implausible in the visual case, for if we can describe a "geometry" of the "subjective visual field" at all, shouldn't it be
more a bounded two-dimensional order than the unbounded three-dimensional structure of the hypothesized physical space? It is a difficult
problem to characterize the psychological notion of "visual space" at all.
But whatever we decide to say about this, the constraints imposed by the
physical geometry of our eyes' retinas should persuade us that the geometry of the visual field can hardly be identified, if it is well defined at all,
with the geometry of the physical space of the world. What I have just
said about visual space holds even more clearly for "tactile space." Poincaré argues that if tactile space could be said to have any dimensionality
at all, it would be the number of muscles whose motion generates the
kinesthetic sensations of touch, rather than the three dimensions of the
space of the visual world.

But if we do not develop the hypothesis about the three-dimensional
Euclidean structure of space from direct apprehension of this structure
in the space of sensation, how do we frame the hypothesis on the basis
of our experience? The answer, according to Poincaré, is this: The sensations that we experience are not random or incoherent. They have an

order and coherence that we can describe by saying that there is a law-like structure to them. For example, consider some sensations due to an object placed in front of our eyes. We change the object and the sensations change. We note that some changes of sensation due to causal manipulation of the object are such that we can restore the original sensation pattern by leaving the object alone and moving our bodies. This is true for rigid motions. By experimentation we can discover the possible rigid motions of objects and their structure, as would be described, for example, by the mathematical group theory of rigid motions.

The hypothesis about the actual geometrical structure of the physical world is an *inference* from the lawlike structure of our sensational experience. We do not perceive space directly as three-dimensional Euclidean; rather, we *infer* that this is its structure from the lawlike systematization that we find we can impose upon the true, immediate sensory data.

*Steps (5) and (6).* A single imaginative parable offered by Poincaré is designed to serve two distinct purposes. First, we are provided with a picture of a world where the sensory data of the inhabitants would naturally lead them to infer that the physical space of their world was not Euclidean, but Lobachevskian. This shows that the kind of sensational experience compatible with a non-Euclidean geometry is perfectly possible, and that we can easily describe it. But, second, we will then see that this "inference" to a non-Euclidean structure for physical space is not the only one that can be made from the data, and that, in fact, one could equally well explain the totality of sensory experience by postulating instead that the space of the world is Euclidean. This feature of Poincaré's "parable" shows us that the naïve empiricist account of geometry, plausible as it may seem in the light of the failure of the aprioristic accounts, is itself inadequate.

With powers actually beyond us we construct in the laboratory the "imaginary world" of Poincaré (Fig. 22). It is a closed Euclidean two-dimensional disk, heated to a constant temperature at the center while the circumference is uniformly cooled to 0° absolute. The temperature gradient along a radius is given by $R^2 - r^2$, where $R$ is the radius of the disk and $r$ is the distance of a point from the center of the disk. We populate the disk with two-dimensional creatures who are interested in determining the intrinsic geometry of their world. We equip them with rigid rods, except that the rods all contract uniformly with diminishing temperatures and in proportion to the drop in temperature, and all the rods have length 0 when their temperature is 0° absolute. We assume ideal thermal contact, so that the temperature of a point on a rod is equal to its location on the disk.

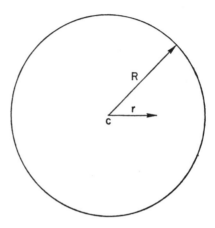

Fig. 22. The world of Poincaré's parable. The center of the disk is kept at temperature $TR^2$. At a distance $r$ from the center, $c$, the temperature is $T(R^2 - r^2)$, where $R$ is the radius of the disk. Consequently the temperature at the edge of the disk is zero. The index of refraction of the medium of transmission of light on the disk at any point $r$ from the center is $1/(R^2 - r^2)$.

The inhabitants are equipped only with light beams and rigid rods whose length at a point is directly proportional to the temperature at the point. Unaware of the temperature distortion of their rods and the bending of their light rays, the inhabitants calculate their world to be a Lobachevskian plane of infinite extent.

The inhabitants of this world proceed to measure distances between points on it so as to determine its intrinsic geometry. They assume, falsely, that a rod remains invariant in its length upon transport. After a short period they come to the following conclusion: Their world is a Lobachevskian plane of infinite extent. For example, geodesics are not straight lines but the geodesics of Lobachevski, and the ratio of circumference to radius of a circle is always greater than $2\pi$.

Suppose we expand the experimental repertoire of the inhabitants by providing them with light rays, but at the same time giving their world a medium of transmission for light whose index of refraction varies as $1/(R^2 - r^2)$. If they triangulate their world with the light rays, assuming that the light travels along geodesics and that the index of refraction of the "vacuum" is constant, they will once again falsely infer that their plane is Lobachevskian, unbounded and infinite.

But now, suppose some clever denizen of the disk realizes that all the experimental data that seemed to lead to the Lobachevskian hypothesis can instead be taken to indicate that the plane is a finite Euclidean disk, equipped with "shrinking" and "light-bending" fields of the kind we have described. How could the issue between his hypothesis and that held by the majority of scientists be resolved? *We* know, utilizing our "superior" standpoint out in Euclidean three-space, that the innovator is correct and the orthodoxy wrong. But how could *they* decide? According to Poincaré, they cannot. They must simply, by *convention, choose* one of the alternative hypotheses.

Now imagine ourselves in three-space performing a similar mapping of the intrinsic geometry of our space. Can't we see that an exactly similar situation could arise? And in this new case, there simply is no "right" theory. Both alternatives are equally correct alernative hypotheses, with only a "conventional" choice to make between them. The parable has shown us that there are describable experimental results that would lead us, making our ordinary assumptions about the behavior of rigid rods upon transporation and about the straight-line passage of light in a vacuum, to infer that our world was non-Euclidean, and that we can conventionally avoid discarding our Euclidean geometry in favor of an alternative, if only we are willing to make sufficient changes elsewhere in our hypotheses about the world.

*Step (8)*.　Step (8) is usually considered the weakest point in Poincaré's argument. What he has not seen is that whereas Euclidean geometry may be in some sense simpler than the non-Euclidean alternatives, in just the same sense of "simpler" the total non-Euclidean-geometry-plus-traditional-remaining-theory hypothesis may be as a whole, simpler than the Euclidean-geometry-plus-revised-remaining-hypotheses theory. So let us ignore step (8).

If we make the necessary changes to carry Poincaré's parable from one about space to one about spacetime, we will see that it contains the core elements of all the basic questions we will have to face about the proper analysis of the epistemic foundations of geometry. In addition, it will throw important light upon empiricist accounts of theory-acceptance in general and will lead ultimately to difficult questions of ontology and metaphysics, as well.

I have been describing Poincaré as a "conventionalist," but I have not made it very clear just what a commitment to a conventionalist doctrine amounts to. I think one can find evidence that Poincaré held at different points doctrines that we will want to maintain as distinct. Since I will later be looking at alternative reactions to the Poincaréan argument in some detail (II,H), I will not yet pursue this line of inquiry. In the next section, when we examine the important replies to Poincaré of Eddington and Reichenbach, we will begin to see what some of these alternative philosophical accounts of the Poincaréan situation are.

## 2. The Empiricist Reply to Poincaré

The physicist Arthur Eddington and the philosophers Moritz Schlick and Hans Reichenbach provide epistemological critiques of geometry that are largely directed to refuting Poincaré's allegation that the selection of a geometric hypothesis can only be a matter of convention. Eddington's views are contained in the "Prologue: What is Geometry?" to his *Space, Time, and Gravitation,* an elementary treatise on the general theory of relativity. The section is quite brief, but the fundamental argument is clearly and precisely laid out. Reichenbach's quite similar views are developed at much greater length, especially in Chapter 1 of his *The Philosophy of Space and Time.* In this section I will summarize their position and go into Reichenbach's development in some detail. Part 3 and Section II,G, to follow will be in the nature of digressions from the main argument, but I shall return to the conflict between Poincaré's position and those expressed in this section shortly in (II,H).

Poincaré's conventionalist doctrine applies to the geometry of space. Eddington and Reichenbach are writing in the context of an already-existing general theory of relativity; consequently, the problem they are concerned with is whether the adoption of a geometry for *spacetime* is a matter of "framing a hypothesis from observational data," as the empiricists would have us believe, or "merely making a conventional choice," as Poincaré alleges. Although Poincaré never contemplates the possibility of non-Euclidean spacetimes, as opposed to non-Euclidean spaces, the application of his doctrine to this new "manifold" of the world is clear.

Eddington's prologue is in the form of a three-way discussion between an experimental physicist, a mathematician, and an exponent of the new theory of general relativity. The experimental physicist gets the role of a rather naïve traditionalist, maintaining the "self-evidence" of Euclidean geometry and the unintelligibility of the alternative world pictures. Naturally enough, the "relativist" has little difficulty in disposing of his arguments. The non-Euclidean picture is consistent. If geometry is meant to characterize the empirically observable inter-relationships among the material objects (rods, clocks, freely-moving particles, light rays) that we would intuitively use to "map" the structure of spacetime, then a set of such observed relationships supportive of a non-Euclidean spacetime structure is just as possible, and conceivable, as the sets that would reaffirm traditional belief in the Euclidean (or, since we are dealing with spacetime, more properly Minkowskian) nature of the manifold.

The debate between the relativist and the mathematical exponent of Poincaréan conventionalism is handled more briefly, but it is sufficiently important to quote at some length. In italicized whole clauses and sentences, the italics are mine:

> *Mathematician:* The view has been widely held that space is neither physical nor metaphysical, but conventional. Here is a passage from Poincaré's *Science and Hypothesis,* which describes this alternative idea of space:
>
> "If Lobatchewsky's geometry is true, the parallax of a very distant star will be finite. If Riemann's is true it will be negative. These are the results which seem within the reach of experiment, and it is hoped that astronomical observations will enable us to decide between the two geometries. But what we call a straight line in astronomy is simply the path of a ray of light. If, therefore, we were to discover negative parallaxes, or to prove that all parallaxes are higher than a certain limit, we should have a choice between two conclusions: we could give up Euclidean geometry, or modify the laws of optics, and suppose that light is not rigorously propagated in a straight line. It is needless to add that everyone would look upon this solution as the more advantageous. Euclidean geometry, therefore, has nothing to fear from fresh experiments."
>
> *Relativist:* Poincaré's brilliant exposition is a great help in understanding the problem now confronting us. He brings out the interdependence of geometrical laws and physical laws which we have to bear in mind continually. We can add on to one set of laws that which we subtract from the other set. I admit that space is conventional—for that matter, *the meaning of every word in the language is conventional.* More-

over, we have actually arrived at the parting of the ways imagined by
Poincaré, though the crucial experiment is not precisely the one he
mentions. But I deliberately adopt the alternative, which, he takes for
granted, everyone would consider less advantageous. I call the space
thus chosen *physical space*, and its geometry *natural geometry*, thus
*admitting that other conventional meanings of space and geometry
are possible*. If it were only a question of the meaning of space—a
rather vague term—these other possibilities might have some advan-
tages. But the meaning assigned to length and distance has to go along
with the meaning assigned to space. Now these are quantities which
the physicist has been accustomed to measure with great accuracy;
and they enter fundamentally into the whole of our experimental
knowledge of the world. We have a knowledge of the so-called extent
of the stellar universe, which, whatever it may amount to in terms of
ultimate reality, is not a mere description of location in a conventional
and arbitrary mathematical space. *Are we to be robbed of the terms
in which we are accustomed to describe that knowledge?*

The law of Boyle states that the pressure of a gas is proportional
to its density. It is found by experiment that this law is only approximately true. A certain mathematical simplicity would be gained by
conventionally redefining *pressure* in such a way that Boyle's law
would be rigorously obeyed. But it would be high-handed to appropriate the word pressure in this way, unless it had been ascertained that
the physicist had no further use for it in its original meaning.

A. S. Eddington, *Space, Time and Gravitation* (Cambridge, 1921),
pp. 9, 10.

The doctrine espoused by Eddington in the voice of the relativist is
only briefly characterized here, but its fundamental themes are clear
and important. Basically the argument is running something like this:
Poincaré has *apparently* shown us that we can construct alternative,
incompatible geometries of space (or spacetime) which are equally
compatible with the same observational data, no matter how much data
we accumulate. The two theories he constructs each have a geometric
part and an additional "physical" part; so, by making suitable modifications in the physical part of the total theory, we can "save" any geometric hypothesis we like from refutation by any amount of experimental data. Now Poincaré thinks that this forces us to the conclusion
that we must simply choose one alternative theory or the other by
making a "conventional" choice. In a sense this is correct, but in a
trivial and not particularly interesting sense. The conventional choice
we must make is neither more nor less than a conventional choice of

using a certain verbal sound or written symbol with a certain *meaning*. The appearance of two "incompatible" total theories is misleading. Properly speaking there is only one theory, although it is written in two different ways. The confusion comes about because, in the *two expressions of the same theory*, words are being used with different meanings.

What the physicist means by "straightest line between two points" is "path of a light ray *in vacuuo* between those points." When we "save" Euclidean geometry by changing the physical theory we are not really saving Euclidean geometry at all. We are changing the usage of such terms as 'geodesic,' 'distance between $P$ and $Q$ along $C$,' etc., in such a way that given these new meanings the old sentences of Euclidean geometry remain true. But they no longer assert the same *propositions*, since the meanings of the terms in the sentences have been changed by the revision in the nongeometric portion of the theory. Poincaré no more establishes the conventional truth of Euclidean geometry than I could establish the truth of the assertion, 'lions have stripes,' by simply changing *my* meaning of the word 'lion' to mean what everybody else means by 'tiger' in English.

Reichenbach presents us with a similar reply to the conventionalist thesis. It does not, I believe, differ significantly in fundamentals from Eddington's briefly-related account; but since Reichenbach has presented us with what has become a convenient (if sometimes misleading) vocabulary for discussing these issues, it will be helpful to offer his version of this reply to Poincaré in a little detail.

If we wish to use a term in science, says Reichenbach, we must first *define* it, or at least, we should realize when we are using it that our assertions have full meaning only if the proper definition for the terms therein is forthcoming upon reflection and analysis. Now definitions can be of quite varied kinds. For example, we can introduce a novel term into a theory by offering a stipulative definition of it in terms of previously accepted meaningful terms of the theory; say, by defining 'bachelor' as 'an unmarried adult male.' But the kind of definition we are concerned with is not quite like this:

> Physical knowledge is characterized by the facts that concepts are not only defined by other concepts, but are also coordinated to real objects. This coordination cannot be replaced by an explanation of meanings, it simply states that *this concept* is coordinated to this *particular thing*. In general this coordination is not arbitrary. Since the concepts are interconnected by testable relations, the coordination may be verified as true or false, if the requirement of uniqueness is added, i.e., the rule

that the same concept must always denote the same object. The method of physics consists in establishing the uniqueness of this coordination, as Schlick has clearly shown. But certain preliminary coordinations must be determined before the method of coordination can be carried through any further; these first coordinations are therefore definitions which we shall call *coordinative definitions.* They are *arbitrary,* like all definitions; on their choice depends the conceptual system which develops with the progress of science.

Wherever metrical relations are to be established, the use of coordinative definitions is conspicuous. If a distance is to be measured, the unit of length has to be determined beforehand by definition. This definition is a coordinative definition.

H. Reichenbach, *The Philosophy of Space and Time* (New York: Dover, 1958), p. 14.

So picking as the meaning of 'one meter' the length of a particular rigid rod in a vault in Paris is a coordinative definition. But, he goes on to argue, choosing to call two objects separated from one another 'the same length,' because one can transport a rigid rod congruent when in coincidence with the first object to the second object and find it is congruent with the second object, is again supplying a coordinative definition. So is taking 'geodesic curve between two points' to be 'path of a light ray *in vacuuo* connecting those points,' and so on.

Given the notion of coordinative definition, let us look at Poincaré's "conventional alternative" theories once again. We save Euclidean geometry in the face of evidence that apparently establishes the non-Euclidean nature of space, or Minkowski spacetime in the face of evidence that apparently establishes that spacetime is Riemannian and not flat, by making certain changes in the remainder of the total physical theory. Like Eddington, Reichenbach would interpret these changes as simply adopting new coordinative definitions for certain important terms of the theory. To refuse to accept the principle that light rays travel null-geodesics in a vacuum is simply to change the coordinative definition for, and hence the meaning of, 'null-geodesic.' To maintain that all "rigid rods" expand or shrink uniformly when transported from place to place, so that they remain congruent to equally transported rods but cease to be congruent with previously congruent rods that were left behind, is simply to adopt a new coordinative definition for the concept 'same length for objects at a spatial separation from one another.' We thought we had two incompatible theories that were equally compatible with the class of possible observational results. We discover that we really have, properly speaking, one theory written in two misleadingly different ways. Since the coordinative definitions for terms in the "two" theories are different, these terms simply have different meanings in the two contexts. The fact that one theory says '*S*' and the other 'Not *S*' no more indicates their true incompatibility than does our disagreeing with a German who thinks ill of "Gift," because to him 'Gift' means a poison and to us 'gift' means a present.

One additional concept of Reichenbach's is that of *universal forces*. The terminology is misleading, but the concept is fruitful. Suppose we wish to construct two "conventionally alternative" geometries of spacetime. They differ in what they apparently say about the structure of spacetime, but how in addition must they differ if they are to remain equally compatible with exactly the same experimental observations?

To discover this we need only reflect back on the ways we can experimentally map out the structure of spacetime, as discussed above (II,D,3). Four kinds of observations were discussed: (1) We can measure spatial separations between events, relative to a particular "state-of-motion" coordinate frame, by using transported rigid rods that are at rest in the designated coordinate frame. (2) We can similarly determine time intervals between events "at a single place," relative to a state-of-motion frame, by using a clock at that place, i.e., at rest in the coordinate frame. (3) We can determine the timelike-geodesics of the spacetime by observing the paths of point masses whose motion is affected only by gravity and inertia. (4) We can determine the null-geodesics of the spacetime by observing the paths of light rays in a vacuum.

So to make a total theory "come out right experimentally" while changing its geometric portion, we would have to introduce a change in the remaining portion of the theory which had the following features: (1) It postulated a "stretching and shrinking" field that preserved congruences between transported rods, but destroyed the congruence of one rod to another "left behind" when the one rod was moved about in space. This is just like Poincaré's temperature field in the two-dimensional parable of (II,F,1). (2) The field would also have to affect congruence in the temporal sense of clocks. Clocks congruent at a place-time would remain congruent when equally moved about to new place-times, but they would no longer "tick at the same rate" with a clock to which they were initially synchronized but which stayed behind. (3) The field would affect the motion of point masses, deflecting them from their "inertial geodesic motion," but this deflection would be independ-

ent of the particular mass or internal constitution of the massive particle. (4) The field must deflect light rays in vacuums from their otherwise null-geodesic path, and the deflection must be independent of the
amplitude or frequency of the light.

With regard to clause (3), the classical gravitational force is a field
of this kind. Some confusion has arisen because Reichenbach refers to
gravity as being just such a field as I am describing, and it is important
to remember that, in relativity, gravity is *not* just a *force,* but something
that affects light rays and the metrical features of spacetime as well.
That is, the field I am attempting to describe must satisfy clauses (1),
(2), and (4), as well as clause (3).

Such a field Reichenbach calls a *universal force.* To save a particular
geometry by postulating a universal force is not to save one part of
physics by making "real changes" in the remainder. It is simply to
change one's coordinative definitions of such notions as 'same length
for objects at a distance,' 'null-geodesic,' etc. The "natural geometry"
of Eddington, the geometry of the usual expression of general relativity,
is the one we obtain by setting the universal forces to zero; i.e., by
"postulating" that there are no such universal forces. To postulate a
nonzero universal force would simply be to change one's coordinative
definitions. The modified theory would not really hypothesize a new
geometry for spacetime at all, it would simply rewrite the old geometric
theory in a potentially misleading way.

One gains clearer understanding of the relationship of one "alternative theory" to the other, says Reichenbach, if he makes the simple distinction between *genuine inductive simplicity* and *mere descriptive simplicity.* In choosing a scientific hypothesis in the light of evidential data,
we must frequently add to the data some methodological principle of
simplicity in order to select out as "preferred" one of the many different
possible hypotheses, all compatible with the specified data.

For example, in attempting to fit a curve to a number of data points,
the scientist will first try a straight line in preference to curves with
more complex algebraic descriptions. This is genuine inductive simplicity at work. It is characterized by the fact that additional data can
force us to reject our simplest hypothesis in favor of one more complex. One additional datum point, for example, may force us to reject
the straight line that we drew in favor of some more complicated curve,
which then becomes the simplest compatible with the extended data.
But no data whatever, no matter how novel or how extensive, can
ever force a decision between one or another of a pair of Poincaréan

alternatives. What this shows us is that, properly speaking, the two alternatives are not really alternatives at all. They are the same theory differently written. They differ from one another only in the form that they use to express a common theory of the world, or in Reichenbach's terms, merely in their *descriptive simplicity*.

This by no means concludes the discussion of these issues. All of (II,H), in fact, will be devoted to a reconsideration of the claims of Poincaré, Eddington, and Reichenbach, and to a discussion of the fundamental epistemological, metaphysical, and semantic issues underlying their disagreement. But in the next section, and in (II,G), that follows, I will digress.

## 3. *The Conventionality of Nonmetric Features of Spacetime*

In (II,B,6) above I noted some important connections between the metrical features of a non-Euclidean space and its overall topological structure. Questions of intrinsic geometry, therefore, are not totally disjoint from questions about the compactness or noncompactness, simple-connectedness versus multiple-connectedness, etc., of the manifold as a whole. As soon as one contemplates the physical possibility of the space or spacetime of the world being non-Euclidean in its metric structure, one immediately begins to speculate about the possibility of its deviating from the Euclidean model in topological features as well. And as soon as one has seen the Poincaréan allegation of conventionality for the metric features, one begins to wonder if a conventionalist account of topology will also go through. To what extent can we claim the attribution of a particular topological form to spacetime, on the basis of the empirical evidence, to be a matter of convention?

The subtle nature of spacetime, as opposed to space-through-time, when combined with these topological questions gives rise to a rather intricate set of questions. These are quite fascinating to the philosopher because anomalous models of the topology of spacetime introduce rather disturbing pictures of the structure of *time*. Variant spatial topologies seem to require at most a little flexibility of the imagination. Variant structures for temporal order, however, introduce impressive stresses into our naïve conceptual scheme. For this reason, I intend to postpone detailed discussion of possible alternative topological structures for spacetime until later (IV,D). At this point I will offer only one concrete instance of "alternative topologies" and do this for a spatial, as opposed to spacetime, structure. The simple instance, however, will give us access to some general issues that arise whenever alternative topological hypotheses are considered, and whenever the question of the conventionality of the alternatives is brought forth.

Consider the two-dimensional space that is the surface of a cylinder. As we have seen above, the intrinsic geometry on this surface, for a sufficiently small region about a point, is indistinguishable to an inhabitant of this surface from a small region about a point on the Euclidean plane (II,B,4). But, in the large, the surfaces certainly differ in their topological structure and this difference is determinable by a two-dimensional inhabitant with no access to the three-dimensional embedding space. Any simple closed curve on the plane can be shrunk continuously to a point, i.e., the plane is simply-connected. But this is certainly not true on the cylinder, as even the geodesic curves that are circles girdling the cylinder cannot be so contracted while kept confined to the cylindrical surface. So we could, being such two-dimensional creatures, determine on the basis of empirical experiment whether the surface to which we were confined was that of a cylinder or that of a plane in its topology, the local equivalence of the two surfaces with regard to metric features notwithstanding.

But is the situation so simple? Considerations clearly stated by Reichenbach show us that it is not. Consider the surface that is a plane divided up into regions by a set of parallel straight lines drawn at constant perpendicular distances from one another. Imagine such a space that has the feature that each such strip on the surface is, at any time, qualitatively identical to any other strip. How could an inhabitant of a surface distinguish between living (1) on a cylindrical surface or (2) on such a repetitious plane? Would not all his observations be exactly the same?

Of course, in order to accommodate the evidence that would naturally lead one to postulate a cylindrical universe for this new picture, one must allow for the changed topology of the world by making suitable modifications elsewhere in one's theory. On the cylindrical model, we can easily imagine an event happening which is like no other event elsewhere in the space at the same time. In the alternative picture, each and every event occurring in the space at a time is accompanied by an infinite number of qualitatively exactly similar events; these events differ from one another only in their *numerical* diversity. In addition, the necessity for a given change in one strip to be accompanied by exactly similar changes in all of the other strips forces us to assume a sort of "causal interdependence for events at a distance from one another." Reichenbach calls this accepting a theory that tolerates causal anomaly. So it is in the principles of individuation of events and in the accepted principles of causal interaction that the plane "repeated-strip" model exacts its price for being made compatible with data that from the other point of view suggest that the structure of the space is a cylinder.

In mathematical terms, the plane of repeated strips is called a *covering space* for the cylinder, and the cylinder is a *quotient space* for the repeated-strip plane. They are related to one another by the many-one mapping, which associates with each member of the infinite set of "qualitatively identical" points on the plane a single point on the cylinder.

Other alternative topologies require different modifications in the remaining physical theory to be made equally compatible with the data. Reichenbach shows, for example, that to try to give an account of data that naturally suggest that one inhabits a two-dimensional spherical surface to the alternative plane topology, we must accept a world in which points at infinity can be reached in a finite time by transporting a finite rigid rod, and in which "events at infinity" can causally affect events at a finite distance. As we have remarked, the possibilities for topological structures and the varieties of conventional alternatives multiply when we consider not only spatially interesting topological structures, but the topology of spacetime itself. I will, indeed, pursue this in some detail in (IV,D); but the material here is enough to make us keep in mind, in the discussion of conventionality which is to follow, that conventionality can be attributed not merely to the metrical features of spacetime but to its topological features as well.

# SOME VARYING VIEWS ON CONVENTIONALISM

The remainer of Chapter II will be devoted to discussing three questions: (1) What could it mean to assert that our knowledge of the actual geometry of the world is merely a matter of convention? (2) What features of geometry and our knowledge of it could lead one to make such a claim? (3) Is the allegation of "conventionality" correct, and does the fact that geometry is conventional in some sense distinguish it in any important way from any other theory about the nature of the physical world?

In this part, I will examine two claims that have been made about conventionalist theses. I believe the claims are incorrect and that their pursuit in depth leads away from the most important issues, issues I will investigate at length in (II, H) below. But since the two positions have been expressed and discussed in some detail in the contemporary literature, it seems appropriate to get these issues out of the way here. The first claim, discussed in Part 1, is that once we have made a distinction between "pure" and "applied" geometry, the epistemological status of the latter ceases to be mysterious in any way. Given this simple distinction, it is claimed, a simple empiricism appears once again as a fully adequate account of geometry. The second claim, discussed in Part 2, is that the "conventionality" of geometry is to be understood as the expression of some important topological features of the spacetime; in particular, its denseness, or its continuity when viewed as a set of points.

## 1. *Pure versus Applied Geometry*

The doctrine of conventionality seems to resurrect the specter of apriorism for the geometry of the world. If, as Poincaré alleges, we can maintain our faith in Euclidean geometry in the face of any experimental results, no matter how intuitively indicative of a non-Euclidean world they may be, isn't Euclidean geometry a priori after all, as the rationalists believed? Those who think that the "pure" versus "applied" distinction is illuminating here will usually maintain this: We can discriminate between *two different theories,* both misleadingly called 'geometry.' Pure geometry is, to be sure, a priori; but in the best empiricist fashion, it tells us nothing about the actual structure of the world. Applied geometry is informative about the world, but like any other such genuinely content-full theory, it is, at best empirically established by induction and hypothesis from the observational data.

What are these two different geometries? To begin to make the distinction clear, we must first restrict our attention to those geometric theories that can be presented in axiomatic form—like traditional Euclidean or the axiomatic Riemannian and Lobachevskian theories. We are, for the moment that is, forgoing interest in such things as Riemann's general analytic geometry of curved spaces, which is not presented in axiomatic form and which may be, for all we know, nonaxiomatizable—current results in logic having shown that the "in principle" finite axiomatizability of a theory is not a self-evident matter, and that there are in fact theories not so axiomatizable.

Once we have a theory in axiomatic form, however, with all the true propositions derivable from some finite set of basic propositions by elementary logic alone, we can then reflect upon two different things that might be called pure geometry: (1) Consider first what we can call *hypothetical geometry.* This is the "theory" we would get by considering all sentences of the form 'If $A$, then $T$,' where $A$ is the conjunction of the axioms of the original geometric theory, and $T$ is any theorem of the original theory. (2) Next consider the set, not of sentences, but of *logical forms of sentences* which we obtain by taking each theorem of the original theory and "disinterpreting" its nonlogical terms; that is, replacing each meaningful predicate or relation symbol of the theory by a meaningless symbolic stand-in, except for the logical terms of the propositions (and, not, if . . . then, all, some, etc.), which are retained unchanged. Let us look at these two pure geometries in turn, ask about their epistemic status, and inquire as to whether reflection upon them throws any real light on the epistemic status of the original geometric theory.

*Hypothetical geometry.* What would the propositions of hypothetical geometry be? They are all sentences of the form 'If $A$, then $T$,' where, given the fact that $T$ is some theorem of the original geometric theory whose axioms are conjoined to form $A$, the consequent of the hypothetical is *logically inferable* from the antecedent; that is, for each of the sentences of hypothetical geometry, $T$ follows as a logical consequence from $A$ taken as premise. The conditionals, then, are simply logical truths, using the simple connection between the validity of a deductive argument and the logical truth of the sentence formed by taking the conjunction of the premises as antecedent and the conclusion as consequent. So the epistemic status of hypothetical geometry is clear, or at least as clear as the epistemic status of logical truths in general. Whatever the grounds for rational belief in logical truths are, surely this is independent of our views about the grounding of our belief in the geometry of the world.

But does this invocation of hypothetical geometry help us with the original problem, the epistemic status of the nonhypothetical geometry we take to describe the world? To be sure, if the geometry in question is of the finitely axiomatized sort, we can reduce questions about belief in the geometry to questions about belief in its fundamental propositions, the axioms. Since the theorems follow by truth-preserving logic from the axioms, once we have established the believability of these latter, the rationality of belief in the former is as clear as the rationality of belief in any conclusion that follows deductively from premises known to be true.

But what does justify our belief in the axioms? It is here that Poincaréan conventionalism and empiricism differ; for the former tells us that this belief is only a matter of convention, immune to the threat of being overthrown by observation, and the latter denies just this point. Invocation of hypothetical geometry won't help us with the basic question, the epistemic status of the nonhypothetical geometry we take to describe the world.

*Pure geometry as "uninterpreted" geometry.* Does the alternative interpretation of pure geometry help any more than the invocation of hypothetical geometry? The issues brought out by a consideration of uninterpreted geometries do come closer to the crux of the epistemological questions, but once again the most fundamental problems remain untouched by the discussion.

The collection of logical forms one obtains by abstracting from a geometric theory to its "mere logical form"—that is, by replacing the sentences of the theory by forms whose only "meaningful" terms are the logical terms, and whose "descriptive" vocabulary has been replaced by place-holding symbols—is a useful object for investigation. The reason is that any purely *logical* features possessed by the original theory (say, its consistency) are preserved in the abstracting process. We can, then, investigate purely logical features of a geometry by abstracting from the geometry to its logical form.

All questions about the relative consistency of the non-Euclidean to the Euclidean axiomatic geometry can be studied in this way. Consider, for example, the proofs of consistency of the non-Euclidean geometrics which work by finding "models" for the non-Euclidean axioms in some fragment of a Euclidean space. This procedure can be viewed as follows: (1) We wish to examine the consistency of the non-Euclidean axioms. (2) We abstract from these axioms to their logical forms. (3) We find a set of theorems of a Euclidean geometry which have the same logical form—although what the theorems assert may be quite different indeed from what the non-Euclidean axioms asserted before disinterpretation. All that matters here is commonality of logical form. (4) We took Euclidean geometry as consistent; hence any set of its theorems we take as consistent. (5) But the consistency of a set of propositions depends not upon what they assert, but solely upon their logical form. (6) Since the axioms of the non-Euclidean theory have their logical form in common with that of the selected set of theorems, we are now, relative to the consistency assumption for the Euclidean geometry, assured of the consistency of the non-Euclidean theory.

What about questions not of *consistency* but rather of *truth?* The logical forms of uninterpreted geometry are not sentences. They are neither true nor false, evident nor nonevident. Such terms simply don't apply to them, only purely logical terms such as 'consistent,' 'satisfiable,' etc. Only the "fully-interpreted" geometric theories have their truth and falsehood, evidentness and nonevidentness, in question. Indeed, the items of uninterpreted geometry are not even "convention-ally" true or false. Conventional truth, whatever it is, is a kind of truth; but logical forms are not in any way true—even conventionally.

Let us return to the problem of the evidentness and truth of ordinary interpreted geometry. Has the excursion into logical forms helped us here? I think not. The issue we are forced to face by Poincaré's arguments is whether the selection of one fully-interpreted geometry over another as the "true" geometry of the world can be rationally guided by observation and experiment, or whether it is instead merely a matter of conventional choice. This issue of the epistemic status of "the geometry of the world" can't be resolved by any amount of reflection on the fact that if you disinterpret geometry, paying attention only to its logical form, the epistemic difficulties disappear. Of course they do. What issue is there in the rationality of belief when there are not even sentences left, i.e., when all questions of truth and falsity have been dissolved in the abstracting process?

Now, I did suggest that reflection upon uninterpreted systems does, in some sense, get us at least a little closer to the real issues than did the contemplation of hypothetical geometries. Why? The fundamental question we will have to face in Section H is that brought to the fore by the differences between Poincaré on the one hand and Eddington and Reichenbach on the other. When we have two total theories, apparently incompatible with one another, but demonstrably equally compatible with any possible class of observational results, should we say with Poincaré that there are two theories between which we must make a conventional choice, or with Eddington and Reichenbach, following an empiricist line on *meaning,* that there is, properly speaking, in this situation only *one* theory differently written in two forms?

Now this is an issue about the meanings of terms in theories. You will not be able to resolve it by simply reflecting upon the fact that if you abstract from a theory to its logical form all questions of the meaning of its descriptive terms disappear. But in the sense that the issue of conventionality is closely tied up with the question of how terms in an interpreted theory *get* their meaning, and just what this meaning *is,* reflections upon the interpreting of the logical form of a theory to generate the full-blooded theory *do* draw us to the crucial issues we will have to face.

## 2. *The Conventionality of the Metric and the Topology of Spacetime*

A fairly extensive body of the recent discussion on the epistemology of geometry has focused on attempts to connect the meaning of conventionality claims with particular topological facts about spacetime. These ideas take their origin from some rather obscure remarks of Riemann which are contained in the same inaugural lecture, cited in the bibliography to this section, in which he outlined the basic features of his generalized analytic geometry of curved spaces. It is my belief that this whole discussion rather misses the most important issues about conventionality, the issues I will discuss at length in (II,H), below. But since much has been said on these points, it seems desirable to briefly discuss this direction of thought here.

First we will need some topological notions. Consider a set of points. Let us put a relational structure on the set $R$ such that for every point $a$ and $b$ in the set, $a$ is $R$ to $b$, or vice versa. We will assume that no point is $R$ to itself; that if $a$ is $R$ to $b$, $b$ is not $R$ to $a$; and that if $a$ is $R$ to $b$, and $b$ is $R$ to $c$, then $a$ is $R$ to $c$. Such structures can differ from one another in significant ways. Suppose, for example, that for each $c$ in the set there is an $a$ such that $c$ is $R$ to $a$, and that there is no $b$ such that $b$ is different from $c$ and $a$ such that $c$ is $R$ to $b$ and $b$ is $R$ to $a$; i.e., assume that there is always a definite element in the set "just below" a given element in the $R$-ordering. For example, the integers ordered by the "is greater than" relation have this structure.

If we instead assume that between any two elements in the $R$-ordering there is always a distinct element, we get a different structure, a so-called *dense structure*. For example, the rational numbers or fractions reduced to lowest terms form a dense ordering under the "greater than" relation. A third structure is obtained if we add the following postulate: Consider any subset of elements of the set such that there is at least one element in the set which is $R$ to every member of the subset. In such a case, we insist that there is a member of the whole set such that (1) it is $R$ to every member of the subset and (2) every member of the whole set which is distinct from this designated member and is $R$ to every member of the subset is also $R$ to the designated member. In the mathematician's terms, we assert that every bounded subset has a *least* upper bound. The real numbers form a structure of this kind under the 'greater than' relation, whereas the rational numbers do not. For example, the set of rational numbers whose squares are less than two has upper bounds, but no least upper bound that is itself rational. On the other hand, the nonrational real number, the square root of two, is a least upper bound for this subset of rationals and the corresponding subset of reals whose squares are less than two.

Now suppose we have a nondense ordered set and we wish to define a "distance" function on pairs of elements of the set. A "natural" way to do this is to count the number of $R$ steps necessary to get from $a$ to $b$ and call this the distance from $a$ to $b$. For dense sets, whether merely dense or "continuous" like the reals, we can't do this because there is no "next" element after a given element in the $R$-ordering, and between any two points there are always an infinite (denumerable or nondenumerable) number of points.

If the space that is the space or spacetime of the world, the argument continues, were nondense, then we could define a metric on it by means of the "nextness" relation that characterizes the topological order of the set. But the spacetime of the world is actually dense, in fact a continuum in its local topological structure, so this means of defining distance between points is impossible. It is this that forces the invocation of transported rigid rods, etc., as the means of specifying the metric. But it is the appearance of these physical "intervening" devices to specify the metric which allows room for Poincaré-type stretching and shrinking fields, and hence for Poincaré-type conventionality arguments. The conventionality of the metric structure of spacetime, then, is to be attributed to its having a dense rather than nondense topological structure.

Let me suggest a few reasons why the denseness of spacetime is neither a necessary nor sufficient grounds for attributing conventionality to its metric structure.

First, the argument is vitiated by the fact that the very contrast between dense and nondense orderings, which seems so clear when one contrasts the integers with the rationals and reals using 'is greater than' for the ordering relationship, becomes difficult to comprehend for multidimensional spaces. If we use the orthodox topological definitions of dimensionality, see (II,B,6) above, we can indeed show that $n$-dimensional continuous Riemannian spaces have the appropriate dimension according to the formal definition. But it isn't at all clear what a three- or four-dimensional nondense space would be. Thinking of, say, the atoms in a crystal lattice won't do, for the dimensionality of this "discrete" space comes about from its embedding in an ordinary continuous space. It is far from clear what a three-dimensional discrete space would be. Take, for example, the important theorem that the dimensionality of a space is preserved by a one-to-one *continuous* mapping of the space into another, a theorem that allows us to characterize dimensionality as a "topologically invariant" feature of a space. What would this mean for discrete spaces?

Second, even in the simple case of a one-dimensional space, i.e., where the topological ordering among the points can be characterized by a single simple *R*-relationship, the connection between topological order and the metric structure is not so simple. True, in this case we *can* define a distance relationship among the points by "counting" the number of intervening points. But other metric structures are possible as well. Any non-negative-valued function on pairs of points, such that if $c$ is between $a$ and $b$ in the ordering then $\text{dist}(a,c) \leq \text{dist}(a,b)$ will give us a new metric on the space. So, while in this rudimentary case we *can* define a metric in terms of the topology, we *need not*. The possibility of alternative metrics and the threat of conventionality still remains. More importantly, we are concerned with the problem of the conventionality of the actual distance between actual points in physical spacetime, and even if the structure were such a simply-ordered discrete space, who is to say that the "natural" topologically-defined metric is the actual distance function? It would be as though we were making the dubious assumption that any two points "next to" one another were a distance apart equal to any other such pair of nearest neighbors.

Third and finally, we should note that while nondenseness, even where intelligible, is not sufficient to characterize nonconventionality for the metric, denseness isn't enough to guarantee conventionality either.

Suppose we believed that the distances between points in space, or the interval between points in spacetime, were immediately and "directly" apprehensible to the senses, with no need for intervening measuring instruments like rods, clocks, light rays, or freely-moving mass points. In this case I submit, the issue of conventionality of the metric structure would never arise, even if the space had all the continuous topological structure we attribute to our actual space or spacetime. The room for "alternative theories equally compatible with all possible observations" comes about not because of some feature of the local topology of the manifold, but because of an "epistemological remoteness" of the metric features of the space. So long as we need to use data confined to observations of a particularly limited kind to gain whatever access we have to the truth of hypotheses about a structure, hypotheses that seem to "outrun" in their content mere assertions about the observations, the possibility of alternative accounts for the data, and thus the possibility of "conventional alternatives," will arise. It is this special feature of the epistemology of geometry which gives rise to the conventionalist critique of a "pure empiricist" position, not any particular topological features of the spacetime, interesting though they may be.

# THE FRAMEWORK OF PHILOSOPHICAL
# PERPLEXITY

## 1. *Poincaré's Parable Revisited*

For the remainder of this Chapter I will be concerned with the most fundamental questions in the epistemology of geometry: (1) On what basic features do the arguments for a conventionalist position rest? (2) How are we to evaluate the two positions in the debate between conventionalists and the defenders of the empiricist position against the conventionalist critique? (3) What implications for the philosophy of science and epistemology in general can we draw from the specific debate about the status of geometric hypotheses?

We can set the stage for the discussion by paying a return visit to Poincaré's two-dimensional world, which he used to establish the conventionalist claim in the first place. Previously we had considered two groups of physical scientists inhabiting this world. Both groups were determined to map the intrinsic geometry of the world, and both had the same evidential basis from which to extrapolate their geometric hypotheses. One group took the evidence to support the thesis that the world was Lobachevskian and of infinite extent. The other claimed instead that the world was the interior of a finite Euclidean disk, and that the appearance of a non-Euclidean structure was due to the presence of distorting fields whose effects on rigid rods, upon transportation, and upon the propagation of light gave rise to the misleading appearance of a non-Euclidean world.

But instead of concentrating on this debate among the physicists, let us instead focus on the parasitical controversy inflicted on the *philosophers* of this world by their reflections on the physicist's quandary. So far I have delineated two schools of philosophy; we shall soon see that there are more than two contending positions that may be defended. One school, the conventionalist, has argued that one can easily construct two alternative total theories of the world which, although they are incompatible with one another, are so constructed that any evidence supporting one theory necessarily supports the other equally well, and any evidence tending to refute the one also tends necessarily to lead to the rejection of the other to an equal degree. But the only rational grounds for accepting or rejecting a physical theory are its conformities and nonconformities with the observational data, so we could not possibly perform any experiment leading us to accept one of these theories as preferable to the other. In such a situation our only recourse is simply to *choose,* by making a decision or adopting a convention, one theory or the other as the "true" picture of the world. But such truth is merely conventional truth, and a careful epistemological critique will keep us from misleadingly confusing such truth-by-convention with the kind of truth that receives its warrant from "genuine" confrontation with the observational facts.

The other school, the defenders of empiricism against the conventionalist critique, replied that the *appearance* of incompatible alternative theories equally compatible or incompatible with any possible experimental results is merely an appearance. Theories are hypotheses adopted to explain the observation facts. A correct doctrine of *meaning,* including a correct understanding of the way in which the terms of a theory obtain their meaning by being associated with experience, shows us that in the case we have been considering there are not, properly speaking, two alternative theories at all. Insofar as the two theories are demonstrably equivalent with respect to predicted experimental outcomes, they are the same theory. The appearance of a multiplicity of theories is due to a simple neglect of the possibility of ambiguity in the use of terms. Once considerations of the *meaning* of terms is brought in, we see that the situation is much simpler than the conventionalist claims. It is, of course, possible to have two incompatible theories that we cannot at a given time decide on between the basis of the observational data yet accumulated. But this situation, common to the frontiers of scientific investigation, shouldn't disturb the empiricist in the least. What is impossible is the existence of two incompatible theories immune to empirical discrimination and test in the face of all possible observations and experiments. Such a "pair" of theories is only a pair of expressions of a single theory, and there is no need to make a choice between them, conventional or otherwise, since there is no choice to be made.

The remainder of Chapter II will be devoted to investigating in a little more detail the features of the scientific situation which give rise to this *philosophical* controversy, the delineation in a little more depth of what the alternative philosophical groups intend to assert along with the presentation of some alternative stances different from these two, and an examination of the way this issue plays a crucial role in the philosophy of science and epistemology in areas outrunning the specific problem of the status of geometry. I won't pretend, however, to offer, a final "resolution" of the philosophical dispute, although I will try to offer what I take to be the best arguments for the contending positions and the most difficult objections they must face.

## 2. *Observational Basis and Theoretical Superstructure*

The claim is frequently made by philosophers of science that for any physical theory a distinguished subclass of the class of propositions entailed by the theory can be selected, which encompasses all of the propositions that follow from the theory which can be said to report results of possible observational and experimental tests. Geometric theories differ from theories in general only in that, within say the context of a presentation of a general-relativistic theory of curved spacetime, we can not only speculate about the existence of such a set of "observational" consequences, but we can also more fully characterize it.

Given the usual presentations of relativistic spacetime theories, what are the features of the world open to observational test? Our "idealized test apparatus" consists of idealized rigid rods capable of transportation throughout the spacetime, idealized clocks, light rays traveling through a vacuum, and point masses traveling unconstrained by any but gravitational "forces." With regard to the rods, we can experimentally determine whether two rods brought into coincidence are the same length. For the clocks, we can determine whether two clocks at the same place are tick-

ing at a uniform rate with regard to each other. And for the light rays
and freely-moving point masses, we can observe what classes of events
constitute their trajectories.

But to fully characterize the geometry of spacetime we need "facts"
that outrun those determinable by direct observation. We need to know,
for example, of four separated points, $a$, $b$, $c$, $d$, whether $a$ is as far from
$b$ as $c$ is from $d$. This requires knowing whether rods at a distance from
one another are congruent (i.e., the same length), not merely knowing
the congruence or noncongruence of coincident rods. Similarly, we need
to know time separations for events at a spatial separation from one another, and, of successive events, $a$, $b$, $c$, $d$, at a place, whether the temporal interval from $a$ to $b$ is the same as that from $c$ to $d$. This means
that we need to be able to establish synchrony for clocks at a spatial
separation from one another, and synchrony of a clock through one period with itself at a later time. If you wish, our observational data using
rods and clocks is all of a purely "local" kind; but it is "global" information that we need to fully determine the metric.

Switching to the use of light rays and free particles, a similar "leap
beyond the observable" is needed to establish the geometry on the basis
of the observations. We need to know what the null-geodesics and timelike-geodesics of the spacetime are. We can observe paths of light rays
and of free particles. Only an additional "assumption"—that light rays
travel null-geodesics when in a vacuum and acted upon only by gravity,
and that free particles in a similar situation follow timelike-geodesics—
can lead us to a knowledge of the geodesics themselves. The observation
is of the behavior of certain material entities, the theory is about the structure of spacetime itself. So two leaps beyond observation are necessary
to go from "all possible observational data" to "the nature of spacetime
itself." The observational basis is the local (coincidence) behavior of
material objects; but the nature of spacetime includes features of a global
nature and features not of the behavior of material objects but of the
structure of the spacetime itself.

As I have pointed out, it is frequently claimed that this outrunning
of the data by the theory, which we have noted in the case of geometry,
is a characteristic of theories in general. So why has conventionalism become an issue with regard to geometry, but not in the case of physical
theory in general? I can only offer a psychological explanation: First,
whereas the assumption of a designated subclass of the consequences of
the theory in question which exhausts all the theory's observational consequences is frequently made, in the case of geometric theories the spe-

cific characterization of this class of consequences is most easily made. It is, in fact, available to the philosopher offering an epistemological critique of the theory by inspection of the *physical* literature. It is as though in the case of geometry the observational-nonobservational distinction among consequences of the theory is worn on the sleeve of the physicists' theoretical speculation. Second, the allegation is frequently made by philosophers that for any full-blown physical theory an alternative theory can be constructed which saves all the same observational consequences. In the case of geometry, however, we can not only speculate about such alternatives but, as the Poincaré parable shows, actually construct them. Once again, geometry differs from theories in general in the forthright, explicit way the conventionalist thesis can be defended.

The step from a theory, which by inductive generalization upon actual observations characterizes the lawlike structure of the observables, to the theory that gives the full specification of the spacetime structure can, in Reichenbach's terms, be made by picking a hypothesis about the universal forces. But we have still not made it clear what such a hypothesis about universal forces amounts to. In particular, is it making a physical assumption about the world, or rather, is it simply adopting a set of coordinative definitions? The next part is in the nature of a temporary diversion from my main concern, but I shall return to this major issue immediately thereafter.

## 3. *Duhemian and Super-Duhemian Alternatives*

There is a well-known argument, of Pierre Duhem, concerning the possibility of crucial experiments in physics. It has sometimes been thought to be crucial to the sort of issue we are now facing. At this point it would be wise to rehearse his thesis and see just how related to, and how distinct from, the argument we have been attending to it is.

Duhem offers an important critique of the notion of a *crucial experiment* in science. Suppose we have a hypothesis, $H$, of general form, "All. . . .", and we wish to confront the hypothesis with an experimental test. Suppose, on the basis of the hypothesis, that we predict an outcome, $E$, for the test, but upon performing the experiment not-$E$ is found to be the result. Can we then affirm that the hypothesis $H$ has been conclusively found to be false?

It might seem, says Duhem, that the situation is just like that in mathematics where general propositions can be refuted by the construction of a single counterexample. For example, the assertion that all continuous functions are nondifferentiable at, at most, a finite number of points is rejected once and for all by Weirstrass's construction of a continuous function on the reals which is nowhere differentiable. But, continues Duhem, the assimilation of the method of crucial experiments in physics to that of counterexamples in mathematics is spurious. For in order to infer an observational result from the hypothesis $H$ we need a vast amount of additional theory, $A$, as well. This theory involves, among other things, the laws governing the behavior of the measuring instruments, etc. The observation of not-$E$ upon performing the experiment informs us only that something is wrong with our total theory, $H$ and $A$; but this failure of the theory to correctly predict the observational facts can be attributed either to the falsity of $H$, or instead, to the falsity of $A$. In other words, if we are motivated strongly enough to retain our belief in $H$ in the face of apparently confuting observational evidence, not-$E$, we can do so at the cost of making sufficiently broad changes in our "auxiliary hypotheses," $A$.

We can broaden Duhem's claim somewhat. In bringing observations to bear on the test of a hypothesis we need not only a set of auxiliary generalizations, $A$, but also a set of beliefs about the particular facts governing the experimental situation as well. For example, in "testing" the Newtonian theory of gravitation by observing the orbits of the planets, we need to assume not only the correctness of some additional laws, say Newton's mechanical laws of general dynamics, but some particular facts about the planets as well. We must assume, for instance, that the planets carry no net electric charge sufficient to introduce electrostatic forces among them which would make our observations unreliable for testing the form of the gravitational force law; because if there were such charges the planets would not, as we assumed, be moving solely under the influence of gravitational forces.

Notice that Duhem is certainly not asserting the skeptical claim that no experiment ever gives us any reason for doubting the truth of any hypothesis. We may very well have good reason, in some specific case, to blame the failure of a total theory to correctly predict the outcome of an experiment on a particular hypothesis $H$, rather than on the auxiliary theory or upon the specific factual assumptions made. His only claim is that such a "refutation by crucial experiment" cannot be assimilated to the mathematical method of counterexample where the existence of the counterexample is taken as logically sufficient to guarantee the falsity of the hypothesis.

Are Poincaréan conventionality arguments just a subspecies of Duhemian claims about refuting experiments? We must see clearly that they are not. Suppose we have been testing a hypothesis, $H$, and have found apparently-refuting experimental data. We recognize some questionable elements in our auxiliary hypotheses, $A$, and are puzzled as to whether we should attribute the unexpected results to the falsity of $H$ or to the mistaken assumption of a false auxiliary premise. Nothing in the Duhemian argument makes it impossible for us to perform a new experiment, designed to discriminate between whether the earlier results were due to the failure of $H$ or to the failure, instead, of the auxiliary proposition. Of course, in this new experiment we will perhaps need new auxiliary premises, and the situation will never be one where the data logically forces a decision upon us, as do counterexamples in mathematics. But that is not the point. Although the alternatives, not-$H$ and $A$, or $H$ and not-$A$, may be equally compatible with some particular experimental outcome, $E$, there is no reason to believe that they will be equally compatible or incompatible with *all possible experimental results*.

It is just this last feature that gives the Poincaréan situation its particular importance. The alternatives, non-Euclidean geometry and no universal forces or Euclidean geometry plus non-null universal forces, are supposed to be equally compatible with *all possible experimental results*. We can say that what Poincaré has given us is a "super-Duhemian" argument. It is this feature that leads Poincaré to his conventionalist claim, and just this feature that leads Eddington and Reichenbach to deny that one really is presented here with genuine scientific alternatives at all.

## 4. The Taxonomy of Controversy

The perplexity we have been led to by the considerations of the epistemology of geometry will be familiar to those acquainted with the history of philosophy since Descartes. Deciding what to say in the face of such examples as the Poincaréan parable is just one instance of the general problem: How should one react, philosophically, in the face of an apparent demonstration of the existence of incompatible hypotheses equally capable of explaining all possible observational results?

Consider, for example, Descartes's well-known invocation of the "malevolent deceiver." Imagine a world governed by an omnipotent being who possesses the power of control over our minds. With malevolent, or perhaps prankish, motivation he fills our previously blank minds with just the sort of sensory data that we ordinarily take to indicate the independently-existing material world causally responsible for our sensations.

Nonetheless, no such material world exists. Were we to live in such a world, and who is to say that we do not, our sensory experience as a whole would be indiscriminable from the sensory experience we actually have. On what rational grounds, then, do we base our disbelief in the existence of such a materially empty world containing a grand deceiver and our belief in the existence of "the external material world"?

Suppose we try to characterize very briefly some possible responses to this familiar argument. Having done so, we can carry over these responses to the case of the epistemology of geometry, to see how they look applied to this particular instance of the general problem. My examination of each possible position in turn will be cursory at best. To try to encompass in a short space the details of all the major philosophical positions that have been proposed, to outline carefully the detailed arguments brought forth to support them, and to rehearse precisely all of the major objections against them, would be to attempt the obviously impossible. All I hope to do is to put some perspective on the issues involved in the epistemology of geometry by offering what amounts to little more than a *classification* of the alternative philosophical options. To a certain extent, some of the deficiencies will be made up later in the book, because the fundamental epistemological issues outlined here, and their metaphysical and semantic consequences, will provide the overall theme of the discussion in later sections. Each of these sections will treat a distinct issue in the philosophy of space and time, and the pursuit of each specific problem will be found, sometimes by different routes, to lead back to some of these same basic concerns.

So in arriving over and over at the same basic philosophical perplexities, some additional details about "ways out" will come to the fore. Let me once again warn, however, that the reader who hopes for a resolution of the perplexities, and not merely some clarification of them and some indications of the particular form they take when the issues are those suggested by theories of space and time, will be disappointed.

Let us now look at some possible "resolutions" of the problem of Descartes's Demon. The treatment will be extremely cursory, but I will go at least a little more into detail when we carry over the schematism into the particular issues of geometric conventionality.

First I must make a preliminary distinction, of overriding importance, between "schools" of philosophy. The Cartesian argument, like that of geometric conventionalists, initially assumes that we can make coherent sense of the notion of "all possible observational consequences" of a theory. So we must first distinguish between those who accept this claim, and those who reject it maintaining that the very concept of "all possible observational consequences of a theory," as contrasted with "all consequences of a theory of whatever kind," is simply not well formulated or even intelligible. Let us call these philosophers believers and disbelievers in an observation basis. Within the class of those who believe in the intelligibility of the notion of an observation basis, we next draw the distinction between those who believe that the total theory is not in any sense "reducible" to its set of observational consequences, and those who believe that, given the distinction of observational consequences from consequences in general, the total theory must be said, in some important sense, to be reducible to its set of observational consequences. Let us call these schools the antireductionist and the reductionist, respectively.

The antireductionists again split into contending schools. First there are those we can call skeptics: The total theories do not reduce to their observational consequences. Yet we can formulate alternative incompatible theories with exactly the same observational consequences. But the only rational grounds for choosing one theory over another to be believed are functions of its observational consequences. We must believe, therefore, that we will frequently be in the position of facing alternative incompatible theories between which we simply have no rational grounds for choice whatsoever.

Next there are the conventionalists. Faced with the same situation as the skeptics, they argue somewhat as follows: We must choose one theory or another, even where we cannot discriminate among the theories in terms of differentiating observational consequences. The only option left for us is to simply choose, by a convention or decision, one of the contending theories from the class equally compatible with the data. We accept one of these theories as true by decision. If this truth is merely "truth by convention," so be it. There is no reason for lamenting this, since it is the best we can do anyway.

Finally there are those we might call apriorists, and they argue like this: Granted that there can be alternative, incompatible theories between which we cannot rationally choose solely on the basis of the conformity or nonconformity of the theory with the experimental data. But choosing which theory to believe in is never a simple matter of examining the conformity of the theory with the data anyway. Considerations of simplicity, systematic power, elegance, etc., are a fundamental feature of any scientific inference from data to hypothesis. In the cases we are considering, such methodological principles must once again be brought into play. Once we see their role clearly we will see that we do have rational grounds for preferring, i.e., believing to be true, one of the many contending theories.

The reductionist position, on the other hand, looks like this: Presented with the set of alternative, *apparently* incompatible theories, all with the same observational consequences, it seems as though we are in the position of having to make a choice, but having to make it on no grounds of rationality whatsoever. But this is merely an appearance. A proper understanding of what theories are, and what the theoretical assertions contained in them *mean,* will show us that all of the "alternative" theories are not proper alternatives at all—they are simply the same theory expressed in different ways. The appearance of multiplicity is simply the confused consequence of failure to note ambiguity of linguistic usage. Insofar as the theories predict the same observational results, they are one and the same theory, and there is no choice to be made.

Exploring any of these positions in depth is, once again, beyond my capabilities here. Let me give a cursory examination to their application in the Cartesian case, however, and then rehearse them one more time, in a little fuller detail, in the case of the debate about the epistemology of geometry.

## THE CARTESIAN DILEMMA

I. Believers in an observation basis: In the context of the Cartesian problem the observation basis is taken to be the immediate content of perception. By means of all too familiar arguments (illusion and hallucination, dependence of perceptual content on the state of the perceiver, etc.), it is maintained that one must distinguish between the objective features of material objects and the immediate contents of sense perception. Only the latter are open to "immediate inspection" by an observer, and assertions about them constitute the body of propositions knowable to him by perception alone.

ANTIREDUCTIONISTS: Proponents of the following three positions, while disagreeing with each other, all agree to the claim that whatever propositions about the existence and nature of the material world are, they are claims that cannot in any sense be "reduced to" or "translated without loss of content" into claims about the content of sensation. The objects and properties of the material world, whatever it is really like, are distinct from the contents and features of any observer's perceptual experience. To go from knowledge about the content of sensation, no matter how broad and how general, to knowledge about the material world and its features is to make an inferential leap.

*Skeptics:* Our empirical knowledge is constrained to knowledge about the immediate contents of perception. Even if we allow inductive generalization from our perceptual data, we obtain at best general knowledge of the lawlike structure of the world of sensory contents. But we can never obtain knowledge of the nature of the material world "beyond the veil of perception." We have no knowledge of this world by immediate perception, and to obtain inferential knowledge of it would require at some point "direct access" into the material world to establish the correct correlation between physical reality and our sensory contents. But again, such direct access is impossible. We must, then, remain in a perennial skeptical withholding of judgment about the nature of the material world, if our body of accepted belief is to be constricted to those beliefs for which we can offer rational justification.

*Conventionalists:* Let us agree that we can offer a variety of alternative hypotheses about the material world, all equally compatible with the possible totalities of sensory data (that the data are caused by an ordinary material world, that they are the result of the psychic manipulations of a malevolent demon, that they are the contents of an ongoing dream, etc.). To accept one of these hypotheses is as rational as to accept another. They are all "equally true." What we need is a correct understanding of *truth,* realizing that it is a matter of convention; but since conventional truth in these matters is the best obtainable, there is no reason to disparage it because of its conventionality. (It isn't clear whether anyone really held this position, but some idealist exponents of "coherence theories of truth" seem to come quite close.)

*Apriorists:* Let us grant that we can have a rational foundation for beliefs about the material world neither on the basis of direct perception nor on the basis of ordinary inductive inference from the perceptual basis. There are a priori arguments, however, that allow us to bridge the gap between the content of perception and material reality. Our belief in these principles of inference is itself founded in no way upon perceptual experience or induction from it. Descartes himself maintained this position, arguing that one could have good (indeed, indubitable) grounds for believing in the existence of an omnipotent, benevolent deity, and that the existence of such a deity guaranteed on a priori grounds the veracity of our empirical data as evidence of a material reality in general. Not, of course, that the evidence of the senses might not sometimes mislead us;

but only that we had as good a reason as possible for believing that such alternative general hypotheses about the origins of our sensory experience as the malevolent demon hypothesis and the eternal dream hypothesis were excludable on the grounds of this a priori argument.

REDUCTIONISTS: The disturbing conclusions one is led to by adopting any of the three preceding positions are unnecessary. All of these positions rest upon a fundamental error, for they all assume that some kind of leap is necessary to go from general knowledge about the contents of perception to knowledge about the material world. But consider any proposition about the material world and its features. What could serve as evidence for the truth of the proposition? Well, nothing but some complex of propositions about the content of the perceivable world of sensory contents. And what use could such a proposition have? Once again, it could serve no useful purpose other than to lead us about in the perceivable world, say to allow us to predict one complex of sensory contents as the likely successor to a complex of a specified kind.

Reflections such as these lead us to assert that what the proposition about the material world *says* is fully analyzable in terms of propositions about the world of sensory contents. Whatever propositions about the material world and its features may seem to say, a clear understanding of what it is for a proposition to have meaning, and what the meaning of a proposition is, will soon assure us that no mysterious inferential leaps are necessary. Once we understand what material object language-assertions say, we see that they are nothing but shorthand for complex assertions about sensations.

Now the body of assertions about sensory contents which even a simple material object proposition abbreviates may be quite complex. Even such a simple claim as, 'There is a table in this room,' may have to be "unpacked" or "translated" into a rather elaborate set of propositions about sensory contents. And some of the sentences of the translation may lead us to realize that not just *actual* sensory contents but *possible* ones count as well. To say that there is a table is to say what sensations an observer *would* have *were* he in a particular sensory position. To invoke such counterfactuals is to enter the realm of possibilia. But the complexity of the translation is not to be confused with the necessity of an inferential leap requiring some sort of justification, less we fall into skepticism or conventionalism; for translations, no matter how complex, simply unpack meanings and require no "justification" for the assertion of their truth.

II. Disbelievers in an observation basis: All of the above positions are the untenable results of an initial epistemological error. We need not be forced to choose between equally implausible lines of argument, that there is an inferential leap from knowledge of sensory contents to knowledge of the material world—with its almost equally intolerable sub-options of skepticism, conventionalism, or apriorism—or the grossly implausible claim that perfectly ordinary simple assertions about the features of the material world are intelligible only as incredibly prolix and complex propositions about the world of sensory contents.

The initial arguments designed to prove to us the "confinement" of our sensory knowledge to that of some mysterious realm of sensory contents are mistaken. Without these arguments, however, we would never have been prompted to distinguish "propositions about the immediate content of sensation" from "propositions about the material world." And without this distinction, the very possibility of "alternative hypotheses all equally capable of saving the empirical data" would never have arisen. The solution to the dilemma of choosing among any of the equally distasteful options explicated under (I) is to realize the implausibility and untenability of some coherent distinction between "consequences of a theory describing possible observations" and "consequences outrunning the realm of the observable." The detailed objections to each of the positions taken under (I) are of little importance compared to this general objection, which, if correct, makes the problem of a choice between the options under (I) appear as what it is: a pseudo-problem introduced on the basis of a fundamental epistemological error at the most basic level.

This sketch gives merely the barest bones of an argument that has been fleshed out in overwhelming detail in the last 400 years of philosophy. Worse, while stating the positions I have done nothing to provide the arguments used to support them. But rather than pursue this most fundamental issue of epistemology in detail, let me carry the schematism over to the case of the epistemology of geometry. Let us try to see how the arguments about the status of geometric beliefs resemble, and differ from, those about "our knowledge of the external world," and in the process look at the most crucial features of a few of the arguments that have been brought for and against each possible position.

## THE GEOMETRIC DILEMMA

I. Believers in an observation basis: The alleged observation basis in the case of geometric theory differs in some important ways from the "immediate content of perception" basis of the Cartesian dilemma. It consists in the kinds of knowledge expressible in propositions that are, first of all, about material objects and their behavior—rods, clocks, light rays, and paths of material particles—and secondly, the expressions of local congruence relations. We are supposed to be able to determine, observationally, whether or not a class of events constitutes the path of a free particle or light ray, and whether or not two rods (clocks) in coincidence are congruent to one another. All other relations among material objects, for example distant congruence among the rigid rods, and all the features of spacetime itself are in the "inferred" realm of theoretical features of the world.

One difference between this observation basis and that discussed earlier is its *publicity*. Whether or not two coincident rods are congruent, etc., are supposed to be questions answerable by observation in an intersubjective way. We do not, at least at this point, have the additional Cartesian perplexity that is the result of making the observation basis the *private* awarenesses of some particular observer, with the resulting tendency to lead to a solipsistic skepticism. Later, in discussing the reductionist approach, we shall see that the geometric case is not as immune to this difficulty as it first appears.

An additional feature distinguishing the present observation basis from that in the Cartesian context is the way it appears to be "extricable" from the theory as normally presented. What I mean is that in the more general philosophical context, the postulation and description of the observation basis is a philosophical construction imposed upon the naïvely accepted realistic theory of the material world. In the case of geometry, however, the consciousness of the problem of the distinction between features of the observation basis and those of the "theoretical superstructure" is again the result of philosophical reflection, but the reflection is much closer to the postulation of the physical theory itself. Perhaps this is only a psychological fact, due to the fact that, whereas the material object theory is a prescientific theory embedded in our ordinary language, the theories of non-Euclidean geometric structure for the world are self-conscious creations of science. Still, there is a sense in which the geometric theories "naturally" characterize their appropriate observation bases in a more direct way than that in which, by philosophical reflection, the "sensory content" observation basis is extracted from the prescientific conception of the material world.

ANTIREDUCTIONIST APPROACHES:    All of these positions agree that it makes sense to ask about such features of the world as the nonlocal congruences of rods and clocks and the structural features of spacetime itself. And, all agree, to ask these questions is not simply to inquire into the general or lawlike features of such things as the local congruence relations among rods and clocks, the paths of particles and light rays, etc. The propositions of the full theory and those of the evidential basis for it are no more reducible the one to the other than the proposition that someone is a murderer is translatable into the set of propositions about the evidence for his commission of the crime. Let me postpone the presentation of arguments for this and for the opposing claim to the discussion of the reductionist position, and move on to an exposition of some features of the various subpositions of the antireductionist position.

*Skepticism:* The skeptic has the virtue of epistemic humility on his side. Why should we assume, after all, that the world must be such that the resources available to human beings are sufficient to fully allow them to determine its nature, at least in principle? If the actual structure of spacetime is beyond reach of human knowledge, why should this surprise us? What a priori reason could we have, after all, to believe that all of the features that constitute the real nature of the world should be accessible to empirical investigation? Why shouldn't there be features of the world which are simply unknowable, in principle, to us?

At this point we should be careful to make a distinction of some importance. The opponents of the skeptic, in particular the reductionist opponents, are not denying the possibility of features of the world being immune to empirical investigation. What they are denying, however, is that we can meaningfully characterize just what these features are. The reductionist hardly denies the claim that there could be some features, "I know not what," of the world of which humans will, in principle, remain totally unaware; but, he claims, the resources for giving meaning to propositions and those for determining the truth or falsity of propositions are fundamentally the same. The objection to skepticism of the kind I am describing is not that it postulates unknowable features of the world, but that it attempts to say just what these features are. It is this that is objectionable, according to the reductionist. If the truth or falsity of a proposition, or at any rate its rational believability or disbelievability, are totally immune to human investigation, then it is misleading to speak of the proposition as being meaningful at all. If there are things we can't know, we can't meaningfully say what they are. I will return to this objection to the skeptic when I discuss the reductionist position.

There is one important difference between skepticism with regard to geometric theory and skepticism with regard to the existence of the external world that I might remark on. Hume has an interesting approach to skepticism. Frequently, he says, we are in the following situation: We believe a certain body of propositions. On philosophical reflection we realize that we have no rational basis for our beliefs, and indeed, that no rational justification for them could be given. Nonetheless, the *psychological* hold of the beliefs on us is sufficiently strong that, think as skeptically with the learned as we will, we cannot do otherwise in our unreflective moments, which constitute the bulk of our lives, than "act with the vulgar"; that is, to unreflectively believe the unjustifiable propositions. We can't help believing in the reality of the external world, as skeptical as we may be in our philosophical study reflecting on a possible Cartesian demon.

In the case of geometry, however, no such psychological necessity seems to hold. Poincaré thought that we would stick to the Euclidean theory, come what empirical data may. But this wasn't because of an ineluctable psychological hold the theory had on us, but because of a self-conscious choice of the descriptively simpler alternative. As a matter of fact, as we know in the light of general relativity, our preference might be for the curved spacetime alternative. But in any case, the choice is ours to make. If we are theoretical skeptics, we need not be psychologically-determined nonskeptical believers in one of the particular alternatives. Perhaps it is this openness of our minds to the various alternatives that makes the tenability of a genuinely skeptical position seem so much more plausible in the case of geometry than it does in the case of belief in the external world in general.

*Conventionalism:* The conventionalist maintains that the difference between the alternative theories is real, that no choice between them can be given a rational basis, either empirical or a priori, and that, since a choice must be made, we might as well make one "by decision." The basic problem with this view is the difficulty of characterizing it in a way that truly distinguishes it from skepticism on the one hand, or reductionism on the other. I think that Poincaré, although claiming that he espouses a conventionalist position, can usually on examination be found to be holding something much closer to what I shall describe as a reductionist view.

Take the problem of characterizing *truth* from a conventionalist point of view, and the accompanying problem of characterizing *reality*. Suppose our choice of a Euclidean as opposed to non-Euclidean world really is merely a decision to be made on our part. If our decision really is arbitrary, what useful purpose is served by calling the theory of our choice true and saying that reality has the features attributed to it by our theory? Shouldn't we either face up to the inevitability of skepticism (we simply can't know which theory is true or what the reality of the world is like) if we really hold to the position that the theories are genuinely incompatible alternatives, or to reductionism (the theories are equally true and attribute the same features to reality) if we slide instead into the reductionist position of taking the theories to be, properly speaking, equivalent? The conventionalist position seems to be too ambiguous to maintain its integrity as an independent position, and on reflection seems to turn rapidly into either skepticism or reductionism.

*Apriorism:* The difficulty with conventionalism that I have just brought out does not infect the aprioristic position. According to the apriorist the alternatives are really incompatible. To be sure, empirical evidence cannot decide for us which of the theories to believe, but this shouldn't disturb us because any theoretical or general belief always requires the invocation of "methodological principles of rational belief," which do not rest upon empirical experience for their acceptability as do principles of rational belief in science.

Before I continue I should carefully delineate what calling this position apriorist means. In particular, it does not mean that the holder of such a position is maintaining that we can a priori determine the nature of the world, nor that the theory of the world we rationally hold is immune to possible refutation by empirical experience. The apriorists I am discussing reserve a crucial role for observation and experiment in directing us to the theory we should believe; unlike the "pure rationalist" who, thinking that the correct description of the world can be inferred by deductive reasoning from self-evident first principles, allows no role for observational data whatsoever.

Some light on this kind of apriorism can be obtained, perhaps, if we reflect upon the role of so-called a priori probabilities in the Bayesian school of decision making. Suppose we must choose one of a number of hypotheses, $H_1, \ldots, H_n$, that exhaust all possible choices and are exclusive choices, i.e., we must adopt some $H_i$ and we cannot adopt two distinct $H_i$'s. According to a standard version of rational decision making, our procedure works like this: First, we know some utilities, i.e., gains and losses that accrue upon adopting $H_i$ when $H_j$ is actually the case. Next, we can perform experiments whose possible outcomes have various probabilities *given* the truth of the various $H_i$'s. Finally, let us suppose that prior to any experiment we can attribute various *probabilities* to the $H_i$'s, our "degree of partial belief" in their truth which we hold independently of any observation or experiment.

With these resources we act as follows: We perform the experiment and observe its outcome. Then, utilizing a theorem of the probability calculus, Baye's Theorem, we compute the a posteriori probabilities of all our hypotheses. We begin with $p(H_i)$'s, the a priori probablities of the $H_i$'s, and $p(O_j/H_i)$'s, the probabilities of the various outcomes relative to the various hypotheses being true. We compute $p(H_i/O_j)$'s, the new probabilities of the hypotheses relative to the experiment, as a matter of fact, resulting in some particular outcome $O_j$. Finally, using the new a posteriori probabilities, we compute an "expected utility" for choosing each $H_i$. This is the sum of the new probability of each $H_k$ times the utility of picking $H_i$ when $H_k$ is the case summed over all the $k$'s. Then we choose the $H_i$ for which this "act of choosing $H_i$" has the maximum expected utility.

Here is a rule for adopting a hypothesis. It takes into account the results of experimental observations in that the choice recommended depends upon the a posteriori probability of each hypothesis, and this depends upon the actual outcome of the performed experiment. But the rule also requires for its applicability an assignment of "probability of truth" to each hypothesis independent of any experimental outcome, the initial a priori probability assigned to the hypothesis.

So the apriorist argues like this: Granted that we cannot adopt a correct geometric theory of the world on the basis of a priori considerations alone, and granted that the empirical data cannot by itself determine for us which geometric hypothesis to adopt. But empirical data can never by itself force the adoption of a hypothesis. In any kind of reasoning from data to theory, a priori considerations of "initial plausibility" of the theories in question must be invoked. So they must be invoked in this case as well. All the alternative theories are equally compatible with the possible observational consequences, but they are not all equally a priori plausible. For a given set of observational consequences, the rational man will believe as true that one of the total theories compatible with these observational consequences whose initial a priori plausibility is highest. This way the hopelessness of skepticism, the unintelligibility of conventionalism, and the antirealism of reductionism are all avoided.

How reasonable is the apriorist's case? Not, perhaps, as convincing as the above makes it appear. To begin with, let us consider a particular case. Suppose a scientific investigator finds himself faced with the kind of observational data that could lead him to postulate a Euclidean world with imposed universal forces, or instead, a non-Euclidean world with no such forces. Which of these alternative theoretical possibilities seems to have greater "a priori probability"? It is hard to predict what the individual reader's response to this is likely to be, but one guesses that the overwhelming majority of readers would find the question unanswerable.

Now those who believe in the importance of a priori probabilities for scientific inference might have two quite distinct views about them. "Subjectivists" are likely to maintain that each investigator has some such a priori probabilities that he assigns to theories antecedent to experiment and observation. There is no criterion of rationality involved here, though, according to them. One has such a priori probabilities, and one believes or disbelieves partly in accordance with them, but there is no standard or rightness or wrongness for one's a priori plausibility assignments, over and above the constraint that the totality of one's such assignments should obey certain constraints of compatibility. From this point of view, if two investigators differ markedly in their a priori probability assignments, there is no question of one investigator being right and the other wrong. They simply differ.

"Objectivists," on the other hand, are likely to maintain that certain standards of rationality govern not only our responses to the results of observation and experiment, but also our initial assumptions of plausibility for theories as well. Since we are ultimately concerned with the question of whether the selection of one of the possible alternative theories is governed by standards of rationality, it is this position that would seem more interesting to us. The subjectivist position comes quite close to being simply conventionalism restated, with some details as to how the conventional choice is made superadded.

But what standards of rationality could govern choices of a priori probabilities when the alternative hypotheses are the kind we have been considering? It is quite clear that the kind of apriorism with which we are here concerned is rather different in kind from the Cartesian "apriorism through theology" we looked at earlier. What one could expect to find is something like this: We must have the alternative theories differing in their a priori plausibility. But, ignoring the comparison of theory with data, the only thing left to work with is the "internal structure" of the theory itself. This suggests that we look for some structural feature of theories, a feature that characterizes them independently of their compatibility with the experimental data and governs their degree of a priori plausibility.

Immediately several features come to mind. The *simplicity* of a theory is frequently alleged to be an important consideration in its scientific acceptability. So is its *generality* or its *explanatory power*. Everything else being equal, i.e., the alternative theories proving equally compatible with the observational data, simpler theories are preferred in scientific practice to more complex theories and more powerful theories to those less great in range and generality.

But now we can see why the apriorist position runs into difficulty in the cases we are considering. The argument is familiar, for it is just Reichenbach's distinction of inductive from descriptive simplicity, rehearsed (II,F,2). Consider the scientist presented with the task of drawing a curve relating two parameters with his data restricted to a finite number of observational "points." An infinity of curves pass through the determined points, but he must select one as appropriate. Here is where simplicity considerations are usually invoked. He should, it is argued— for want of any additional grounds for different action—choose the simplest curve through the points. Now already a number of different methodological problems appear. First there is the problem of characterizing just what simplicity is. A fully general "measure" of the simplicity of a theory seems a methodological pipe dream. Worse, the simplicity of a theory does not seem to be something preserved under transformations of the expression of the theory into a different but logically equivalent form. For example, a curve that appears simple given Cartesian orthogonal coordinates might have quite a complex algebraic description in terms of polar coordinates, but it is still the same curve. If simplicity is invoked as a rule of acceptability, the acceptability of a theory seems to depend on which of the logically equivalent means of expression we choose for it.

Again, simplicity is objected to since the connection between *simplicity* and *truth* seems so dubious. What the scientist wants, the argument goes, is the *correct* curve. But what kind of reasoning can we offer to convince someone that the simplest hypothesis compatible with the data is most likely to be the true hypothesis? And if no argument can establish such a connection, what reasons do we have in the first place for invoking simplicity in our mechanism for choosing hypotheses?

What is important for us, however, is that even if we accept the rationality of utilizing simplicity considerations in inductive inference, we still do not have the kind of a priori principle of selectivity we need for our case. The reason is this: Consider two curves of differing simplicity which the scientist is considering as candidates for his hypothesis. Both curves do go through all the presently-available data points, otherwise we would not still be considering them as candidates for the "true" theory. But the possibility of rejecting at least one of these curves on the basis of further experimentation remains open, for the curves do not make the same predictions for all possible experimental tests. It is this feature that makes the rule, "Adopt the simplest hypothesis compatible with the present data," seem somewhat more innocuous than might first appear. For even if we do make this choice, we are not stuck with it, in the sense that ongoing experimentation can "test" our choice and, conceivably, reject it in favor of some more complex hypothesis.

But it is just this possibility of further experimentation, discriminating between the alternative hypotheses, which is lacking in the case we have been considering; for the alternative total theories we have in mind are supposed to be identical in their totality of observational consequences. So if they differ in simplicity at all, it is not the same kind of difference that discriminates, for example, the straight line through two points from any of the parabolas that also pass through the points. For any straight line and any parabola there are at least some points on one but not on the other, but for our alternative hypotheses all the "points" confirming one confirm the other as well. The two hypotheses, as Reichenbach says, differ only in their *descriptive* simplicity, not in their *inductive* simplicity. And this difference appears more like that between the two *descriptions* of a straight line, one in Cartesian and the other in polar coordinates, than like that between a straight line and a parabola.

If the apriorist is to make out his case, it will have to be in a manner that does not consist of the observation that simplicity considerations are always necessary for inductive inference, and so there is no reason not to invoke them in the question of deciding between the alternative hypotheses here. For the use of simplicity as an a priori rule for selecting among hypotheses that are merely equally compatible with *present* observational data is quite a different matter from invoking it as a standard for selecting among hypotheses that have *all possible* observational consequences in common.

REDUCTIONISTS: I think that reflection upon the difficulties of the various antireductionist positions we have just considered will make considerably clearer and more intense the "pressures" toward adopting a reductionist position. If, as the reductionist claims, the various alternative theories under consideration can all be shown to be, properly speaking, a single theory expressed in differing ways, all the problems of motivating and rationalizing a choice vanish at once.

One can view the reductionist position as the result of expanding upon the following theme: The difficulties faced by the skeptic, conventionalist, and apriorist are the result of their belief that one can formulate alternative hypotheses that have identical observational consequences. But this is an illusion. To frame a hypothesis is to formulate a set of propositions. Propositions are not "marks on paper" or "sounds in the air," but meaningful assertions. We must reflect not only upon how we *test* a meaningful proposition for truth or falsity once we have it in mind, but how we formulate genuinely meaningful propositions in the first place. Once we have done this, we realize that the source of meaning is the same as the source of experimental adjudication—observation and experience.

A proper theory of meaning will have as one of its most fundamental consequences the awareness that the meaning of a scientific proposition is exhausted by the experimental or observational consequences derivable from it. If this is so, two theories with identical observational consequences can be seen to be identical in their meaning, whatever variation in their overt linguistic appearance there may be. To speak of adjudicating *between* these theories is as misleading as if one were to try to decide whether Newtonian mechanics expressed in English or Newtonian mechanics expressed in German was the correct dynamical theory. One can, to be sure, experimentally test the correctness of Newtonian mechanics, but one tests this theory, not a particular linguistic expression of it. Appearances to the contrary, the same thing holds in our case of "alternative" geometric theories. Euclidean geometry combined with a physics that postulates certain universal forces is the same theory as the corresponding observationally equivalent non-Euclidean theory with no universal forces. There is no deciding between them to be done.

This doctrine about meaning, and the accompanying thesis of the reducibility of theories to their observational consequences, is at least as old as Berkeley's rejection of "substance" as a meaningless metaphysical term; the detailed construction of reductionist programs has provided a great bulk of such various philosophical "programs" as phenomenalism, behaviorism, and logical positivism. Obviously I will not be able to do remote justice to all of the claims that have been made. Instead I will attempt to characterize just a few of the most fundamental features of the reductionist approach, emphasizing, as usual, not the general philosophical approach, but rather the way in which the controversy over geometry seems to serve as an illuminating paradigm of the more general philosophical issues. I will make some remarks about three issues: (1) Some comments about the doctrine of meaning fundamental to the reductionist claim. (2) Some consequences of the reductionist claim with regard to metaphysical and ontological issues. (3) Some observations about the most fundamental objections to the reductionist claim that one might raise.

(1) The reductionist requires a doctrine of meaning which has the consequence that two theories with exactly the same observational consequences are said by the doctrine to have the same meaning. Actually, this is a rather weak constraint upon a theory of meaning. It requires only that the notion of 'same meaning' be made clear, and this only when it is *theories* whose meaning is in question. Most reductionists go much further than this, however. They usually elaborate a doctrine that, first of all, attributes meaning to *terms,* not merely to propositions or theories, and secondly, attempts to say what the meaning of a term is, not merely to tell us when two terms have the same meaning. This more elaborate theory has a number of consequences that people have found objectionable. Let me very briefly indicate what such a fuller theory of meaning usually looks like, expand on the difficulties it is frequently alleged to have, and then make some comments upon the possibility of a "weaker" theory sufficient to ground the reductionist claims about "same meaning" for theories, but to which the standard objections are perhaps not applicable.

The standard theory usually begins with a distinction between observational and theoretical *terms.* This is a bifurcation of the vocabulary of the theory or, perhaps, its set of concepts. Observational terms are said to get their meaning by being associated directly with some element of sensory experience, say in the manner that 'red' is claimed to receive its meaning in our language by being associated with reddish sense-data. Given the meanings of the basic terms of the observational vocabulary, meanings are assigned to other terms by means of definitions. The defined term appears alone on the left-hand side of the definition, and on the right-hand side appears a meaning-assigning phrase containing only terms of the observational vocabulary or previously defined terms. These are Reichenbach's coordinative definitions—for example: 'null-geodesic' = (def) 'path of a light ray in a vacuum.' The theory is hierarchical in the sense that there is a natural order by which terms over and above the observational level are introduced.

One standard objection to this theory is the difficulty in practice of reconstructing these definitions by examining already existing theories with their already meaningful theoretical vocabulary. Frequently the following claim is made: For the "theoretical terms" of actual theories, no "definition" of each term in terms of the observational vocabulary alone can be given. But if the hierarchical theory of meaning were correct, we should be able to reach a definition of each theroetical term in terms of the observational vocabulary alone by following a chain of definitions whose ultimate result has the single theoretical term on the left-hand side and a meaning-assigning phrase, using only the pure observational vocabulary, on the right-hand side. The reason we cannot find such definitions, it is claimed, is that the network of theoretical terms is frequently introduced in a holistic way. To define one such term requires reference to the other theoretical terms of the theory, and no term-by-term elimination to the observational vocabulary is in fact possible.

The plausibility of this objection varies from case to case. In our geometric case, such "explicit" definitions of the theoretical terms seem fairly easy to obtain, witness the definition of null-geodesic above. In broader reductions, however, like that of "material object language" to "pure sensory-content language" the possibility of such term-by-term translations of the theoretical vocabulary does seem more doubtful. Try, for example, to fully "unpack" the meaning of 'There is a desk in this room' in terms of what "pure sensations" one does or would have in situations themselves characterized entirely in "pure sensory-content" language. What we should note here, though, is the extent to which this particular "term-by-term definition" theory of the meaning of theoretical terms outruns the initial claim, which was simply that, taken as a whole, theories have the same meaning if they have the same total sets of observational consequences.

Another alleged difficulty of the "term-by-term explicit definition" theory is this: Consider any of the propositions that give the explicit definition of a theoretical term in terms of the observational vocabulary. Since such propositions are definitive of the meaning of the term, the assertion of the propositions can only amount to asserting an analytic truth. Thus if 'null-geodesic' means 'path of a light ray in vacuum,' then to assert that all light rays in vacuums follow null-geodesics is to assert a proposition whose truth depends at most upon logic and the meanings of terms alone. Such propositions then are not revisable. At least they are as irrevocably true as the truths of pure logic. To deny such a proposition is simply to change the meaning of the theoretical term involved.

Now many of the philosophers we have been concerned with would have no objection to this consequence at all. For example, Reichenbach would welcome it as an explicit realization of the role played in science by coordinative definitions. According to him, some of the propositions of any theory must be analytic, for otherwise the theoretical terms would have no meaning attributed to them, and the possibility of meaningful synthetic propositions using these same terms would be blocked. Some propositions containing theoretical terms must be analytic, according to him, in order that any meaningful synthetic propositions using these terms exist at all.

Others, W. V. Quine for example, whose philosophical views in general are not totally out of sympathy with at least portions of the reductionist claim, would be strongly averse to the way the particular theory we have been discussing "localizes" the meaning-assigning features of a theory to some particular subset of its propositions, taking the remaining propositions as unabashedly "purely synthetic." Their doctrine is a more holistic one, and they argue that the best one could do is to look at a theory as a whole and examine its totality of observational consequences. To go further and distinguish the propositions of the theory into the two classes, analytic and synthetic, is to try to impose on our knowledge of theories an unjustifiable theory of meaning. The theory is there, and so are its observational consequences, but there is no need to try to sort the propositions of the theory into "meaning-attributing" and "fact-stating" ones; for realistically, all the propositions of the theory are on a par in that they all function as strands in a network that functions to bring together the observational consequences that "tie the network down at its periphery."

Some light can be thrown on these questions by asking how various philosophers would respond to changes in accepted scientific theory. Suppose we move from one theory to another with different observational consequences. All would agree that this constituted a genuine theoretical shift. Suppose we switch from one theory to another whose total set of observational consequences is identical to the former theory. The minimum condition we must impose to call a philosopher a reductionist is that he take the new theory as being, properly speaking, merely a novel formulation of, or perhaps a novel expression of, the same theory as was the older set of propositions. Philosophers like Reichenbach would insist that this "shift" be explicable as the change in meaning of some specific terms of the theories, and that one could localize the shift in meanings by locating which analytic coordinative definitions of the older theory are now denied truth in the new theory. The more holistically-inclined philosophers, on the other hand, would deny the possibility of localizing the meaning changes to specific terms, just as they deny the possibility of locating meaning-attribution in the theory to some specific propositions. What we must be clear about here, though, is that one could quite easily accept the most crucial reductionist claim, that same observational consequences imply the same theory, without accepting the particular theory of meaning-attribution familiar from Reichenbach and similar philosophers' works.

(2) I think it is important at this point to realize two truths: (*a*) Adopting a theory of meaning which takes the meaning of theoretical terms to accrue to these terms by means of the role they play in a theory that has observational consequences that can be expressed in an observational vocabulary whose meanings are, in some sense, "primitive," does not *in general* commit one to a doctrine that the theoretical entities and properties referred to or expressed by the terms in the theoretical vocabulary are somehow "unreal" or "convenient fictions." (*b*) On the other hand, the reductionist doctrine, as it is usually formulated, whatever particular theory about the meaning of theoretical terms accompanies it, does seem to have some such "instrumentalistic" or "fictionalistic" consequences with regard to the theory about the reality of theoretical entities or properties.

To see that a doctrine about the accrual of meaning to theoretical terms does not in itself commit one to the unreality of theoretical entities and features, consider a theory that looks something like this: Suppose we consider a theory with a number of observational terms and one theoretical term, say 'is electrically charged.' The theory will tell us about connections between being electrically charged and having various observational characteristics. Take the propositions of the theory and conjoin them into a single sentence. Consider the new sentence we get by replacing the term 'is electrically charged' wherever it appears by a property variable, 'is $F$,' and putting the expression 'There is a property $F$ such that . . .' in front of the now single-sentence theory. This is a modified version of replacing a theory by its so-called Ramsey sentence. The claim is, among other things, that what 'is electrically charged' *means* can be determined by looking at the Ramsey sentence. It means 'having a property $F$ such that, etc. . . .' So, in this sense, meaning accrues to the predicate 'is electrically charged' by the total role it plays in a theory whose other terms, being observational, have their meaning "primitively given."

But does such a theory deny that there *is* a theoretical property, that of being electrically charged, which has just the features attributed to it by the theory? Of course not; for the existence of such a theoretical property with these features is just what the modified Ramsey sentence asserts—"There is a property $F$ such that, etc. . . ." So a doctrine that claims merely that meaning accrues to theoretical terms through their role in a theory that contains observational terms does not in any way commit one to the claim that the entities and properties referred to or expressed by the theoretical terms are unreal, convenient fictions, or anything of the sort.

How would an exponent of the Ramsey sentence type of reductionism view our present dilemma? He would say this: We have two incompatible theories with the same observational consequences. What each of these theories really says can be seen by replacing each theory by its appropriate Ramsey sentence. You get these by (1) conjoining all the axioms of the theory; (2) replacing each theoretical term in the theory by a variable; and (3) putting an existential quantifier for each new variable in front of the new one-sentence theory.

Now each Ramsey sentence will have the same observational consequences as the theory it replaces. But, while the older theories were incompatible with one another, the new Ramsey sentence theories will not be. Of course, they may not be identical either. Each of the new theories will be devoid of theoretical terms, so the reduction in *meaning* of the theoretical to the observational terms will have been accomplished. The new theories will still postulate the existence of theoretical entities and properties as explanatory of the observational facts. But the new explanations will no longer contradict one another.

The reasonableness of replacing theories by their Ramsey sentences, the consequences of doing so, and the ability of this kind of reductionism to solve the dilemma of apparently incompatible theories all "saving the phenomena" are all questions still requiring much exploration. The reader should remember, though, while thinking about the more radical kind of reductionism which we shall next explore, that there is this more modest kind of reductionism which, while claiming that theories reduce in their meanings to theories stated purely in observational terms, allows the theories the legitimacy of their ontological claims about the reality of theoretical entities and properties.

But what about the more usual reductionist doctrine? I think that we can see that it does seem to have deep ontological consequences. Consider the two alternative theories with identical observational consequences which have constituted our paradigm pair: Euclidean geometry plus universal forces or non-Euclidean geometry with the universal forces set to zero. Take the propositions from these two theories respectively which assert (1) geodesics are straight lines and (2) geodesics are not straight lines. According to the first alternative (1) is the case, according to the second alternative (2). But according to the reductionist theory, the two alternative theories are not really alternatives at all, but merely alternative expressions of the same theory. To believe that there "really are" geodesics and that they "really are straight lines" or "really are not straight lines" seems to many philosophers grossly incompatible with the reductionist view about the equivalence of the two alternative theories, since one asserts some feature as holding of geodesics and the other denies it.

The way out of this dilemma for this kind of reductionist seems obvious. It is to reject the "realist" interpretation for the theoretical entities and properties referred to by the theories. There *really* are, according to this view, the entities and properties talked about in the body of propositions which constitutes the *common* set of observational consequences of the two theories. Since all of the alternative theories say the same things about these observational entities and properties, we can adopt a "realistic" attitude toward them without fear of being put in the apparently absurd position of maintaining that certain entities exist, maintaining that they have certain features, and yet maintaining that two theories are equally true which assert quite incompatible things about these entities.

I think we can see that whatever detailed theory of meaning we adopt as explicative of the reductionist position, this version of the reductionist doctrine cannot remain simply an epistemological and semantic theory. It has inevitable metaphysical and ontological consequences as well. Since the bulk of Chapter III and significant portions of Chapters IV and V are devoted to just these basic issues of ontology, I will forgo pursuing this issue any further here.

(3) The reductionist position as we have just seen it explicated has many obvious virtues. The dilemmas of the antireductionist are simply dissolved. At some cost, to be sure, since we must now abandon a naïvely realistic view of theoretical entities and properties; but this seems a small price to pay. Since the theory does not seem to require the particular term-by-term theory of meaning which usually accompanies it, many of the usual objections to reductionist positions, in terms of the difficulty of finding explicit definitions for particular theoretical terms and in terms of the illegitimate splitting of propositions into analytic and synthetic, fail to be convincing objections to the general reductionist position. They are interesting objections only to some of its particular forms.

Before proceeding it is important to once again emphasize the relation between reductionism and skepticism. The reductionist does not maintain that he has refuted "problematic skepticism" in general. Nothing that he says is incompatible with the claim that there may be features of the world totally beyond the possibility of human knowledge. What is impossible however, according to the reductionist, is the formulation of *meaningful* propositions that cannot be confirmed or disconfirmed empirically (unless, of course, they are merely analytic). The reductionist does not claim that the limits of our knowledge necessarily constitute the limits of the world, but only that the limits of our knowledge constitute the limits of what we can meaningfully assert or deny to be the case.

The most problematic feature of reductionist accounts, I believe, is one that is independent of the detailed formulation of the position; for it rests upon a questioning of the nature and status of the so-called observation basis. In the geometric case we have been considering, the specification of the observation basis is fairly clear, since it is "read off," as it were, from the presentation of the geometric theory. We obviate the problems faced by all of the antireductionist positions by the twofold process of (1) adopting a theory of meaning which attributes direct meaning only to terms descriptive of things and features in the observation basis, and allowing meaning to the theoretical structure only insofar as it ties together observational propositions and (2) differentiating between the "reality" of the observational world and the "mere convenient fictionality" of the theoretical entities and features.

But should we stop with an observation basis of, say, coincident congruence among rods and clocks and observable trajectories of light rays and free particles? Do these features of the world, in other words, constitute the genuinely irreducible observational basis? Once one has started on the job of reducing the meaning of theories, ultimately, to the meanings of their observational terms, and the reality of the world to the reality of the observational entities and their observational features, it seems very important to carry out this reduction as fully as possible. To carry out "one stage" of the reduction, reducing a portion of a theory to some subsection of it but not continuing until all reductions have been made to the "ultimate" observational vocabulary and "ultimate" observational class of things and properties, seems like trying to set up housekeeping halfway down philosophy's notoriously slippery slope. The only true equilibrium, one feels, is to be found in a reduction to the lowest and irreducible level.

But once one has acceded to this point, the difficulties are manifold and all too familiar. The only truly observable features of the world soon come to be identified with "pure sensory contents." If one takes these as "ideas in the mind," following Berkeley, one soon is faced with all the traditional difficulties of solipsism: The only world I can truly be sure of the existence of is the world of my internal consciousness, and the only meaningful assertions I can frame are those about its features; if their is a genuine external world I can never have good reason for believing that this is so, nor can I even intelligibly assert propositions about it, irrespective of the limitations on my knowability of their truth value.

If, instead, the sensory contents are the atoms of a "neutral monist" or "phenomenalist" picture, where mind as well as matter is taken to be a logical construction out of sensory contents, although we are not afflicted with the solipsism of the idealists we are certainly afflicted with all of the rather well-known difficulties of coherently characterizing the phenomenalist position.

In any case, it seems that the price to be paid for avoiding the difficulties of the antireductionist position is rather higher than we anticipated. It is not only the reality of spacetime features themselves which go, eliminated in terms of the real local relations among the idealized material objects that we use to map the spacetime, it is the whole of the "external world" in its usual sense which must be sacrificed, leaving us with a world whose only genuine realities are immediate sensory contents and their observable interrelations. This is not the place to argue the virtues and vices of traditional philosophical idealism and phenomenalism. I can only show how quickly these "deep" questions come to the surface once an initial reductionist stance is tolerated.

Reductionism requires, if not the "reduction" of the whole world to the world of the immediately apprehended, at least such a reduction for all in the world whose nature can be meaningfully discussed by us. A program that starts by denying real existence to a small fragment of the naïve realist's external world seems innocuous enough. But it soon leads us to conclude that nearly all our ontological commitments to the reality of the material world must be taken as *façon de parler,* properly intelligible only in terms of a commitment to the somewhat spare reality of the world of immediate contents of sensation. At this point the ontological prejudice in favor of those features of the world which, from the naïve realist's point of view, merely happen to be apprehensible directly by the human perceptual apparatus becomes, for many, unbearable.

II. Denying the observation basis: Having seen the uncomfortable positions we are led to both by antireductionist and reductionist fundamental premises, it should no longer come as a surprise that many philosophers have attempted to undercut the whole controversy by arguing that antireductionists and reductionists suffer from the common mistake of assuming the intelligibility of the notion of observation basis in the first place. If the distinction between propositions verifiable by direct observation and those establishable only by theoretical connection to the observational level is denied to begin with, the very questions about how to treat the "nonobservational" propositions epistemologically and metaphysically will never arise. Given a dichotomy resulting in perplexity, a quite natural response is to deny the intelligibility and coherence of the dichotomy in the first place.

The details of various attempts to cast doubt upon the observation-basis/theoretical-superstructure dichotomy will be beyond our scope. However, there are two distinct approaches, each designed to attack different versions of the observation basis thesis. If the observation basis is taken to be some set of entities and features in the "external" physical world, the attack usually rests on the gradualness of the transition in the world from observable to unobservable features, arguing that this actually slow transition is grossly misconstrued by the philosophical theories that make the distinction one that is hard and fast enough to constitute a distinction "in principle." If the distinction is instead between immediate sensory contents as observables and all physical features as unobservables, the claim is more likely to be one that attempts to show that the postulation of a world of immediately observable sensory contents is simply unintelligible given the assumption that all physical things and features are forever immune to direct observation.

What I would like to do, instead of presenting these critiques of the observational/theoretical distinction in detail, and offering the best arguments I can find for and against their soundness, is to offer just one argument designed to show that there must be something coherent in the postulation of the dichotomy, no matter how persuasive the critical objections may be to detailed unpackings of the observational/theoretical distinction. I am arguing, in other words, that as painful as the adoption of either the reductionist or antireductionist positions may be, the pressures toward their common assumption, that one can coherently characterize the notion of the observational consequences of a theory as a proper subset of its total consequences, are very hard to resist.

My argument is fairly simple. In the case of "alternative geometric theories" we feel philosophically perplexed. Faced with all the alternative philosophical attitudes possible to the results of Poincaréan parables, we feel strong tensions in trying to decide just what the parable would be taken to show. But there are other cases of alternative theories saving the data in physics where the response of the scientific community is clear and unequivocal. It is, in fact, to adopt a straightforward reductionist position, maintaining that what we have before us is an unquestionable case of two alternative presentations of the same theory and not two genuinely alternative theories at all. But if the reductionist interpretation of the situation is clearly the correct one, the assumptions of the reductionist position, in particular the assumption that one can clearly differentiate the observational from the nonobservational consequences of the theory, must also be justified.

So the structure of my argument is really like this: If the denial of the possibility of characterizing the observation basis of a theory is utilized to undercut both the reductionist and antireductionist positions we have been examining, it can only be at the price of maintaining that a scientific position universally adopted by the scientific community actually rests upon an incoherent notion. Naturally, this will only convince one of the intelligibility of the assumption to which the rejector of the notion of an observation basis denies intelligibility, to the extent that one is convinced that the scientists are correct in maintaining the position they unquestioningly and universally hold.

The scientific example I have in mind comes from the development of quantum mechanics, in particular from the universally accepted position that the Heisenberg and Schrödinger versions of the theory constitute not two alternative theories between which we must choose, but rather two alternative *representations* of one and the same theory.

In the quantum-mechanical framework one "prepares" systems by some physical operation. One leaves the system alone for a while and later makes observations upon it. The outcomes of various experiments performed at later times are attributed certain probabilities by the theory, which probabilities depend upon the preparation of the system, its internal constitution, and the experiment performed. Into the mathematical apparatus of the theory go two symbols characterizing the state of a system at a time and the observable to be measured by a particular experiment. The probabilities that an experiment at a time on a specific system will result in the various possible outcomes of the experiment are a mathematical function of the state of the system at the time and the "state" of the observable at that time. In the Schrödinger representation, the state of a system is initially fixed by the preparation of the system. But as time elapses the state changes according to a dynamical law and the internal constitution of the system. The mathematical representatives of the observables are operators, constant in time. So the changing probabilities of outcomes as the system dynamically evolves are attributed to its time-dependent state. In the Heisenberg picture the state of a system is characterized by its preparation, and is represented by a temporally unchanging mathematical form, the unchanging state-vector for the prepared system. The operators for each observable, on the other hand, are time-dependent mathematical structures, and the changing probabilities of outcomes are now captured in the dynamical evolution of the operators.

Since the evolution of the probabilities predicted by both the Heisenberg and Schrödinger representations are identical, depending in both cases upon the preparation of the system, its internal constitution, and the experiment to be performed, it is universally accepted that the Schrödinger and Heisenberg theories constitute merely alternative representations of the same theory. The moral is clear. "States" and "observables" are not themselves directly inspectable quantities. They are, rather, constituents of the theoretical superstructure of the theory. So the fact that states are time-dependent in the one picture and time-independent in the other is irrelevant. The correlations between preparations and probabilities of experimental outcomes is the same in both accounts, mediated in the same way by the dynamics determined by the internal constitution of the system. The unquestioning adoption of the view that the Schrödinger and Heisenberg representations are merely representations of the same theory is a clear indication that, at least in this case, the intelligibility of the observation-basis/theoretical-superstructure distinction is assumed by all. And if it makes sense here, why should it be denied intelligibility for physical theories in general?

Of course the observation basis of quantum theory is not the philosopher's ordinary basis of "directly perceivable entities and qualities," for the correlations that are preserved by the alternative representations may be among such "theoretical" properties as the magnetic moment of an electron. Yet the analogy is sufficiently strong, I think, and the phenomena "saved" in the switch from one representation to another sufficiently like the observational phenomena saved in the switch from one Poincaré alternative to another, to see the strength of this argument for making an at least plausible case for the ultimate characterizability of an observation basis.

If the above argument is correct, the attempt to outflank the dilemmas faced by reductionists and antireductionists alike by denying their common assumption of the intelligibility of the observational/theoretical distinction is far too facile. It appears that we shall have to take the questions raised by the Poincaré parable and the philosophical responses to it seriously. We shall have to worry about what constitutes the observation basis of theories, how we are to interpret the relation of theoretical superstructure to observational basis, how we should respond to the questions raised by the possibility of apparently incompatible theories sharing a common observation basis, and what consequences for our ontology and our semantics we are to draw from this epistemological puzzle.

So far our response to these issues has been inconclusive. And, alas, so it shall remain. But the remainder of the book will perhaps enlighten us a bit further, for we shall investigate in some detail additional ways these general philosophical issues impinge upon and are impinged upon by specific questions of physical theory.

# BIBLIOGRAPHY FOR CHAPTER II

## SECTIONS A AND B

PART 1

The reader interested in seeing Euclid's version of Euclidean geometry should see:
> Euclid, *The Elements*, Book 1, edited by T. Heath.


PART 2

The clearest treatment of Hilbert's study of the axiomatization of geometry is:
> Coxeter, H., *Non-Euclidean Geometry*.*

For a brief and elementary discussion of the independence of the parallel postulate, see:
> DeLong, H., *A Profile of Mathematical Logic*, chap. 2.

The brilliant book:
> Hilbert, D., and Cohn-Vossen, S., *Geometry and the Imagination*,

contains in its fourth chapter, primarily devoted to differential geometry, a clear, nontechnical development of important aspects of the non-Euclidean axiomatic geometries.

PART 3

The clearest treatment of Hilbert's study of the axiomatization of geometry is still Hilbert's orginial lectures, collected in:

Hilbert, D., *Foundations of Geometry.*

PART 4

A clear treatment of Gauss's theory of surfaces is contained in the first two chapters of:

Laugwitz, D., *Differential and Riemannian Geometry.**

PART 5

The Laugwitz book cited immediately above goes on to develop general Riemannian differential geometry from the classical standpoint. A comprehensive advanced treatment of general Riemannian geometry from the older point of view is:

Eisenhart, L., *Riemannian Geometry.***

More modern (and more readable) is:

Willmore, T., *An Introduction to Differential Geometry,***

which develops the older methods in some detail and also introduces the reader to some aspects of the more modern approach such as global problems, exterior algebra, coordinate-free characterizations of a space, etc.

PART 6

For affine spaces and spaces with nonintegrable metrics, see:

Weyl, H., *Space, Time and Matter.***

Affine spaces are also treated, along with the general theory of differential manifolds, in:

Auslander, L., and Mackenzie, R., *Introduction to Differential Manifolds.***

The book:

Spivak, M., *A Comprehensive Introduction to Differential Geometry,***

will, when it appears, be the best available source for an introduction to differential geometry and its generalizations from the modern point of view. Volume 2 of this work will also give a comprehensive survey of the classical works in differential geometry from the modern standpoint, thereby bringing together for the reader the older and newer approaches.

A useful brief survey of the basic concepts of differential geometry, useful to the beginner finding his way through the literature is the article "Riemannian Geometry," by Anderson, J., in:

Chiu, H., and Hoffman, W., *Gravitation and Relativity.**

Numerous excellent treatments of general topology exist. The interested reader might try:

Kelly, J., *General Topology*,**

or:

Hocking, J., and Young, G., *Topology*,**

or:

Alexandroff, P., *Elementary Concepts of Topology*.*

## SECTION C

PART 1

Minkowski's original paper "Space and Time," reprinted in
    Einstein, A. et al., *The Principle of Relativity*,
gives a clear exposition of Minkowski spacetime.
    A nontechnical development of the fundamental notions also appears in:
    Taylor, E., and Wheeler, J., *Spacetime Physics*.
In addition, any of the standard textbooks on relativity such as:
    Bergmann, P., *Introduction to the Theory of Relativity*,**
    Møller, C., *The Theory of Relativity*,**

or:

    Synge, J., *Relativity: the Special Theory*,**
give a clear discussion of the structure of Minkowski spacetime.

PART 2

An elementary discussion of Riemannian relativistic spacetime can be found
in:
    Eddington, A., *Space, Time and Gravitation*.
A condensed but fairly technical treatment is in:
    Einstein, A., *The Meaning of Relativity*.**
A straightforward development of Riemannian geometry as applied to general relativity can ge found in:
    Adler, R., Bazin, M., and Schiffer, M., *Introduction to General Relativity*.**
A somewhat more difficult, but more "modern," approach is available in:
    Anderson, J., *Principles of Relativity Physics*.**
Additional treatments will be found, of course, in any of the standard texts on general relativity, such as Synge's cited in the bibliography to (I,D,2).

## SECTION D

PART 1

Clifford's speculations can be found in his:
    Clifford, W., *The Common Sense of the Exact Sciences*.
See also the references to Clifford cited in:

Graves, J., *The Conceptual Foundations of Contemporary Relativity Theory*,

as well as the detailed examination of Clifford's views to be found in this text.

## PART 2

Numerous excellent treatises on general relativity exist. For example:

Synge, J., *Relativity: the General Theory*,**

and the works of Møller, Bergmann, Anderson, and Adler, Bazin, and Schiffer cited in the bibliography for (II,C,1) and (II,C,2). Chapter 3 of Graves cited above covers many of the features of general relativity with close concern for their philosophical relevance.

## PART 3

The problem of mapping the spacetime is reviewed in

Graves, J., *Conceptual Foundations*, chap. 3, sec. 10.

The problem is treated discursively and at some length in

Reichenbach, H., *Axiomatization of the Theory of Relativity*.

In addition, see the article "Gravitation as Geometry—I," by Marzke, R., and Wheeler, J., in:

Chiu, H., and Hoffman, W., *Gravitation and Relativity*.

## SECTION E

### PART 1

For the Greek use of geometry as the paradigm of a priori science, see the passages on knowledge as "recollection" in:

Plato, *The Meno*.

Aristotle's theory of science, including his treatment of the difference between beliefs based upon induction and those founded upon deduction from self-evident first principles can be found in:

Aristotle, *The Posterior Analytics*.

The reader unfamiliar with the rationalist mode in philosophy might want to look at:

Descartes, *The Meditations*,

or:

Spinoza, *Principia Ethica*.

The latter of these two works is an attempt at a comprehensive metaphysics and ethics modeled on the axiomatic treatment of geometry found in Euclid.

### PART 2

Kant's doctrine about time and space can be found in

Kant, I., *The Critique of Pure Reason*, "The Transcendental Aesthetic."

The translation by Kemp-Smith, N., is highly recommended. See also:

> Kant, I., *The Prolegomena to Any Future Metaphysics.*

An intensive exegesis of Kant's doctrine is available in:

> Paton, H. J., *Kant's Metaphysics of Experience.*

A contemporary study of the Kantian approach is:

> Bennett, J., *Kant's Analytic.*

For Russell's modified Kantian doctrine see:

> Russell, B., *An Essay on the Foundations of Geometry.*

## PART 3

The reader unfamiliar with the classical empiricist position would do well to read:

> Hume, D., *An Inquiry into Human Understanding,*

or his more extensive work:

> Hume, D., *Treatise of Human Nature.*

A contemporary statement of the empiricist position, as expounded by the logical positivists, and highly readable is:

> Ayer, A., *Language, Truth and Logic.*

## SECTION F

### PART 1

Poincaré's three seminal essays on the epistemology of geometry can be found together in:

> Poincaré, H., *Science and Hypothesis.*

### PART 2

Eddington's insightful critique of Poincaré is to be found in the prologue to:

> Eddington, A., *Space, Time and Gravitation.*

Reichenbach's extensive development of the line first espoused by Eddington is in:

> Reichenbach, H., *The Philosophy of Space and Time,* chap. 1.

The reader interested in Reichenbach's general approach should also refer to:

> Reichenbach, H., *Experience and Prediction,*

and his:

> Reichenbach, H., *Philosophical Foundations of Quantum Mechanics.*

This last book, while devoted to problems in quantum mechanics rather than to problems in space and time, contains some of the clearest expositions of Reichenbach's general philosophical line.

A brief and lucid treatment of geometry from the Reichenbachian point of view is contained in:

> Carnap, R., *Philosophical Foundations of Physics,* part 3.

In the course of developing his own non-Reichenbachian analysis of conventionality, Grünbaum treats the Eddington-Reichenbach views at some lengths in:

Grünbaum, A., *Philosophical Problems of Space and Time*, part 1.
This last work also contains a very extensive bibliography of works on the philosophy of space and time up to 1963.

## PART 3

On the conventionality of topological features of space see:
Reichenbach, H., *The Philosophy of Space and Time*, sec. 12.


## SECTION G

### PART 1

For the allegation that many of the issues of conventionality can be made clearer by a reference to the distinction between pure and applied geometries see:

Nagel, E., *The Structure of Science*, chap. 8,
and also Hempel, C., "Geometry and Empirical Science," reprinted in Feigl, H., and Sellars, W., *Readings in Philosophical Analysis*.

For a critique of the relevance of the pure-applied distinction to the problem at hand see:

Sklar, L., "The Conventionality of Geometry," *American Philosophical, Quarterly*, Monograph Series, vol. 3.

### PART 2

The suggestion that conventionality is related to the denseness of the manifold is first made in:

Riemann, B., "On the Hypotheses which Lie at the Foundation of Geometry," reprinted in Smith, D., *Source Book in Mathematics*.
This thesis is developed at length in:

Grünbaum, A., *Philosophical Problems of Space and Time*, part 1.
A lengthy, detailed, and very incisive critique of the thesis is presented in:

Putnam, H., "An Examination of Grünbaum's Philosophy of Geometry," in Baumrin, B., ed., *Delaware Seminar in the Philosophy of Science*, vol. 2.
Another presentation of Grünbaum's position, reprinted from Volume Two of the *Minnesota Studies in the Philosophy of Science*, along with an extensive reply to Putnam, appears in:

Grünbaum, A., *Geometry and Chronometry in Philosophical Perspective*.
See also:

Sklar, L., "The Conventionality of Geometry."

## SECTION H

### PARTS 1 AND 2

There is, of course, a vast array of literature on the problem of the characterization of theories in terms of observational basis and theoretical superstructure. For a start, the interested reader might consult:

Hempel, C., "The Theoretician's Dilemma," in Feigl, H., Scriven, M., and Maxwell, G., eds., *Minnesota Studies in the Philosophy of Science*, vol. 2.

Scheffler, I., *The Anatomy of Inquiry*, part 2.

For a critique of the observational/theoretical distinction, a good source is:

Hanson, N., *Patterns of Discovery*, chap. 1.

### PART 3

For Duhem's thesis see:

Duhem, P., *The Aim and Structure of Physical Theory*, chap. 6.

For a discussion of the way in which the conventionalist argument in geometry differs from the original Duhemian argument, see:

Grünbaum, A., *Philosophical Problems of Space and Time*, chap. 4.

### PART 4

To even survey the literature on the "veil of perception" and the "justification of our knowledge of the external world," would be to read a great portion of Western philosophy, at least since the seventeenth century. The reader unacquainted with this philosophical theme might look at some of the following, however:

Descartes, R., *The Meditations*,

for a clear statement of the problem, the puzzle of the malevolent demon, and a proposed apriorist resolution to the problem;

Berkeley, G., *Three Dialogues*,

Hume, D., *A Treatise of Human Nature*,

and:

Price, H., *Perception*,

for the development of the phenomenalist reductionist account.

For a critical examination of the underlying assumption of a "sense datum" observation basis, see:

Austin, J., *Sense and Sensibilia*.

For a survey of the reappearance of these issues in the context of the philosophical scrutiny of scientific theories, the references cited to (II,H,2) will provide a good introduction.

For an introduction to the Bayesian approach to theory acceptance see:

Jeffrey, R., *Logic of Decision*.

For a critique of term-by-term meaning reduction in science, see the Hempel item cited in the bibliography to (II,H,2).

For a discussion of the alleged nonviability of the positivist's analytic/synthetic distinction, see:

Quine, W., "Two Dogmas of Empiricism," in his *From a Logical Point of View*;

and:

Putnam, H., "Analytic and Synthetic," in Volume 3 of the *Minnesota Studies in the Philosophy of Science*, edited by Feigl, H., and Maxwell. G.

On the Ramsey sentence approach to the meaning of theoretical terms, see:

Ramsey, F., *The Foundations of Mathematics and Other Logical Essays*.

Also, consult the Hempel item cited in the bibliography to (II,H,2), and:

Lewis, D., "How to Define Theoretical Terms," *Journal of Philosophy* 67 (July 9, 1970), and:

English, J., "Underdetermination: Craig and Ramsey," *Journal of Philosophy* 70 (Aug. 16, 1973).

# *ABSOLUTE MOTION*
# *AND SUBSTANTIVAL SPACETIME*

# SOME PRELIMINARY REMARKS

The problem I shall probe in Chapter III has an origin and development as curious as any in philosophy or science. The so-called issue of absolute space has exercised physicists and philosophers for four centuries, and not just any physicists and philosophers. The roster of names of those who have attended to this problem seems like a roll call of scientific genius in the Western world—Newton, Leibniz, Huyghens, Berkeley, Mach, Einstein, and Reichenbach are just a few of the greatest figures who have confronted this issue.

The problem itself has a flavor that is unique. No single issue has brought together in such an intimate way the experimental results of the observational scientist, the speculative resources of the scientific theoretician, and the critical acumen of the philosophers. Yet the problem has been sufficiently subtle, with the basic threads sufficiently entangled, that no other problem seems to have led as many men of unquestionable genius to say as many things that seem incoherent or grossly implausible in retrospect.

At the end of Chapter II, in the consideration of some of the consequences of the reductionist approach to the epistemology of geometry, we saw questions of ontology coming to the fore. If, as the reductionist claims, the total meaningful content of a geometric plus physical theory is contained in its observational consequences, and if "alternative" theories with identical observational consequences seem to attribute incompatible features to the structure of spacetime itself, the conclusion we seem led to is that the attribution of "reality" and "real features" to spacetime itself should be taken merely as a conveniently fictional way of talking. Following through the reductionist claim, then, we seem to end up with the conclusion that genuine reality should be attributed only to the observable things of the world and their observable features. The propositions of the observation basis are to be taken at face value. If they say that something exists and that it has some specific features, and if we believe them to be true, then we should take ourselves to be committed to a world with such entities and features. But the propositions of the theoretical superstructure are not to be read in this "naïve" way. Rather, they are to be taken as shorthand for complex assertions of the kind suitable to the observation basis. Or, more holistically, the whole of the theory is to be viewed as merely a convenient systematization of its totality of observational consequences.

The traditional metaphysical issue about space is whether we should view space itself as an entity possessed of a structure, or whether we should view spatial language with its prima facie reference to spatial entities and their features as merely a misleading way of talking about ordinary material objects (things, light rays, etc.) and a systematic structure of relations they bear to one another. It seems pretty clear, to begin with, that the positions taken in this debate and the arguments used to support them are going to be based, in a way now familiar, upon underlying epistemological considerations.

If the whole foundation of the debate were this consideration, however, this chapter would be merely an intensive further probing into issues already outlined in Chapter II. The reason the whole pattern of discussion takes on a novel aspect is because of a most curious change in the metaphysical debate due to some important arguments of Isaac Newton's. Basically Newton argues this: It might seem that the issue of the substantival, or merely relational, nature of space hinged solely on philosophical argument. I can show, however, that there are *experimental or observational facts* that resolve the debate. In fact, they resolve it in favor of the substantivalist or, in any case, one who shares with him a rejection of the relationist position as totally inadequate.

Newton's argument changes the whole atmosphere of the philosophical debate. If he is correct, we cannot resolve the ontological issue of the "reality" of space by applying to our common-sense knowledge of the world the critical apparatus of the philosopher. We must, in addition, take account of particular fundamental features of the best available contemporary science; and as this science develops and changes, we must perhaps change our metaphysics accordingly. And so, it has been alleged, electromagnetic theory, special relativity, general relativity, and perhaps other actual or possible scientific theories must all be examined in detail in order to correctly adjudicate the issue of the reality of space.

Now Newton's argument is actually multistaged. First there is an argument from the observational facts to the necessity of postulating absolute motion of a specific kind, absolute acceleration. Then there is an argument from the existence of absolute accelerations to the inadequacy of the relationist theory of space. Finally, there is the question of going from a refutation of relationism to the establishment of the alternative "correct" theory. Each stage of the argument, as we shall see, rests upon various additional premises and principles. Some of these are explicit in Newton's argument and some implicit, waiting for us to draw them out. This pattern of a complex multiple argument resting at least partly on hidden assumptions is repeated in the arguments of Newton's critics and successors, so the job of untangling the various strands in the debate is nontrivial, to say the least.

In broad outline, the investigation looks like this: First I will try to state the fundamental claims of both substantivalists and relationists, sticking to the basic philosophical positions and the philosophical arguments for them. Then I shall try to unpack as clearly as possible Newton's scientific argument for absolute space and the critical objections to it. After this I will pursue the changes in scientific theorizing through the four centuries succeeding Newton's work, at each stage examining some features of novel scientific theories and some reasons why physicists and philosophers thought these scientific results relevant to the metaphysical issue. Finally I shall return to "pure philosophy," seeing if I can in a manner similar to the final portions of Chapter II lay out in some schematic order the various metaphysical positions and the underlying arguments for them.

In anticipation I might remark that one of the deepest problems to face is Newton's argument for a doctrine of the reality of space. The argument proceeds through an attempted empirical demonstration of the existence of absolute motion. At each stage one must reflect on at least two basic questions: (1) Does the theory in question really support an acceptance or rejection of the existence of absolute motion? (2) How should we take the existence or nonexistence of absolute motion as decisive in determining our metaphysical attude toward the reality of space?

There is one important complication one should always be aware of.

Newton believed in both the reality of space and in "absolute time." As we shall see, one cannot, even prerelativistically, ignore temporal features of the world in trying to come to a conclusion about the reality of space. Naturally, once one has moved to the relativistic theories, it is spacetime whose reality or "absoluteness" is fundamentally at issue. This shift from the prerelativistic to the relativistic doctrines about space and time will bring with it interesting changes in the structure of the argument. But, as we shall also see, the mere invocation of the relativistic approach will not solve the basic philosophical puzzles, although it will certainly put the attempted resolution of these philosophical problems in a novel context.

# THE TRADITIONAL ONTOLOGICAL DISPUTE

## 1. *Substantival Views*

The first position I shall examine can be called the substantivalist. Actually, this designation is somewhat misleading, since it is not just the postulation of the independent reality of space as a kind of substance which I include under this title, but the postulation of an independently existing temporal structure or, relativistically, an independently existing spacetime as well. The common element in each of these views is the claim that the structure in question can be said to exist and to have specified features independently of the existence of any ordinary material objects, where the latter phrase is taken to include even such extraordinary material objects as rays of light, physical fields (electromagnetic fields, for example), etc. According to a Newtonian substantivalist, for example, even if there were no matter in the universe whatever, there would still be space with its standard three-dimensional Euclidean structure, and there would still be "instants" of time which together form a temporal order. Finally, there would still be those ordered pairs of places in space and instants in time which constituted event locations—even if there were no events at all.

Within the Newtonian framework, there is a clear analogy between space and ordinary material objects. Space is an object, although one of a quite idiosyncratic sort. It is infinite in extent, Euclidean and three-dimensional in structure, and it persists through time—as do ordinary material objects—but its persistence is characterized by its total unchangingness through time. It cannot be immediately apprehended by the senses, any more than can a magnetic field; but like the field, as will be seen, it can be indirectly inferred from observable phenomena by a legitimate scientific inference. The analogy of the temporal structure of instants to an object is a little less clear, obviously, for the temporal structure cannot be said to persist through time, as can ordinary material objects of substantival space. Nonetheless, the "substantivality" of the position is clear enough in that the claim is made that the temporal instants and their structure would exist whether events occurred at any of these instants or not.

The points of substantival space have distance between them, for the set of points has the full Euclidean structure. The events of the temporal structure have temporal separation between them. One can indeed use material objects to *measure* these features of the space and time themselves, but one must never take these, perhaps inaccurate, measures of spatial and temporal intervals for the intervals themselves. Measuring rods can expand and shrink and can move about in substantival space unbeknown to us, and clocks can change their rate of ticking, but "absolute" space retains its unchanging structure and the intervals of time their real temporal magnitude.

What is the relation of matter to the substantival space and the substantival time? The answer is usually stated in a "container" or "arena" view. Objects are "in" space and events take place "in" time. Now we mustn't push the metaphor of containments too far, for objects are not in space the way a present is in a box. Rather, the meaning is unpacked something like this: Consider the relation of coincidence defined on idealized point-unextended objects. No two point material objects can be coincident. But every point material object, out of which extended material objects are constituted as wholes and made up of their parts, is coincident with some point of substantival space. So the set of point material objects which constitutes an ordinary extended material object can be said to be in space in the sense that the object as a whole is coincident with a set of points of the substantival space.

Analogously for time, utilize the notion of temporal coincidence or simultaneity. Many concrete events may be simultaneous with one another, and every material event (the death of Caesar, for example) is simultaneous with some instant of time.

As one moves from the substantival metaphysics underlying the orthodox Newtonian picture to that appropriate to a relativistic physics, one can see that details of the theory can change markedly but the underlying theme of the independent existence and structure of space and time—or rather spacetime now—remains the same.

For special relativity the appropriate structure is that of Minkowski spacetime. Here the elementary constituents are no longer points and instants, but rather event locations. They have no definite spatial separation or definite temporal separation since these exist only relative to a particular state of motion chosen as reference frame (II,C,1). But they do, of course, have their invariant interval between them taken in pairs. Now as Minkowski spacetime is the container, it is events that are the contents. The ideal unextended material entity is the concrete event, and the extended material entity is the "world-history," consisting of the "history" of an object taken as the set of happenings that constitute its life. But the rejection of objects for histories, and of space and time for Minkowski spacetime, still leaves much unchanged from this point of view. Spacetime has its independent existence and structure, as space and time did before. So, for example, a spacetime totally empty of all events is perfectly intelligible in this view. And, as before, the spacetime remains totally immune in its structure to any changes due to the emplacement in it of any material events or their interactions. As in the Newtonian theory, spacetime is still the passive arena or container of the world. The notion of container is now to be unpacked, of course, in terms of the claim that every material unextended event (say, an idealized collision of idealized point masses) is coincident with some event of the spacetime.

Shortly (III,D,3), I shall examine another spacetime, neo-Newtonian spacetime. Once again, its postulation is compatible with a metaphysical substantivalist position. Event locations form the basic constituents of this spacetime, as they do of Minkowski's. But between any pairs of events a definite, invariant temporal separation exists although an invariant spatial separation does not. The spacetime has additional invariant structure as well, whose details I shall leave to later. It will suffice for now to point out that it shares with Minkowski spacetime the possibility of being given the interpretation of an arena of events and, like Minkowski and Newtonian spacetime, the feature that the structure of the arena is independent of what takes place within it.

When I get to general relativity the picture is slightly more complicated. But, as we shall see, not in such a way as to make the substantivalist account prima facie untenable. As I described it in (II,C,2), the spacetime appropriate to general relativity is that of the Riemannian generalization of the four-dimensional pseudo-Euclidean Minkowski spacetime. Once again event locations are the basic constituents and, as in Minkowski spacetime, the invariant structure of the spacetime is characterized by the intervals along curves between pairs of event locations. But now, of course, the spacetime can be "curved," i.e., can have intrinsic metric structure different from the "flat" structure of the Minkowski spacetime.

There is one crucial difference between the spacetime of general relativity and those we have just looked at. As I have noted (II,D,2), the structure of spacetime is lawlike-connected by the field equations with the distribution of nongravitational mass energy contained in the spacetime. In this way, the spacetime of general relativity does not have the "total immunity to causal effect" that was shared by the spacetimes appropriate to Newtonian mechanics and to special relativity. I have already noted that to speak of the mass-energy distribution as causing modifications in the spacetime might be misleading, and it is better simply to assert a lawlike connection between the two structures—the intrinsic geometry of the spacetime and the distribution of nongravitational mass-energy contained in it. But in any case, in the general-relativistic picture one has for the first time the possibility of *different* spacetime structures serving as arenas for the action in the world.

I might note at this point that general relativity, despite maintaining a lawlike connection between the structure of spacetime itself and the structure of the mass-energy distributed in it, also allows for the possibility of spacetimes totally devoid of any material happenings. In fact, as we shall see, the theory is compatible with the possibility of a variety of nonequivalent spacetimes with varying metric structure each being possible spacetimes of a world empty of nongravitational mass-energy.

What arguments can be given in favor of the substantivalist position? The dialectic of the argument between substantivalists and their opponents is curious, but it is one not unfamiliar to philosophers from other disputes. It runs something like this: The substantivalist position seems the "natural" position in that it is the position one "reads off" the discourse of ordinary language and ordinary science. One talks in ordinary discourse about the space in a room, or the empty space between galaxies. One ordinarily thinks of objects as being "in" space and events "in" time. Scientifically, one talks about the structures of space and of time, or of spacetime. In other words, it looks as though a substantival view—a view of space, time, or spacetime as real, existent, and independent in their existence—is presumed in everyday ordinary and scientific discourse.

Now the antisubstantivalist is the one on whom rests the burden of proof. It is he who must show that the apparent commitment to the existence of spacetime itself is merely an illusion. The argument usually takes the form of trying to show that one can make intelligible sense of all the ordinary and scientific discourse that seems to commit one to the existence of spacetime, without postulating any such entity as spacetime, and further, of trying to show that the postulation of spacetime, besides being otiose, is philosophically illegitimate.

As I have remarked, the pattern of argument is not unique to this situation. Consider the debate over the existence of numbers as abstract objects, for example. In ordinary arithmetic we seem to talk about numbers and their interrelations. The mathematical Platonist takes this talk seriously. He argues that one certainly cannot reject all arithmetic discourse as nonsense, that arithmetic discourse commits us to talking about numbers as abstract objects, and that hence some abstract objects, the numbers, exist. The arithmetic nominalist is then faced with the twofold task of showing (1) that he can make intelligible sense of all of ordinary arithmetic with its apparent reference to numbers as abstract objects without postulating any such abstract entities, and (2) that the postulation of such entities is philosophically abhorrent.

The only unique feature of the debate over the reality of spacetime comes about through the rather peculiar reply to the antisubstantivalist initiated by Newton. For in this case the substantivalist replies not just with a philosophical critique of the antisubstantivalist's arguments, but with an attempt to buttress the substantivalist's position by a *scientific* argument resting upon allegedly crucial observational facts.

There is one variant on the substantivalist position which will become important later on and which I might anticipate at this point. So far the substantivalist claim has been only that among the entities of the world are space, time, or spacetime. The views I have been looking at all take it for granted that one has not exhausted the contents of the world in referring to spacetime, for, of course, there are also the ordinary material objects which are its contents as well. The more extreme view I would like to note now, however, is that all there is is spacetime. From this point of view, the ordinary material contents of the world should be viewed as "pieces" of spacetime itself.

The view is not brand new. But its scientific plausibility is limited to the period after the discovery of general relativity. Plato perhaps, in the *Timaeus*, took *chora* or space as the "matter" of the world, and there are some features of Descartes's "plenum" view of space which again suggest the idea of space as the total "stuff" of the world. As noted earlier (II,D,1), one of the earliest speculations about the possibility of a non-Euclidean world, that of W. K. Clifford, identified the ordinary matter of the world as small, mobile regions of highly curved space. Also noted (II,D,2), was Einstein's speculation about matter as singular regions of spacetime, in his attempt to derive the dynamical law of motion of general relativity from the field equations alone.

The fullest development of the program to identify ordinary matter with the spacetime itself is the program of geometrodynamics of the physicist John Wheeler and his collaborators. This program is, in part, a systematic attempt to make unnecessary the postulation of any entities and features over and above spacetime and its structure in the full description of the world, ordinary material things and their histories included. I will not expound at all upon this here, since I will sketch some important features of this program below (III,E,5). At this point I will make only the following preliminary observation: The identification of all of the material world with the structured world of spacetime is not to be interpreted as the linguistic trick of simply replacing objects by the region of spacetime they occupy and some novel "objectifying feature"—say replacing 'There is a desk in the (X,T) region,' by 'The (X,T) region desks.' The *scientific* program of reducing matter to spacetime is rather more on the order of the scientific program of reducing material objects to arrays of their microscopic constituents or identifying light waves with electromagnetic radiation. In the reduction, the assertion of the existence of a material object at some spacetime location is to be shown reducible to the assertion of some spacetime feature holding in the spacetime region, say its having a certain intrinsic curvature over the region. It is only general relativity with its postulation of a "dynamic" variable curved spacetime which makes this plausible, and we shall later see how this program might be attempted. I need only note now that the existence of such programs gives us two distinct substantivalist positions: (1) the rather modest position that demands only that spacetime be accepted as one of the many real existents constituting the world and (2) the more ambitious program which asks one to believe not simply that spacetime exists and is real but that it is the only existent of the world.

## 2. Relationist Theories of Spacetime

The view I shall put up against the substantivalist positions just outlined is that of relationism. According to the relationist, the postulation of space, time, or spacetime as entities existing in their own right with structures properly attributed to them in reality is simply a confusion. It is just one more case of philosophers being misled by the surface appearance of ordinary and scientific language. In the world "really" are material entities—material objects if one has a prerelativistic point of view—and material events in both the prerelativistic and relativistic versions of the relationist position.

Now the relationist is not, of course, asserting that all talk of spatiality and temporality is nonsense. The claim is only that, properly understood, all spatial and temporal assertions should be seen not as attributing features to space or time or spacetime, but rather as attributing some spatial, temporal, or spatiotemporal relations to material objects. To think of, say, space as an *object* is to think that because the relation of brotherhood can hold between male siblings, there must be in the world not only people and their relations but some mysterious entity, the brotherhoodness, as well.

As usual, and as should be familiar from my remarks in Chapter II about the ideal nature of the observation basis entities postulated in critiques of the epistemology of geometry, the basic entities of the world among which spatial, temporal, or spatiotemporal relations hold are taken as a kind of idealized entities. Spatial relations are among objects and the objects usually chosen are ideal, unextended point objects. Ordinary extended objects are taken as constituted out of these as wholes constituted of parts. The necessity for this is clear, since the theory hopes to account for the spatial features *internal* to ordinary material objects, as well as the external spatial relations they bear to one another, and the only way to be sure that all of these internal features will be accounted for in the theory is to take as basic relata those pieces of the objects having themselves no internal spatial structure—the unextended point pieces.

For temporal and spatiotemporal relata the idealization is that of the instantaneous event and, in the case of spatiotemporal relata, events that are spatially unextended as well as instantaneous, i.e., temporally unextended.

Insofar as talk about spatiality is meaningful, the relationist alleges, it is talk about the spatial relations born to one another by the point material objects of the idealization. To view it as talk about "the structure of space itself" is to wallow in metaphysical confusion. Insofar as temporal talk is intelligible, it is talk about the temporal relations instantaneous events bear to one another. And insofar as spatiotemporal talk is meaningful, it is talk about the spatiotemporal relations born to one another by the ideal unextended and instantaneous concrete events that occupy the spacetime world on the substantival view. Properly speaking, the relata occupy the appropriate spacetimes only in the sense that they bear spatiotemporal relations to one another. Any other sense of "occupation" is absurd, for there is no spacetime itself to be occupied. Relations like being coincident spatially or being simultaneous temporally or having a given invariant spatiotemporal interval between them are relations material points, events, or unextended events can have only to one another. They cannot have these relations to points of space, instants of time, or event locations in spacetime, for there are no such things over and above the material "occupants" of them. And certainly any talk of relations born to each other by points of spacetime itself is pure nonsense, for there are no such "points."

Frequently relationists have tried to characterize in some interesting way the relations that constitute the spatial, temporal, or spatiotemporal relations. That is, members of any given pair of existents may bear many different relations to one another. For example, my brother bears to me not only the spatial relation of being approximately 250 air miles away from me, but the relation of being my brother. The usual assumption is that there is a special related "family" of relations that constitute the spatial family, another constituting the temporal family, etc., and that these can be simply picked out from relations in general by some important distinguishing features. Consider for example the following important quote from Leibniz:

> I hold space to be something merely relative, as time is; that is, I hold it to be an order of coexistences, as time is an order of successions. For space denotes, in terms of possibility, an order of things which exist at the same time, considered as existing together, without inquiring into their manner of existing.
>
> H. G. Alexander, editor, *The Leibniz-Clarke Correspondence* (Manchester, 1956), pp. 25–26.

I will have a great deal to say immediately below about the role of the "in terms of possibility" rider in this "definition." For now, however, simply note Leibniz's attempt to pick out the spatial relations from all other possible relations. They are distinguished from the temporal ones in that they are relations among things at an instant, whereas temporal relations are among things (actually events) that are not taken as coexisting at the same time. And they are distinguished from relations in general by their externality. If my brother ceased to be my brother, it would mean, presumably, some change in his "manner of existing." But he can move spatially relative to me and be internally the same sort of existent. Actually, such attempts to define the "spatiality" of spatial relations are usually taken as ineffective if not unintelligible, so I shall simply give to the relationist his designated family of relations: spatial, temporal, or spatiotemporal, as the need may be.

It was sometimes maintained, especially by Newtonian opponents of the Liebnizian relationists, that the relational theory was incapable of providing an adequate account of the *quantitative* relations among things, even if it could account for their qualitative relations. That is, while '$a$ is spatially coincident with $b$' could be distinguished by the relationist from '$a$ is not coincident with $b$,' he could not account for the difference between '$a$ is one meter from $b$' and '$a$ is two meters from $b$.' But this is an ineffective rebuttal. Consider the way, from the substantivalist's point of view, spacetime is mapped by the use of material measuring instruments, as rehearsed in (II,D,3). The same resources of transported measuring rods and clocks, paths of light rays and free particles, etc., are available to the relationist as to the substantivalist. All such procedures require only the observation of spatial, temporal, or spatiotemporal relations among the material measuring instruments. The substantivalist interprets the results as establishing the metric structure of spacetime, the relationist as establishing only the metrical spatiotemporal relationships among things and events. But the resources for generating metrical relationships and distinguishing one from another are clearly available to the relationist, contra the Newtonians' claims.

At this point it is extremely important to take account of the "in terms of possibility" clause in the passage of Leibniz quoted above, for the invocation of possibility is a very important feature both of the relationist account and of the *philosophical* arguments of the substantivalists against it. The full Leibnizian doctrine of space and time, as it appears in the context of his total metaphysics, is extraordinarily complex. It is also puzzling and very idiosyncratic. He believed not only that space was merely a set of relations among material things, but that, properly speaking, there weren't really material things or spatiotemporal relations among them. We won't be paying any attention to this "deeper" Leibnizian metaphysics, only to his remarks in his correspondence with the Newtonian, Samuel Clarke; for here he seems, by and large, to ignore his overall metaphysics in favor of the much more easily comprehended, and frequently held, relationist position.

Why speak of space as a family of relations considered "in terms of possibility"? The answer seems pretty clear. Ordinary and scientific language countenances speaking of unoccupied spatial locations, i.e., points of space where no material object is located. But if space is nothing but the set of spatial relations point material objects bear to one another at an instant, how can the notion of unoccupied location be made intelligible? There simply is nothing there to relate, spatially or otherwise, to other material objects. One solution is to deny the possibility of unoccupied positions; this is at the source of much of the seventeenth-century skepticism about the possibility of a genuine vacuum which became an object of amusement to later generations of scientists.

The alternative to this is to countenance possible but nonactual spatial relations. When I speak about the unoccupied space inside the vacuum jar, that is, I am talking about spatial relations an object *could* have to the bell jar (or to other material objects) but which no object actually *does* have to the other actual objects of the world. By a similar invocation of possible but nonactualized relations the possibility of countenancing instants at which no events occur, or spacetime locations that are not occupied by any unextended instantaneous event, becomes clear. This invocation of possible but nonactual relations to allow intelligibility to empty-space language in the relationist view should appear very familiar to philosophers. It resembles in every detail the invocation of material objects as "permanent *possibilities* of sensation" by the phenomenalists, to enable them to countenance talk about material objects that are, at a given time, unobserved. To speak of there being an unobserved tree in the yard is to talk, in shorthand, about what a perceiver *would* perceive *were* he in the appropriate perceptual situation. To talk about the empty space in a container is to talk about the spatial relations an object would have were it situated, contrary to the actual facts, inside the container.

But I shouldn't leave this invocation of possibilia in the relationists' scheme without a couple of important comments:

1. The invocation of possible, as opposed to actual, relations provides the relationist with the resources he needs to be able to analyze in his view the meaning of assertions about unoccupied spatial locations, temporal instants, or event locations. But will it allow him to contemplate the possibility of spaces totally empty of any material objects, or times or spacetimes in which nothing whatever occurs?

We must be a little careful here. One can specify a particular empty place in a world with at least some actual material objects in it by specifying what spatial relations to the material objects an object at that place would have. In a totally empty world, however, there are no actual reference points and hence no possible specifications of individual points. One can't, even in thought, view the empty space as a set of locations waiting to be filled, for nothing "individuates" one location from another. Nevertheless, it doesn't seem that the relationist is barred from saying something like this: Suppose there are no material objects. One can still make general assertions about what the structure of the spatial relations among material objects *would* be if there were any objects. To ask where an object would be in the space were it in existence is nonsense, since there are no "points waiting to be filled." But to ask whether space *is* Euclidean, meaning "Is it the case that if there *were* material objects their spatial interrelations *would* obey the laws of Euclidean geometry?" seems to be legitimate. If one has already allowed possibilia, with the language of counterfactual and subjunctive assertion, into one's framework, then some sense to the notion of totally empty space (or time or spacetime) seems to be available. One cannot meaningfully ask where things would be were there any, but one can ask what the structure of the set of spatial relations would be were there any things to be so related.

2. The introduction of "possibility talk" into the relationist scheme has been taken by many substantivalists to be a gross defect in the position. The fundamental argument runs something like this: What does it ever mean to ask, "What would be the case?" In ordinary scientific discourse the introduction of such "subjunctive discourse" is always grounded on an underlying nonsubjunctive theory. To say, for example, that a piece of salt *would* dissolve *were* it placed in water, is quite coherent. It is coherent because we believe the salt to have an actual underlying microstructure that satisfies condition $F$ in the law

"All things that have an *F* microstructure dissolve when they *are* placed in water." The dispositional term 'soluble' which we predicate of the salt and whose meaning is unpacked by the subjunctive 'would dissolve were it placed in water' is merely a "place-holder" in an overall theory. The use of the term, and the accompanying subjunctive language, is legitimatized by the underlying theory of the *actual* features of the situation. As a slogan one might adopt, "Reference to possibility is intelligible only when legitimatized by belief in an underlying actuality."

But look at the relationist use of possibility language. What would the spatial relations be of an object at a place, if there were an object where there actually is not? The substantivalist is satisfied that he knows, since the unoccupied space of the region is still actual space and he has a theory about the structure of this space, occupied or not. But what is the underlying reality whose structure legitimatizes the rash use of counterfactual assertions about what spatial relations would be held were objects distributed other than they are for the relationist? In the case of the use of subjunctives to make intelligible talk about empty places in a world that has at least some actual material things in it, perhaps the relationist would reply that his subjunctive talk is grounded not on the belief in an underlying actual space, but rather on the general truths about actual relations he has inferred from experience. Since, he says, the structure of the actual relations among actual objects obeys say, Euclidean geometry, I can make assumptions about what would be the case were objects where they are not, and it is this "generalizing from generalizations" that grounds my assertions—for example, that the unoccupied regions of space are also Euclidean.

For his assertions about totally empty space, however, he is a bit worse off. What is the *sense* of saying that the spatial relations among objects would be describable by Euclidean geometry were there any objects, unless such a statement is grounded on the assertion that there is *actually* an existing Euclidean space waiting to be occupied. Here one doesn't even have a set of general truths about actual relations among actual material objects to draw on. It is important to note that in this argument the substantivalist is not claiming that the relationist suffers from an *epistemic* difficulty in his talk about totally empty universes. He is not saying that the relationist would have no good reason to believe the possible but nonactual relations to be Euclidean; he is saying, instead, that such an assertion simply makes no sense from the relationist point of view. At this point, of course, the hard-nosed relationist can always simply deny the intelligibility of talking about totally empty space, just as his even harder-nosed colleague who is unhappy with subjunctives in any situation can simply deny the intelligibility even of empty places in a world some of whose points are occupied.

So far I have merely been examining some crucial components of the reductionist viewpoint. But what are the arguments for his position? As I have already noted, both substantivalist and relationist usually agree that the substantivalist metaphysics is the one that is naïvely read off from ordinary and scientific language. Usually, also, they agree that this puts the burden of proof on the relationist. It is his metaphysics that is novel and must be defended against the substantivalist position by argument. Of course, once the debate begins both sides feel the necessity of marshaling *reasons* for accepting their positions.

The primary source for arguments supporting the relationist position is Leibniz, especially in his correspondence with Clarke. Leibniz offers not one, but a whole battery of "principles" whose truth he affirms and which he claims lead one inevitably to a relationist theory of space and time. The arguments frequently resemble one another, but they are different in crucial respects—different, in fact, in ways that have not always been sufficiently noticed. Since one can extract all of the standard justifications of relationism from a careful consideration of Leibniz, I shall spend quite an extensive amount of time rehearsing his arguments. In the process, I will even have to make detours to see how some of these arguments function in roles different from that of defending the relationist standpoint, for these other roles will be crucial to us in other contexts and it is important to see how the arguments used fit in to the general Leibnizian scheme.

Let me first distinguish Leibniz's arguments into three classes: (1) the epistemic or verificationist, (2) the pure metaphysical, and (3) the metaphysical-causal. Within the second category, as we shall see, the possibilities exfoliate in a way that makes keeping track of the issues not too easy.

1. The epistemic argument is one that will be very familiar to the reader from Chapter II. It goes something like this: The meaningful assertion of the existence of some entity or feature of the world requires that the presence or absence of that entity or feature, or a change in that feature, have some observational consequences. To affirm the existence of features of the world with no detectable consequences is not to espouse some kind of meaningful skepticism, but rather to affirm the intelligibility of the unintelligible. Now suppose

space itself were a substance. It then would make sense to ask what the position of the whole of the material world in space is, how fast the world as a whole is moving with respect to this substantival space, etc. But we can only observationally determine the spatial relations of material objects to one another, the motions of material objects relative to one another, etc. There are no observations that could conceivably determine the position of the world as a whole in substantival space, nor its velocity with respect to substantival space, etc., assuming of course that, for example, in changing its position in substantival space the internal spatial relations of material objects relative to one another remain constant. So we now see that: (*a*) Belief in substantival space requires the intelligibility of assertions about the position of the world as a whole in substantival space and its motion with respect to substantival space; but (*b*) such assertions are clearly meaningless by the verificationist principle. Therefore, there can be no such thing as substantival space.

Here is a typical Leibniz version of the argument:

> Motion does not indeed depend upon being observed; but it does depend upon being possible to be observed. There is no motion, when there is no change to be observed. And when there is no change that can be observed, there is no change at all. The contrary opinion is grounded upon the assumption of a real absolute space. . . .
> H. G. Alexander, *op. cit.*, p. 74.

Now we have already seen some of the responses one might take to such verificationist arguments in the discussion of the reductionist and antireductionist positions (II,H,4). I will not rehearse these here but only remark that the change in direction of the controversey over the substantival or relational nature of space initiated by Newton comes about when Newton argues that one can, contra Leibniz, observationally determine at least some kinds of motion of objects relative to substantival space itself.

2. What I call Leibniz's "pure metaphysical" arguments all rest upon versions of the so-called principle of the identity of indiscernibles. We shall see that there are many versions of this principle of varying believability and varying applicability to particular issues.

Let me start off crudely:

P.1.    *Let* a *and* b *be objects. Suppose* a *has all and only the properties that* b *has. Then* a *is identical to* b.

Well, of course. One property $a$ has is 'being identical to $a$.' And $b$ has the property 'being identical to $b$.' So if $a$ has all the properties that $b$ has, then $a$ has the property 'being identical to $b$' so $a$ *is* identical to $b$. In this form the principle is certainly true, but totally useless. For whenever the identity or diversity of $a$ and $b$ is in question, the person who denies their identity will simply affirm that there is at least one property of $b$ not possessed by $a$, being identical to $b$, and of course vice versa.

Here is a subtler version:

P.2.    *Suppose* a *has all and only the "purely qualitative" properties that* b *has. Then* a *is identical to* b.

Now the principle might be making a nontrivial assertion. But there are two difficulties: (1) It is extremely difficult to know what 'purely qualitative' means. (2) Under its usual interpretations the principle appears not to be a logical truth, and may not even be empirically true.

The adjectival phrase "purely qualitative" is supposed to discriminate between those properties, like being red or being five feet tall, that don't "make reference to particular objects," as do properties like being Richard Nixon or being the brother of Mao Tse-tung. As even a little reflection will assure the reader, trying to make the distinction sufficiently clear for philosophical contentment is no mean trick. I don't intend to pursue the issue here.

Suppose, though, we do know what 'purely qualitative property' means. Does the principle seem to be a logical truth? The usual reply is to offer counterexamples to it in the sense of constructing possible (but not actual) worlds where the principle doesn't hold. If it is a logical truth it should hold in any possible world, so if there is a possible world where it is false then it is not a logical truth, even if it is a truth. The usual counterexample is a world consisting of nothing but two billiard balls that are internally qualitatively identical (i.e., exactly the same color, mass, shape, etc.) situated, say, one meter apart. Now if there is no substantival space, all the properties that are purely qualitative and hold of one billiard ball, hold of the other—its shape, size, color, mass, and the relational fact of its being one meter away from another such billiard ball. Even if there is substantival space, the billiard balls differ only in that one is coincident with some particular points of the space and the other billiard ball with different points. But that, it is said, is not a purely qualitative difference.

If the construction is really the construction of a possible world, then it seems that P.2 is certainly not a logical truth. Of course it still may be an empirical truth about the actual world—who can tell? But even if it is, can it do the job we might have in mind of providing a rejection of the substantival view of space which proves that there could not be such a thing in any possible world? No.

I am not finished with P.2 in the least. The principle can be read as saying that within a given possible world there cannot be two individuals who have all their qualitative properties in common. I have already noted above (II,F,3) that the issue of the conventionality of the topology of a world hinges, at least in part, on the possibility of "identifying" or "disidentifying" things, or events or locations in the world. That is, to make a new topology "save the phenomena" where an old and different one worked before, one thing we need is to say of what was taken to be *one* event or thing that "it" is actually a *multiplicity* of events or things, or conversely, to identify the members of a formerly numerous set as the unique member of a unit set. Obviously P. 2 is relevant here, since accepting it forces one to a preference for unity over multiplicity wherever possible.

Now we shall see later (IV,D,3) that the pursuit of this issue leads to even greater complications in the interpretation of Leibniz's principle. In (IV,D,3) I will have a chance to carry out these methodological investigations in the context of the important question of their applicability to the problem of the conventionality of the topology of spacetime in the context of general relativity. Since general relativity with its intrinsically curved spacetime suggests the real possibility of unusual topologies to the manifold of the actual world, the question is not of purely speculative interest. I will reserve further discussion of P.2 for that section.

I wish to emphasize here that there is a third version of the principle of the identity of indiscernibles, and it is this version that a Leibnizian might bring to bear in his attempted refutation of the existence of substantival space.

P.3.      *Suppose we have possible worlds A and B such that they are the same with regard to every purely qualitative feature. Then A is the same possible world as B.*

Notice, now, that it is the identity of indiscernible possible worlds which is being proposed, not the identity within some possible world of indiscernible individuals.

First, I think that we can easily see that providing a counterexample to P.3 is not as simple as providing one to P.2, and may in fact be impossible, leading one to suspect that P.3 may very well be logically true. To even get the discussion under way I must assume that we can make sense of the "purely qualitative" notion as applied to features of possible worlds and, indeed, that the very notion of "alternative possible worlds" is intelligible to us. I will not follow up these issues in detail, for to do so would plunge us into the most problematic aspects of metaphysics one can imagine. I will instead offer some reasons for believing that P.3 might be logically true, even if P.2 is not, by assuming some intuitive sense of possible-world and purely-qualitative features on the part of the reader and showing how the attempt to construct a simple counterexample to P.3 fails.

To construct a counterexample to P.3 I must describe two possible worlds, $A$ and $B$, such that: (1) $A$ and $B$ have every qualitative feature in common, (2) yet $A$ is a different possible world than $B$. Let me try. Suppose our only two qualitative properties are $F$ and $G$, and the individuals of possible world $A$ can be denoted $a$ and $b$. Let $A$ be the possible world in which $a$ has $F$ and not $G$, and $b$ has $G$ but not $F$. Consider the possible world, $B$, in which $b$ has $F$ and not $G$, and $a$ has $G$ but not $F$. Isn't $B$ qualitatively just like $A$? In both there is one thing with $F$ but not $G$, and one thing with $G$ but not $F$. But isn't $A$ a different possible world than $B$ since the thing with $F$ but not $G$ is $a$ in $A$ but $b$ in $B$?

But this won't do to refute P.3. I have called the thing that is $F$ but not $G$ in $A$, $a$; and I have called the thing that is $F$ but not $G$ in $B$, $b$. Similarly for the thing that is $G$ but not $F$ in $A$ and $B$. But is the $a$ of $A$ the same individual as the $a$ of $B$? Or, rather, is not the $a$ of $A$ the same individual as the $b$ of $B$? We have gotten into the problem of the "identity of individuals across possible worlds," a problem that besets those trying to construct a semantic theory for modal logic. Now this is an area of notorious confusion and controversy. Usually what is done is this: To decide which individual in possible world $A$ is identical with a specific individual in possible world $B$ we select a specific set of properties which we call *essential*. The thing in $B$ which is identical to $a$ in $A$ is that thing in $B$ which has all its essential properties in common with those possessed by $a$ in $A$. In the possible world in which Blücher does not arrive in time for the battle of Waterloo, who is Napoleon? Well, he is the individual who commands the French forces, was born in Corsica, etc. It doesn't matter that in this possible world that man is a victorious general of the Battle of Waterloo, since we don't count Napoleon's being defeated at Waterloo in the actual world as one of his essential properties.

But now in the possible worlds $A$ and $B$, $a$ in $B$ has *all* the purely qualitative features $b$ had in $A$, and $b$ in $B$ has all those $a$ had in $A$. So if we are to make any sense of "transworld identification" at all, $a$ in $A$ is identical to $b$ in $B$, and $b$ in $A$ is identical to $a$ in $B$. So $A$ is a world where something is $F$ but not $G$. The same thing is $F$ but not $G$ in $B$, although we labeled it differently. Similarly for $b$ in $A$. So $A$ and $B$ are actually the same possible worlds and I haven't constructed a counterexample to P.3 at all.

Now I am not affirming that coherent sense can be made of all of this in the ultimate sense of philosophical intelligibility one might demand. All I have claimed is that: (1) P.3 is a different principle than P.2. (2) The usual counterexamples constructed to make one doubt the logical truth of P.2 fail to carry any force against P.3. (3) The typical kind of counterexample one might think to construct to cast doubt on P.3 won't do since it assumes a dubious principle of "identification across possible worlds." (4) When one adopts a plausible rule for identifying individuals across possible worlds, it seems if anything to back up P.3.

Now there are other accounts of what it is for an object in one possible world to be the same as an object in some other possible world. These accounts do not rest upon the "commonality of essential features" definition I have used above. If one adopts such an account, it may become possible for there to be two distinct possible worlds that differ not at all in their purely qualitative features. I wouldn't like to pursue this difficult problem of metaphysics here. I think it will be sufficient to argue that this last version of Leibniz's principle is certainly the *most* plausible version of it, even if not ultimately acceptable, and then go on to show that it is this version of the principle which the defender of the relationist account needs for his metaphysical attack on substantival space.

How does a Leibnizian use P.3? Suppose substantival space exists. Consider the two possible worlds that consist of (1) the actual world and (2) the actual world displaced five feet to the north in substantival space. It seems as though there are two possible worlds being talked about here, but the worlds are identical with respect to every purely qualitative feature. So they are, by P.3, the same possible world. So substantival space does not exist. Again, consider a world in which the universe of material things is, as a whole, at rest in substantival space and another in which it is all moving uniformly in substantival space with a velocity of five feet per second in the northward direction. If substantival space exists, these are different possible worlds. But they are qualitatively identical and, hence, by P.3, actually the same possible world. So substantival space does not exist.

In viewing these "applications" of P.3 to a particular argument the observant reader will probably begin to see how the postponed difficulties involved in never really explaining what a "purely qualitative property" is make Leibniz's case seem quite weak. The substantivalist will presumably reply to these arguments like this: Suppose, for the moment, I accept P.3. It is usable as a refutation of the existence of substantival space only if I accept your assumption that position in substantival space and motion with respect to substantival space are not purely qualitative features of the material things of the world. Otherwise, the possible worlds described are qualitatively distinct and, even given P.3, individuable. So your use of P.3 to reject substantival space is simply question-begging, since it rests upon the assumption that position in and motion with respect to substantival space are not purely qualitative features of things, and these assumptions are unacceptable, to say the least, given that no clear meaning to 'purely qualitative' has ever been proposed.

I think that this is correct to the extent that it makes it clear to us that the metaphysical version of the principle of the identity of indiscernibles is unclear, to say the least. I think that many would argue that by the time one is finished unpacking the meaning of 'purely qualitative property' one is going to end up with a version of P.3 indistinguishable from the epistemically grounded verifiability principle we saw in (1) above and in the description of the reductionist views about theories in (II,H,4). It is interesting to note, as we shall see in (III,C,1) below, that it is Leibniz's epistemic version of the argument to which Newton's reply is primarily directed.

3. Leibniz's third argument, what I have called his metaphysical-causal argument, looks like this:

P.4.    *Suppose a theory allows us to distinguish two distinct states of the world, A and B. But suppose that it is impossible, in principle, to discover a causal reason why A exists rather than B or vice versa. Then the theory must be rejected.*

His application of this "principle of sufficient reason" to the rejection of substantival space is curious, to say the least: Suppose substantival space existed. Then the world, as a whole, would have some position in it. But when God created the world, nothing could have served him as sufficient reason to place the world at one position in substantival space rather than in some other. So position in substantival space cannot exist, and so there cannot be such a thing as substantival space.

If P.3 suffered from unexplained metaphysics, this argument surely suffers from the invocation of theology. Keeping the theological framework, one can see a number of replies: Why can't God just arbitrarily place the material world anywhere in substantival space he chooses? This amounts, pretty much, to rejecting P.4. Who are we to know God's reasons? Just because we can't see a reason why the material world should be at one place in substantival space rather than another, who are we to believe that we have plumbed fully God's reasoning powers or motivational choices?

Dropping the theology, Leibniz's argument looks even worse. Suppose we consider the world at a time. It has, according to the believer in substantival space, some position, as a whole, in substantival space. Why this particular position? What is wrong with this answer: Because yesterday it was, as a whole, in this position and no forces arose to move it. So there is a sufficient reason for the present place of the material world in substantival space—its previous position and the forces that have acted in the meantime. After all, why is Jupiter where it is today relative to the Earth? Answer: Because of the positions and velocities they had yesterday and the forces acting upon them in the intermediate period. In other words, if we take "giving a sufficient reason" for a fact as doing what scientists ordinarily do—invoking initial conditions as well as laws to explain a present fact in terms of past facts and operative forces of change—then we can give a sufficient reason for the "absolute" place of the material world in substantival space.

Now Leibniz would be very unhappy with this notion of "giving a sufficient reason," since he believed that, at some ultimate level, the explanation of contingencies in terms of other contingencies would have to be replaced by a demonstration that the "contingency" of the state of the world was mere illusion. With God's knowledge and wisdom we would see that the way the world is is the way it has to be. "Has to be" not relative to some other contingency, but "has to be" period. But could we ever find such an explanation of the world's present location in substantival space? No, he would say. But now it appears that Leibniz's argument against substantival space rests not simply upon

P.4, the principle of sufficient reason, but upon a particular "rationalist" model of scientific explanation, in which all facts must ultimately be shown to be the case of necessity. So the argument rests not only upon a somewhat dubious methodological principle, P.4, but upon an account of scientific explanation which most present-day philosophers and scientists would reject out of hand.

I think we have seen enough to know that Leibniz's most persuasive argument is one that has the familiar feature of resting ultimately upon epistemological considerations. Even if one rejects P.2 as implausible, the use of P.3 as question-begging, and the use of P.4 as resting upon a highly dubious account of what it is to give reasons for the world's being the way it is, one is still left the invocation of P.1: What sense does it make to speak of the world's position in or motion with respect to substantival space, when there is no conceivable way of observationally determining what this position or state of motion is?

I will now turn to Newton's reply, a reply designed to support the doctrine of substantival space against just this epistemological critique. At the very end of Chapter III I will return to some of the purely philosophical concerns that have just exercised us.

# NEWTON AND HIS CRITICS

## 1. *Newton's Argument for Substantival Space*

Newton attacks the relationist position just where the arguments for it are strongest. We have seen that Leibniz's most persuasive argument against the substantival view is that observations are confined to those we can make on material things (in general, including light rays etc.), their properties and relations. No relation of these material things to the alleged substantival space has any possible observational consequences. But assertions about features of the world which have no possible observational consequences are illegitimate, or, as more modern "positivists" would assert, meaningless. So we cannot scientifically tolerate the postulation of a substantival space—existing in its own right and with its own features, conceivable even as empty of material objects, and serving as the container or arena of the material world.

Newton says this: Many of the relations that material objects bear to space itself have, it is true, no observational consequences. However, the *acceleration* of a material object relative to space itself does have observational consequences. From these observational consequences one can infer, in the same way thoreticians frequently infer unobservable entities from their observable consequences, to the existence of space itself. Once we know that space itself exists, we may infer the existence of other relations that material objects bear to it, for example their position in it and the velocity of their motion relative to it, even if the variation in these relations has no empirical, observable effects.

Two thought experiments Newton proposes are crucial. First, take a bucket of water suspended by a rope in one's laboratory. Twist the rope and let it unwind. The bucket begins to spin. Initially the bucket and the water in it are at rest with respect to one another, and the surface of the water is flat. Soon the bucket is in rather rapid rotation with respect to the water, and yet the water remains flat. Finally one observes the water again at rest with respect to the bucket, the two together spinning with respect to the laboratory, and now the surface of the water has become concave, the water rising up the sides of the bucket.

The concavity of the surface of the water can only be explained by the existence of forces acting on the water, forces sufficient to distort it from its normal equilibrium flat surface which it has when at rest in the earth's gravitational field. Clearly these forces arise because of the motion of the water, for it is the spinning motion that is the inevitable feature concomitant to the existence of such forces. But it can't be the motion of the water with respect to the bucket which induces the forces, for when bucket and water are in maximum relative motion the surface is flat, and it becomes concave only when the water is once again at rest with respect to the bucket.

Now one might think that such forces are the result of the relative motion of water and the laboratory, or perhaps the relative motion of water to the earth. But when one realizes the pervasiveness of such centrifugal forces throughout the observable astronomical universe (it is the same force that "keeps the planets from falling into the sun"), despite the widely variant material surroundings of these objects, one quickly sees that the motion responsible for the forces cannot be taken as motion relative to any *ordinary* material object.

Newton's second thought experiment is designed to show that the motion responsible for the forces can't be taken as motion relative to any material object at all. Imagine, he says, two weights tied together by a string, inhabiting an otherwise empty universe. Two possible situations are imaginable. The weights might be in rotation about their common center, or they might not be. If they are not this is observationally indicated by the absence of tension in the rope joining them. If they are in rotation, however, the rope will experience a tension due to just the same forces, which have their origin in the motion of the objects, that caused the water to slosh up the sides of the bucket. But now, surely, the motion responsible for the forces cannot be called motion relative to any material object. For there are no other objects in the world than the weights and their rope, and in both cases, non-rotation and rotation, the weights are at rest with respect to themselves.

It is not only circular motion that gives rise to such forces. Linear accelerations will do as well, as any passenger in a swiftly starting or stopping train can tell. He can observe the acceleration of the vehicle without ever looking out the window by observing the forces exerted on the objects in the vehicle by their accelerations.

So accelerations give rise to *observable* forces. And these accelerations are not accelerations relative to ordinary material objects. But all accelerations are accelerations relative to *something*, and so must these be. The something is clearly space itself. The ubiquity of the connection between acceleration and forces and its omnitemporality are now easily explained. No matter where and when experiments are performed, they are performed in space itself. And space, substantival space, is everywhere alike, unchanging in time and the same in every direction. No wonder, then, that similar rotations and accelerations give rise to similar forces, when the forces are due to the acceleration of the object in question relative to substantival space which is everywhere, everywhen, and in every direction the same.

Now once we know that substantival space exists, just as the naïve realist thought, as the container of all material objects, we realize that many relations to it of material objects exist as well. There is for example the position of a material object in space and the velocity of the material object with respect to space. Of course these relations of material objects to space have no observable consequences. But they must exist, since substantival space must exist as the reference object for those accelerations that give rise to the observable forces. If the verificationist principle insists that all real features of the world be observable or at least have observable consequences, so much the worse for the verificationist principle.

Newton's argument fell on the philosophical disputants concerned with the substantivality of space like a bomb, changing the whole pattern of the debate from that point forward. It would no longer do to rehearse the standard philosophical pros and cons of substantival versus relationist positions, for any adequate theory of space had now to answer Newton's apparently simple and conclusive argument for the substantivalist position. Actually, Newton's argument is far from simple.

At each stage it relies upon "hidden premises." The discovery and criticism of these premises has been one of the major occupations of defenders of the relationist position since Newton.

But the debate could not long continue, even at the level of the traditional philosophical dispute as modified by Newton's novel scientific argument for substantival space. For Newton's argument rests upon a particular scientific theory, his own dynamics. And as scientists changed their accepted "best available theories," the role the contemporary science played in the debate over the substantival nature of space changed as well. At least some scientists thought, in fact, that changes in science would eventually resolve the issue once and for all. That is, that the philosophical debate into which Newton added his scientific considerations could finally be dispensed with entirely if only enough good science were applied. As we shall see, they were wrong.

Let me begin by noting a few of Newton's hidden assumptions, assumptions necessary to argue from the data (orbits of planets, tension in ropes, sloshing water in spinning buckets, passengers thrust back in their seats as airplanes accelerate, etc.) to the existence of substantival space. Of course in doing this we are using massive doses of hindsight generated by the work of philosophical critics of Newton and advances in science since his time. Let me break up the class of "Newtonian hidden premises" into four classes:

1. First there is the rather enormous inductive leap made by Newton when he assumes that the lawlike connection between accelerated motions and inertial forces holds at every place, at every time, and in the same manner always. Of course any scientific inference to a lawlike theory from the observational data requires the kind of inductive leap that leads from particular facts to generalizations, but it does seem as though Newton is making a very bold conjecture indeed on the basis of the evidence available to him. On the other hand, the scope of the evidence available to Newton, and even more available to us, is quite broad. Astronomical observation is certainly sufficient to indicate that the existence of phenomena such as centrifugal forces induced by rotations is very widespread in space and time.

Interestingly enough, none of Newton's critics really objected to this stage in the argument, for all were agreed that inductive generalization constituted legitimate scientific inference, and all were as persuaded as Newton of the universality of the dynamical effects from which he argues. Only with the postulation of general relativity has it become respectable to speculate upon the possibility that the connection we observe between accelerated motions and inertial forces may indeed take on different forms at sufficiently remote times and places.

2. Newton's leap from the observed behavior of objects in the actual world, as instanced by the bucket thought experiment, to the behavior of objects in a possible world quite other than the actual, as instanced in his claimed behavior of weights on a string in the nonactual universe in which all other matter is imagined out of existence, is a leap that cannot be assimilated to the inductive generalization of (1). One can easily imagine someone agreeing with Newton that in the world as it is actually constituted, rotational and other accelerated motions actually will give rise to inertial forces, but at the same time claiming that such forces will not be present whatever the state of motion of the test system in the imagined world totally devoid of other matter, or even claiming that the notion of accelerated motion in such a world is incoherent.

One cannot, without disposing of a great deal of accepted scientific language, reject "reasoning to behavior in possible worlds on the basis of observations in the actual world" altogether, although some remarks of Ernst Mach's would have one so dispose of all possible-world inference as illegitimate. For example, what would become of scientific inference about "ideal objects" such as ideal gases, ideal point masses, etc., if such talk is ruled out of court altogether? There are no ideal gases in the actual world, but this has never stopped scientists from describing what their behavior would be like if there were any.

Nonetheless, it is important to note the role played by this inferential step in Newton's argument and the opening for criticism it allows. It is only the behavior of the weights in the otherwise empty universe, in particular the remaining association of some motions of them with tension in the connecting string, that allows Newton the claim that whatever the "real accelerations" are with respect to, it cannot be some material entity that provides the reference frame. As I have noted, Mach rejects Newton's inferential leap from the actual to this imagined world, arguing that it is scientifically unacceptable. In the discussion of Mach's views below (III,D,2) we shall see that there is room for important discussion here, even if one does not go so far as Mach to reject "inference to behavior in possible worlds other than the actual" altogether.

3. Newton associates the existence of the inertial forces with accelerated motion. He then affirms that accelerated motion must be

motion that is in acceleration relative to some entity. Since it can't be an ordinary material entity, it must be space itself.

But must motion, of any kind, be motion with respect to some *object?* Newton simply assumes that this is so. There is a curious paradox here. Newton, attempting to establish the substantivalist view of space in contrast to the relationist position, assumes automatically that all motion is relational. To say of an object, *a*, that it is in a certain state of motion is incomplete. For motion of *a* is always motion with respect to some other object, call it *b*. 'Has velocity *r*' and 'has acceleration *s*' are, Newton assumes, relational or two-place predicates. And the "holes" in the predicates must be filled by terms referring to objects for a meaningful assertion to result. If the bucket is in accelerated motion, it must be in accelerated motion with respect to some *object*. If ordinary material objects won't do, then it must be some extraordinary object, substantival space. Newton is antirelationist in the sense that he believes that one cannot "reduce" the assertion of the existence of substantival space to the assertion of the existence of a family of spatial relations among material objects, but he is a relationist in the sense that he believes position, velocity, and acceleration all to be relational notions. An object has position, velocity, and acceleration only with respect to some other object, even if this other object is the nonmaterial substantival space itself.

It is curious also that many anti-Newtonian positions reject just this Newtonian assumption. Some of these positions remain substantivalist in a sense, and some are attempts at restoring the pure relationist position against Newton's critical attack. The former assume, with Newton, that the spatial notions of position, velocity, and acceleration are, indeed, relational. But they deny the necessity of always "filling in" the second place in the relation with a term denoting an *object,* in Newton's sense. As will be seen below, (III,D,3), (III,D,4), and (III,E), the neo-Newtonian spacetime, special-relativistic, and general-relativistic views of "absolute acceleration" all break with Newton at this point. They all share, at least on the surface, the Newtonian view that the "arena" of the material happenings of the world has its own reality over and above the reality of the material things in it. But all deny the existence of substantival space in Newton's sense. Newtonian substantival space is a sort of "super thing" similar to a material object that can persist through time. Neo-Newtonian spacetime is not like the things of the world, and in the relativistic theories even the ordinary notion of a material thing that persists through time is rejected. Still, in all of these theories, given their intuitive realist interpretation, absolute accelerations are relative to some reference entity, but the entity is no longer the "object" of Newton's theory.

The use made of this inferential step of Newton's by the defenders of the pure relationist theories is most peculiar. They admit that Newton is correct in associating inertial forces with a kind of motion, real accelerations. But they deny that one need take 'is absolutely accelerated' as attributing a relation of some object to any other entity. That is '*a* is in absolutely accelerated motion' becomes a complete assertion, not needing to be filled out by answering the question: "In accelerated motion relative to what?" So that to defend a relationist ontology of space or spacetime, one must resort to denying that what appears to be a relational concept, being accelerated, really is one. It is a simple, nonrelational property. I will show in the next section how Berkeley tried something like this approach to defend relationism against Newton, and I will return to this question at the end of Chapter III, in (III,F), where I examine in some detail the question of whether the substantivalist and relational theories can both be made at least physically compatible with all the physical theories, requiring philosophical principles for deciding between them.

4. On reflection, Newton's association of inertial forces with accelerated motions with respect to substantival space seems very queer indeed. How is this association established? It is *not* established by independently noticing the existence of the forces and the existence of acceleration with respect to substantival space and then noting a lawlike connection between these two phenomena, for one cannot in any way *independently* observe the absolute acceleration of the object with respect to substantival space. The only grounds for assuming that such an acceleration exists, and that there is an entity relative to which the acceleration is an acceleration, is in fact the existence of the forces! There appears to be a curious circularity here. The only evidence for the existence of the absolute accelerations is the forces generated in objects under certain states of motion, and we then hypothesize the absolute accelerations as being explanatory of the forces.

One wants to be a little cautious here. Surely there is nothing in principle wrong with (1) observing a phenomenon, (2) on the basis of the phenomenon postulating some underlying unobservable theoretical feature of the world, and (3) then postulating the theoretical feature as being *explanatory* of the observed phenomenon. Isn't that the way of theoretical explanation in general: the data giving us inductive reason for accepting the theory, and the theory then being put forward to explain the data?

Yet there is something peculiar in this case. What suggests that *accelerations* be invoked to explain the inertial forces in the first place? The answer seems to be that we see the forces coming into being only in association with observed accelerations *relative to ordinary material objects*. The sloshing of the water is associated with the spinning of the bucket relative to the laboratory, the earth, or perhaps, "the fixed stars." So the observational facts that fix our attention on acceleration as being causally responsible for the inertial forces is an observed association of the forces with observable accelerations relative to material objects. But one then goes on to disassociate the forces from such "ordinary" accelerations, by Newton's argument, and to associate it with some unobservable "absolute" acceleration instead—unobservable, that is, except insofar as it gives rise to the inertial forces.

The curious structure of the reasoning here suggests that one look a bit more closely at just what the association of inertial forces with states of motion amounts to observationally. Suppose we forget about absolute motions and reflect upon the general truths associating inertial forces with relative motions of material objects with respect to one another. What generalizations can we draw based upon ordinary generalizing inductive reasoning, without ever mentioning substantival space or acceleration with respect to it?

Suppose we have a material reference frame which has the feature that objects at rest in it suffer no inertial forces. We can observe that objects in uniform motion with respect to this reference frame, or equivalently objects at rest in reference frames in uniform relative motion with respect to our original reference frame, also suffer no inertial forces. In addition, there is the fact that if an object suffers inertial forces when at rest in a given material reference frame (like a passenger strapped in an accelerating airplane), then objects of the same mass fixed in any reference frame in uniform motion with respect to the initial frame will suffer identical forces. A similar proposition holds with regard to the accelerations that are not linear, but due to rotation of the reference frame. It is these facts that allow Newton to account for all inertial forces by (1) blaming them causally on accelerations alone (and not upon their particular "absolute velocities," etc.) and (2) making all such absolute accelerations with respect to a single "rigid" object, substantival space. If the equivalence of material reference frames in uniform motion with respect to one another did not exist, then the Newtonian explanation would fall through, since there would be objects with the same acceleration with respect to substantival space and the same mass but which suffered different inertial forces.

So an observable connection of inertial forces with observable states of motion does exist. It is not the connection of inertial forces with the not-independently-observable acceleration with respect to substantival space, but instead the facts summed up in the claim that dynamics obeys Galilean relativity, i.e., that with regard to purely mechanical phenomena, the same laws will describe the interaction of matter without the postulation of extra inertial forces in any two reference frames that are in uniform motion with respect to each other. As we shall see, this is what many take as the "hard" evidence behind Newton's claim, and it is this that any adequate theory of inertial forces must account for. Even if one rejects Newton's association of inertial forces with unobservable absolute accelerations, one must still account somehow for this important lawlike association of inertial forces with observable states of relative motion.

One additional fact is sometimes taken to be empirically determinable by observation—and of crucial importance. This is that some reference frames identifiable by association with particular material object systems appear more or less inertial. That is, objects at rest in these frames or in uniform motion with respect to them seem to suffer no inertial forces. The surface of the earth is not such a reference frame, as the existence of the Coriolis forces generating the well-known cyclonic and anticyclonic motion of the atmosphere attests, but a reference frame fixed "in the fixed stars" seems, observationally, to be pretty close to inertial. I shall return to this point below (III,D,2).

Though I have said a great deal about Newton's doctrine of substantival space, I have said nothing about his parallel claim that there is an absolute time that must be postulated to account adequately for the observable data. I will not discuss this in detail here but will have some things to say about this in Chapter IV where I take up the question of the relation between accepted physical theory and philosophical views of time. It will be sufficient here for me to note that a notion of the real time between events, as opposed to the time as measured by some clock or other, is crucial for Newton.

Real accelerated motions must be distinguished from merely apparent accelerations, for only real accelerations given rise to inertial forces. For example, from the viewpoint of an accelerating car the world outside appears to be accelerated in the direction opposite to that in which the car appears accelerated relative to observers stationed on the road. But one acceleration, that of the car, is real and the other "merely relative," as is attested by the fact that it is the passengers in the car who suffer the inertial forces. When the car accelerates, their coffee sloshes in its cup, not the coffee of the outside observers.

Now I can cook up a clock that, if I use it to record time intervals, will make a uniform motion appear accelerated. For acceleration is change of position per unit time per unit time, and if I keep changing the rate of the clock the chart of distance versus "time" will appear the curved path characteristic of accelerated motion and not the straight line characteristic of uniform motion. But one can tell, having once picked an inertial frame as reference, if an object is really accelerated relative to that frame. Do things attached to the object suffer inertial forces? A clock whose use leads us to characterize a motion as accelerated relative to an inertial frame, when the nonexistence of forces tells us that it is really in uniform motion, is a clock that fails to *correctly* record the time interval between events. In this sense, there is an "absolute" time scale, and clocks are good or bad as they are in conformity with this scale or not.

## 2. Some Early Critics of Newton

The truly exciting replies to Newton's argument arose with the radical changes in physics from the middle of the nineteenth century onward, despite the fact, as we have seen, that philosophical criticisms and replies were always in principle available to one who examined Newton's hidden premises in sufficient detail. I will reserve the exposition of the later responses to Newton to (III,D) and (III,E) below, since they are each important enough to warrant their own detailed presentation. In this section I will only make a few brief remarks about three early responses to Newton by antisubstantivalists: the responses of Leibniz, Huyghens, and Berkeley.

Leibniz agrees that Newton has shown a true distinction between merely relative motion and "real motion," for the latter results in the inertial forces while the former does not. But he hopes to account for this distinction not in terms of the object to which the motion is relative, but in terms of the cause of the motion. In any motion of two material objects relative to one another, he says, at least insofar as the motion is a relative acceleration, at least one of the objects must be *moved,* i.e., have forces applied to it. He speaks of the body in "true motion" as having "the cause of the change in it."

> I grant that there is a difference in an absolute true motion of a body, and a mere relative change in its situation with respect to another body. For when the immediate cause of change is in the body, that body is truly in motion; and then the situation of other bodies, with respect to it, will be changed consequently, though the cause of the change be not in them.
>
> H. G. Alexander, *op. cit.,* p. 74.

He goes on to characterize real motion as a subspecies of relative motion, real motion being relative motion in which the cause of the relative motion is in the body truly moved. He is on to something here, but there are great difficulties. First of all, there can be, according to Newton, absolute motions when there are no relative motions at all. That is, if all the matter in the universe were in rotation, say, there would still be inertial forces generated, despite the fact that there were no relative motions of the material parts to one another at all.

The notion of a cause "being in the object moved" is sufficiently vague to be near useless. In addition, one seems to be able to imagine situations where there are inertial forces but no causal forces inducing the motion at all. Again imagine a uniformly rotating universe of material objects. Now imagine it to have been always in rotation, the rotation not caused by any imposed torque at all, but simply eternally sustained by the conservation of angular momentum. There would still be inertial forces present on any object at rest in this coordinate frame.

In one place the physicist Christian Huyghens attempts to show that the motions that give rise to inertial forces can be characterized in terms of the features of merely relative motion of material objects with respect to one another, obviating any need to postulate substantival space as the reference object of absolute accelerations. Imagine a rotating disk, he says. Consider two points on opposite sides of the center. While rigidly attached to one another, they are in motion with respect to one another, their velocities pointing in opposite directions. It is this relative motion that can be credited as the source of the inertial forces.

But the argument is confused. First of all, it is totally irrelevant to the existence of inertial forces generated by linear accelerated motions. Second, the appearance of "velocities in opposite directions" for the points of the disk is generated by fixing one's reference frame to one with respect to which the disk is rotating. If we picked a reference frame fixed on the disk, all the points of the disk would be at rest in this frame, and there would be no relative motion of them with respect to each other at all. Once again we seem to need a special, nonmaterial reference frame to account for the fact that the disk is *really* in accelerated motion, and Newton would, of course, fix that reference frame in substantival space. In other passages Huyghens seems to be anticipating, vaguely, what I shall describe in (III,D,3) as neo-Newtonian spacetime, but I shall forgo discussing his anticipation of this since I will devote some detailed attention to the contemporary sophisticated version of this approach.

Berkeley makes many deprecatory remarks about Newton's doctrine of substantival space, but fails to come up with any adequate alternative positive doctrine. Like Leibniz he tries to account for absolute motion as a species of relative motion, arguing that the motion is absolute when it is relative and when forces are being applied to the object to keep it in motion. But his position is vitiated, once again, by the possibility of absolute motions that are not motions relative to any material objects at all, and by the fact that there are absolute motions of objects to which no forces are being applied at all. For example, the steadily rotating disk has inertial forces acting throughout it. Indeed, if it is not strong enough, the centrifugal forces will tear it apart. Yet once it is spinning at a constant angular velocity, no forces or torques need be applied to keep it in motion.

There is another argument of Berkeley's in *De Motu* designed to account for centrifugal forces but, besides being irrelevant to Coriolis forces and the inertial forces generated by linear accelerations, it is so confused and question-begging as to be not worth detailed consideration. On the other hand, Berkeley is in other places extraordinarily prescient, offering remarks which interestingly anticipate Mach's later account.

# THE EVOLUTION OF PHYSICS AND THE PROBLEM OF SUBSTANTIVAL SPACE

In this section I shall examine the impact on the question of the substantivality of space of four scientific theories. One of these, the electromagnetic theory of the nineteenth century, constitutes a now partly-rejected important scientific advance. Another, Mach's theory of absolute motion, constitutes a speculative attempt at a scientific resolution of the problem of "absolute acceleration" in terms favorable to the relationist theory of space; a speculation that, as we shall see, never achieved scientific fulfillment as a full-fledged theory. A third, the neo-Newtonian theory of spacetime, constitutes the construction of an account of spacetime designed to better "fit" Newton's own dynamical theory at a time when Newton's dynamics was no longer accepted as correct. As such, the invention of neo-Newtonian spacetime constituted something more of a philosophical clarification of Newtonian mechanics than a genuine novel scientific theory of the world. The fourth theory, special relativity, is, of course, one of the major scientific achievements of all time. In Chapter IV I will examine the foundations of this theory with some care. Here I will be concerned with the theory only to the extent that its adoption forces on one changes in his views about absolute motion and the substantivality of spacetime. As is shown, the adoption of the theory has very little effect in resolving either the basic questions about the "absoluteness" of some motions or those about the substantivality of the manifold itself; although, naturally, the kind of manifold with which it deals differs radically from that of either Newton or the neo-Newtonians.

## 1. *The Electromagnetic Theory of the Nineteenth Century*

As a result of his systematization of the theory of electricity and magnetism in the later part of the nineteenth century, James Maxwell predicted, as a consequence of the general laws governing electricity, magnetism, and their interaction, the existence of electromagnetic radiation, radiation that could exist in otherwise empty space independently of any material objects.

One important consequence of Maxwell's theory is that this electromagnetic radiation, if there is any, would have to travel through empty space with a definite velocity. The velocity is calculable from the measurement of two completely "nonkinematic" parameters, the dialectric constant and the magnetic susceptibility of the vacuum. When the appropriate measurements are performed and the expected velocity of electromagnetic radiation calculated, the predicted value for this velocity turns out to be the well-known velocity of light in a vacuum. Immediately upon discovering this fact, Maxwell made the bold conjecture, since confirmed by the evidence as well as a hypothesis can be, that light is identical with a species of electromagnetic radiation.

Actually, this isn't quite what Maxwell said. Contrary to the present metaphysical attitude of scientists, which treats radiation as a "substance" on a par with material objects although qualitatively different from ordinary matter, Maxwell viewed both light and electromagnetic radiation as "disturbances" in some all-pervasive medium, the aether. What he actually says is that the identical values of the predicted velocity of propagation of electromagnetic radiation and the observed velocity of light lead him to postulate that the electromagnetic aether and the optical aether are one and the same medium of transmission of wave phenomena.

What is important for our purposes is this: The theory predicts a specific value for the velocity of light in a vacuum, a value independent of the direction of propagation of the light. Now according to all prerelativistic theories of space and time, a wave can be seen as propagating through an isotropic medium with the same velocity in every direction, given that the velocity of the wave is independent of its source, as is the velocity of light, only by an observer who is at rest in the medium. An observer in motion through the medium will see the wave propagating away from him in his direction of motion at a slower velocity than the wave propagating opposite to his direction of motion, since he is catching up to the one and receding from the other. So by observing the velocity of light in varying directions in different reference frames, we should be able to determine the reference frame whose state of motion is that of being at rest in the aether.

As a matter of fact we cannot so determine this rest frame. It is this fact that leads, ultimately, to the rejection of the aether theory of propagation and to the adoption of the theory of special relativity. Here, however, I am going to assume that, contrary to fact, the experiments designed to determine the frame of reference at rest in the aether are successful, and that we have determined which state of motion of an observer can be characterized as being at rest in the aether. What should the effect of such fictional experimental results be on the doctrine of absolute motion and substantival space?

Many nineteenth-century scientists seemed to feel that such positive experimental results would go a long way toward eliminating many relationist objections to the doctrine of substantival space. Why? Well, one of the primary objections of the relationists to Newton's doctrine was that the postulation of substantival space committed one to the existence of phenomena with no observational consequences whatever, a gross violation of verificationist canons of science. Whereas acceleration with respect to substantival space was detectable by means of the inertial forces caused by the acceleration, the absolute position of an object in substantival space and the velocity of an object relative to substantival space have no mechanical consequences at all.

It was generally assumed that one could establish the velocity of a coordinate frame with respect to substantival space by simply determining the velocity of the frame in question relative to a material frame found to be at rest in the aether by means of some optical experiment as I have described above. So, if the aether theory is correct, we could now find an empirical correlate to velocity with respect to substantival space, and at least some of the relationists' objection to the Newtonian theory would be undercut.

The whole doctrine rests upon a chain of implicit assumptions that are dubious, to say the least. First of all, it is simply assumed that the rest frame with respect to the aether would be a frame in which no inertial forces were generated on an object at rest in that frame, i.e., it was simply assumed that the aether would constitute an inertial frame.

Surely this is a necessary assumption, for substantival space is obviously an inertial frame according to the Newtonian theory. What would have been the result had the rest frame in the aether been found, and the aether turned out to be in rapid rotation, is hard to say. As far as I can tell, no scientists ever even considered this a possibility.

But even if the aether turned out to be an inertial frame, what right had anyone to assume that this frame was at rest in substantival space, much less that the aether could be *identified* with substantival space? An aether that was in rapid but uniform motion with respect to substantival space would equally well constitute an aether that formed an inertial reference system. Again, this possibility never seemed to be noticed by the scientific community.

On can see, in fact, that the assumption that the aether, supposing it to have been found to constitute an inertial frame, was identical to or even at rest in substantival space, rather than a different "entity" in rapid motion through substantival space, is just the kind of assumption with no possibility of observational test which is abhorrent to the verificationist exponent of the relationist theory of space. From his point of view, discovery of the aether frame constitutes no ground whatever for believing that one had discovered the frame at rest in substantival space and no ground whatever for believing that "empirical sense" had been given to the notion of 'absolute velocity.'

Even if one accepts an identification of the aether with substantival space, the verificationist is still likely to be dissatisfied. Consider the notion of position in substantival space. Would the discovery of the rest frame in substantival space be sufficient to establish the empirical meaningfulness of this feature of things, a feature they are alleged by the Newtonian theory to have with no observational consequences following from their having it?

The answer to this question is a little more complex than one might think. If one can determine absolute velocities, can one determine absolute positions? In one sense we can. Find an object whose velocity with respect to substantival space is zero. Identify the point in substantival space at which it is located by simply calling it 'the position of the reference material object.' Now find the positions of other material objects relative to the reference material object. This gives the "absolute" position of these objects, and if they too are at rest with respect to substantival space, they too can serve as "markers" labeling specific points in substantival space.

Isn't this enough to give empirical meaning to the notion of position in substantival space? Leibniz would say, "No," I believe. Suppose the whole world had been made by God such that the whole of the material universe were elsewhere in substantival space than where it actually is. Would this have any empirical consequences? No, not even if motion with respect to substantival space, even uniform motion, does have empirical testability. Again, would there be any way of empirically refuting the assertion that at some instant the whole of the material universe instantaneously changed its position in substantival space, retaining, of course, the relative positions of material objects among themselves constant? Once again the answer is negative, even if ordinary motion with respect to substantival space has empirical results.

The negative answers to these questions would, I think, be sufficient for a verificationist as strict as Leibniz to maintain that even an identification of the aether frame with the rest frame in substantival space is insufficient to make the notion of substantival space scientifically legitimate; for while this would be sufficient to establish good empirical meaning to assertions about the velocity of objects with respect to substantival space, it would not, as the arguments of the preceding paragraph show, be sufficient to give "good empirical meaning" to assertions about the positions of objects in substantival space. It is clear that a strict verificationist would be satisfied with the legitimacy of the concept of substantival space only if some observational means were provided to allow us to label individual points in this space, independently of their being occupied by any material objects. Even being able to find objects empirically determinable to be at rest will not suffice for him, for while this does allow us to tell when an object has changed its place in substantival space, it doesn't allow us to say which point in substantival space a given material object occupies. Simply saying, "It occupies the place it occupies" won't do for a Leibnizian verificationist.

But since the results of the experimental attempts to find the reference frame at rest in the aether led, contrary to expectation, to the conclusion that one could not determine which frame this was by any empirical means, the issues I have been discussing become moot, and I won't pursue them any further.

## 2. Mach's Theory

Ernst Mach was simultaneously a working theoretical physicist and a positivist philosopher. He made, in his *Science of Mechanics,* a heroic attempt to replace Newton's theory by an alternative theory. The alter-

native is supposed to be adequate to account for the inertial forces Newton took as the primary data supporting his doctrine of substantival
space, but to lack any such "metaphysical" elements that infected, according to Mach, the Newtonian scheme.

His proposal rests on three elements: (1) A philosophical critique of
the Newtonian theory, (2) an important "observational fact," and (3)
a speculation about a possible future scientific theory.

Many of Mach's criticisms of Newton are the familiar verificationist
objections we have already seen. One criticism of Mach's is of particular
importance, however. He certainly admits the presence of "inertial
forces" in the actual world: how could one reject Newton's bucket
thought experiment when what it represents is a vast accumulation of
empirical experience? But he rejects Newton's inference as to how the
weights on the string would behave in a universe devoid of other matter
as totally unwarranted. At least, according to Mach, it is an unjustifiable
leap to argue from inductively established generalizations about the behavior of objects in the actual matter-occupied world to a proposition
about how they *would* behave in a world as radically different from our
own as to be devoid of the enormous masses of the astronomical universe. Some of Mach's comments about the universe being "given to us
only once, complete with fixed stars" seem to imply, in fact, that it is
not just unwarranted to make assertions about such possible but nonactual worlds but, since science is the attempt to infer the general behavior of the actual world and is summed up in the laws it discovers,
that such assertions about possible worlds radically different in kind from
the actual world are meaningless, rather than simply unsupported by the
empirical evidence. In any case, Mach refuses to accept Newton's claim
that one could distinguish, in the world devoid of other matter, the two
situations of the weights rotating and the weights nonrotating. Newton's
claim that the former case would lead to observable tension in the rope
connecting the weights is, according to Mach, simply unwarranted.

The observational fact Mach uses is that the fixed stars constitute an
inertial frame. The experimental evidence for this is simple enough. Fix
a gyrocompass on earth pointing in a given direction. As the earth rotates, the gyrocompass will point in different directions relative to a
frame fixed in the earth. After a period of about 24 hours, the compass
will point once again in its initial direction. This fixes the "inertial" period of the earth's rotation. Next fix the "astronomical" period of rotation by observing a star at the zenith and seeing how long it takes for the
same star to reappear overhead. If we make compensation for the fact

that the earth is moving in its orbit around the sun as well as rotating, then the two periods turn out to be the same. How would Newton explain this? The only answer can be that the stars we use to fix the astronomical period are themselves a nonrotating frame, for otherwise the gyrocompass and repeated zenith times would not agree. So, Newton concludes, as a matter of fact, the system of distant stars (or perhaps better, galaxies, since the Milky Way galaxy is itself known to rotate) form an inertial framework.

Mach's speculative scientific theory takes this "contingent fact" of the inertiality of the fixed stars in Newton to be evidence for quite a different conclusion. According to Mach, the system of fixed stars *must* form an inertial framework, since it is acceleration relative to the fixed stars, and not acceleration relative to some mysterious substantival space, which is causally responsible for the existence of inertial forces. Each experiment of Newton's which gives rise to inertial forces is counted by Newton as evidence of acceleration relative to substantival space. But since the fixed stars are themselves an inertial framework, one might equally well describe each such motion as an acceleration relative to the system of fixed stars, or, if you wish, an acceleration relative to the average smeared-out mass of the universe. So, we may blame the existence of the inertial effects on forces generated on the test object by these distant masses due to the relative acceleration. Just as masses generate static gravitational forces on each other, dependent only upon their mass and their separation, so they generate inertial forces on each other. The magnitude and direction of the force exerted is dependent upon the relative acceleration of the material bodies. But nowhere in this scheme is there a necessity to postulate some mysterious, nonmaterial entity, substantival space.

Mach can be seen as agreeing with Newton that the inertial forces must be blamed upon an acceleration of the test object, and also agreeing with Newton's "relational" doctrine of motion that all accelerations are accelerations relative to some *thing*. Where Newton fails, he argues, is in his attempt to show that no material object could be the proper reference frame for the absolute accelerations to be relative to. According to Mach, all that Newton's arguments show is that no *local* variation in the state of the surrounding material objects will affect the inertial forces felt by a test system. But nothing in Newton shows that one cannot account for these forces by referring to the acceleration of the test system relative to the far distant, enoromus masses of the astronomically-observable material world.

The greatest difficulty with Mach's proposal is this: To account for the existence of inertial effects he must postulate the existence of a force that acts between all material objects which is (1) dependent for its magnitude and direction on the relative acceleration of the objects, (2) highly dependent on the masses of the objects, so that the enormous masses of the far distant objects can make their effect predominate over any local variation in the material surroundings of the test system, and (3) of enormously long range, so that the effect of the very distant masses is not overwhelmed by the smaller but far closer masses of the local surroundings of the object. When we spin the bucket relative to the fixed stars the water sloshes, and when the earth rotates its atmosphere suffers Coriolis forces. The effect on the test systems of the farthest galaxies outweighs the effect of the earth, sun, or even the Milky Way galaxy, as far as inertial forces are concerned.

Mach simply postulated that such forces exist. The forces with which he was familiar, gravitational and electromagnetic, plainly would not do. They are far too short-range or else have the wrong form, depending only on separation and relative velocity and not on relative acceleration. As we shall see below (III,E), Einstein hoped to find such Machian forces in his relativistic theory of gravitation, general relativity. But it appears that general relativity is not the ultimate relationist theory for which Mach hoped.

I will discuss this problem in some detail in (III,E) below, but I might here just touch upon the following question: Given Mach's assertion that we are presented with the universe just once, complete with the system of fixed stars, could one ever possibly test to see whether Mach or Newton was actually correct? Don't both theories have exactly the same observational consequences? Well, no. Even if we agree with Mach that one can never perform Newton's "spinning weights in an empty universe" experiment, and, hence, never use its results as a crucial test between the Newtonian and the Machian theory, the structure of the Machian theory still seems such that some kinds of experimental tests for or against it, relative to the Newtonian theory, seems possible. According to Mach, the inertial forces are generated by the relative acceleration of a mass with respect to other masses. So one should be able to modify the inertial forces suffered by an object by modifying the structure of its surrounding masses. Granted that the enormous mass of the universe of fixed stars, combined with the long range of the force, will always fix the predominate structure of the inertial effects, still a sufficiently gross change in the relation of the test system to its surroundings should introduce, according to a Machian theory, some change no matter how small in the relation between acceleration relative to the fixed stars and the inertial forces that result. According to Newton, the inertial effects are the result of acceleration relative to a totally unmodifiable substantival space, and nothing we do can ever change their lawlike association with acceleration. So the possibility of a crucial experiment is not ruled out entirely. In addition, it may be that a general theory that we believe for independent empirical reasons will have theoretical consequences that differentially support the Machian or the Newtonian theory. We shall see how this direction of investigation goes when I ask whether general relativity supports a Machian or Newtonian world-view in (III,E) below.

I might note at this point that in some versions of his theory Mach denies any distance dependence of his new force at all, thereby making the test between Mach and Newton, which depends upon local variation in the inertial forces with local variation in the distribution of the surrounding matter, impossible. But for the future, I will stick with the version of the Machian theory which allows at least a weak dependence of the force upon the separation of the interacting objects.


## 3. Neo-Newtonian Spacetime

In this section and in the next one we will see how the notions of absolute position, velocity, and acceleration fare in the light of two doctrines about the structure of spacetime whose very nature was quite unimagined by Newton or his critics throughout the seventeenth through nineteenth centuries. Both of these accounts do assume, at least on the surface, the reality of spacetime as an "entity" over and above the material things that may happen to be located in it. Once again one must reflect upon the fact that the critical examination of Newton is two-staged: (1) To what extent was Newton correct in thinking that the facts required the postulation of absolute motions, as opposed to motions merely relative to other material objects? (2) To what extent was Newton correct in believing that the existence of absolute motions requires the postulation of a substantival space, or of spacetime in general, as an independent existent?

Both the theory of neo-Newtonian spacetime and the spacetime theory of special relativity postulate spacetime structures radically different from Newton's "substantival space persisting through time." And the notion of absolute acceleration becomes quite different from Newton's in these accounts. While the accounts agree with Newton that absolute acceleration exists, and that an object absolutely accelerated is accelerated relative to some entity, both deny this "reference entity" the status of a *thing* possessed by Newton's substantival space.

Both theories would give some comfort to a verificationist, for in both of them some of the features of the world which Newton postulates to exist but to have no observable consequences are denied existence. But neither theory would fully satisfy a pure relationist, for in both cases the theories, at least superficially, speak of a spacetime whose existence is postulated to be independent of the existence of any material objects whatever and whose features are postulated to be real and independent of the material objects that happen to occupy the spacetime arena. In (III,F) I shall once again examine in detail the relation between a theory's view about absolute motions and its metaphysical commitment to a substantival spacetime.

The structure of neo-Newtonian spacetime looks like this: The basic entities of which the spacetime is constituted, as a whole is constituted of its parts, are event locations, which as before we will call events, hoping that the reader will not confuse events as locations with events as possible or actual happenings at the locations. Between any pair of events there is a relation that generates the temporal separation of the events. As in Newton's theory, this is an absolute notion. Any two events have a definite temporal separation, which may be zero, and their separation in time is not relative to any particular reference frame, state of motion, or whatever.

A class of events that are simultaneous forms a space. It is assumed as in Newtonian spacetime that the relation of simultaneity, of having a temporal separation of zero, is an equivalence relation: every event is simultaneous with itself; if $a$ is simultaneous with $b$ then $b$ is simultaneous with $a$; and if $a$ is simultaneous with $b$ and $b$ with $c$, then $a$ is simultaneous with $c$. One can, therefore, divide up the class of all events into "equivalence classes" under the relation of simultaneity, i.e., into classes that are disjoint (have no members in common) and when taken together exhaust the class of all events. Each such equivalence class of simultaneous events is called a space.

The structure of each space is assumed to be that of Euclidean three-space. So far the theory parallels Newton's in every detail. But where neo-Newtonian spacetime differs from that of Newtonian is in the way that the spaces are "glued together" to form a spacetime. In Newton's theory of space-through-time, we can keep track of any given spatial point through time. That is, at $t_1$ we can ask of a point of the space at

$t_1$ whether it is identical with some point of the space at $t_0$. In neo-Newtonian spacetime this is impossible. The points of the spaces are events. For simultaneous events $a$ and $b$ we can ask whether they are the same event—two events at different times are, obviously, different events—but we cannot ask, as one can in Newton's theory, whether the two events occur at the same place in substantival space. In Newton's theory we can view places in substantival space as classes of events that occur at different times but are related to one another by the equivalence relation of being spatially coincident. In neo-Newtonian spacetime, the notion of spatial coincidence holds only for events that are simultaneous. There is no spatial coincidence or spatial noncoincidence for events that occur at different times.

Now in the Newtonian theory it makes sense to ask for the average velocity of a material particle between two events in its history. In the neo-Newtonian theory it is still coherent, of course, to ask for the velocity of a particle between two events in its history, if it is the velocity relative to some particular material frame which is in question. For we can ask whether the position of the particle relative to some other material particle at the later time is the same as or different from its relative position at the earlier time. But since we cannot ask whether the absolute position at the later time is even the same as or different from its absolution position at the earlier time, the notion of "absolute velocity" for a particle is just not defined in the theory. Since the spaces fit together to form spacetime in a way that forbids inquiry into the sameness or difference of position for events at different times, there is simply no such thing as "the" change of position of a particle through time, and, hence, no such thing as "the" velocity of a particle through an interval of time. And since there is no such thing as "the" average velocity of a particle through an interval of time, there is no such thing as "the" instantaneous velocity of a particle at a time, since this notion is defined as a limit of a series of average velocities. So neo-Newtonian spacetime satisfies at least one verificationist objection to the Newtonian theory. There is no longer in this view an absolute velocity of a particle existing as a real but totally unobservable feature of the world.

But if absolute velocity simply does not exist in the neo-Newtonian view, how can there be such a thing as absolute acceleration in this account? The answer is simple. We simply build in enough additional structure to the spacetime to allow for the definition of noninertial motion. It works like this: So far the only relation nonsimultaneous events have to one another is their temporal separation. I now introduce a new three-place relation, that of three nonsimultaneous events being inertial, or if we wish, the relation of $c$ being inertial relative to $a$ and $b$. I assume that there is such a relation. The empirical correlate of this relation is clear. Suppose that we have three events, $a$, $b$, and $c$, none of which is simultaneous with one of the others. How can we test to see if the third is inertial relative to the first two? We look to see if there is a possible path of a particle such that: (1) Three events in the particle's history are located at $a$, $b$, and $c$, and (2) the particle is at rest in an inertial frame, i.e., in a frame such that any system at rest in that frame has no inertial forces acting upon it. Essentially, collections of events, all of which are nonsimultaneous with any other event in the class, constitute inertial classes of events if they are all related to one another by being locations of events in the history of some particle moving free of forces, moving "inertially."

With this additional structure built into the spacetime, it is easy to see how absolute accelerations are both definable and measurable. What is the average acceleration of a particle over an interval of its trajectory? Take the particle at the beginning of the interval. Find a reference frame that is (1) inertial and (2) such that at the initial time the test particle is at rest in this inertial frame. At the end of the interval, find the new inertial frame in which the particle is at rest. Now find the *relative* velocity of the second frame relative to the first at the end of the time interval. We don't know what the absolute velocity of the first frame is; indeed, there is no such thing as its absolute velocity. But we do know that the first inertial frame has had no *change* in its velocity throughout the interval, since it is an inertial frame and these, by definition, fix the meaning of "frame that suffers no velocity change over an interval." So the relative velocity of the second frame to the first at the end of the interval gives the "absolute velocity change" of the particle over the interval, since the particle was at rest in the first frame at the beginning of the interval and at rest in the second at the end. Now take this absolute velocity change and divide it by the temporal separation between the events at the beginning and end of the intervals. This is the "average absolute acceleration" of the particle over the interval. By the usual limiting process we generate the notion of instantaneous absolute acceleration.

In this neo-Newtonian view, then, absolute accelerations are both real and empirically measurable. Absolute velocities are not measurable, but that is because they simply do not exist. It is clear that neo-Newtonian spacetime is somewhat more appropriate to the facts that led Newton to postulate substantival space-through-time as his model of spacetime. It makes fewer unnecessary postulations about the structure of spacetime, in the sense that it puts into the structure of spacetime only those features that have empirical consequences.

Well, not quite. To be sure, the old verificationist objection to Newtonian spacetime, that it postulated the existence of absolute velocities with no observable consequences, is now vitiated. But the very hard-nosed verificationist is still likely to be dissatisfied. Consider the world at a given moment in time. According to the neo-Newtonian theory there is a set of spacetime locations which are simultaneous and constitute the space of the world at this time. The material objects of the world are situated at various "places" in this space. But would it make any observational difference, we can hear Leibniz object, if the world were situated at the time at some other location in "space itself," all spatial relations of material objects relative to one another kept unchanged? No it would not. Once again this extreme version of verificationism is contravened, as indeed it must be by any spacetime theory that postulates a spacetime with an existence over and above the existence of the material objects "in" the spacetime and yet denies any direct observability of the spacetime locations themselves.

## 4. *The Spacetime of Special Relativity and Absolute Acceleration*

I have already described the spacetime appropriate to special relativity, Minkowski spacetime, in some detail in (II,C,1), and here I will only rehearse its most crucial features. My concern in this section is not to motivate the adoption of this spacetime model, I leave that to Chapter IV below, but only to see how the notion of absolute motion fares in the light of this new spacetime theory.

As in the case of Newtonian and neo-Newtonian spacetime, the spacetime of special relativity is postulated, at least on the surface level, as an independent entity having its own existence and its own structure and, as in the other cases, being totally unaffected by the presence of whatever events happen to take place in its arena. It goes without saying, I think, that the theory is going to fall prey to just those objections of the extreme verificationist to neo-Newtonian spacetime which I brought out at the end of the last section. In this case, however, the objection will be stated slightly differently. In Minkowski spacetime there is no such thing as "the space at a time," since, as we saw in (II,C,1) the notion of simultaneity of events or event locations is no longer invariant. The hard-nosed verificationist argument would now have this form: Since it would make no empirical difference if all the material events of the world had taken place at event locations other than those at which they "actually" occurred, all spatiotemporal relations among the material events being preserved as they are, the notion of spacetime location of an event, as opposed to the notion of actual or possible material event itself, is illegitimate.

Some important features to note about Minkowski spacetime for present purposes are these: (1) Unlike Newtonian or neo-Newtonian spacetime, the notion of an absolute time separation between event locations is not well defined. Events have a temporal separation only relative to a particular state of motion of an observer. (2) As in neo-Newtonian spacetime, and as opposed to Newtonian spacetime, the notion of the spatial separation of nonsimultaneous event locations is not well defined. There is simply no such thing as "the distance" between events that are nonsimultaneous. Taking into account the relativity of simultaneity the picture looks like this: Independent of the specification of a reference frame in a particular state of motion we can say of two events neither that they are simultaneous nor that they are spatially coincident. If the events are such that relative to some state of motion of an observer they are simultaneous (as we shall see in Chapter IV only certain pairs of events have this feature), then we can ask for the spatial separation of these events relative to this particular observer. If the events are such that they are nonsimultaneous relative to every observer, we can still define a notion of spatial separation for them relative to an observer in a particular state of motion. Once again, however, the "absolute temporal separation" and the "absolute spatial separation" of events are not well-defined notions in Minkowski spacetime.

Not surprisingly, the notion of the "absolute velocity" of a material particle is also not well defined according to the theory. There simply is no such feature of the particle's history. Curiously enough, there is a notion of absolute velocity appropriate to light waves in a vacuum and, indeed, for any signals that travel at the velocity of light (gravitational radiation, for example, if there is any such thing). For according to the theory: (1) The notion of a reference frame's being inertial is well defined and (2) the velocity of light and similar signals in vacuum is a constant relative to all such inertial systems and the same in each of these systems. Why this is so will be seen in Chapter IV.

Is absolute acceleration a well-defined notion in Minkowski spacetime? As I noted above, the notion of inertial frame is well defined in Minkowski spacetime, just as it is in neo-Newtonian and Newtonian spacetime. Naturally, then, so is the notion of absolute acceleration.

Consider the geodesics of Minkowski spacetime. If a one-dimensional set of events consists of events at what I called (II,C,1) *spacelike separation* from one another, i.e., if the events are not connectible by any material causal signal, then the geodesics will consist of events that are simultaneous and on a straight line in space relative to some observer. If the events are all at *lightlike* separation from one another, i.e., if a light ray could pass through the set of events, then the set of events constitutes a geodesic path which is the path of a possible light ray. If the one-dimensional set of events consists of events that all have *timelike* separation from one another, i.e., if they constitute a possible path of some material particle, then they will constitute a geodesic in the spacetime if and only if a particle in inertial motion could pass through all the events in question. The Minkowski spacetime theory, and the special theory of relativity on which it is based, assumes the possibility of observationally distinguishing inertial motions of particles from noninertial motions.

Now the status of absolute acceleration in Minkowski spacetime is different from that in neo-Newtonian spacetime. In both cases it is meaningful to ask of a set of events whether it constitutes the path of a possible inertial motion. In the neo-Newtonian theory, however, we can go further, for once we have the notion of inertial motion and temporal separation of events we can, as we have seen, define an invariant notion of magnitude of absolute acceleration. In special relativity the "ordinary" absolute acceleration is no longer invariant. Since the temporal separation of events is no longer an invariant notion, there will no longer be a notion of the magnitude of ordinary absolute acceleration definable in the theory, but only the notion of the acceleration of a particle relative to a particular inertial frame chosen as reference. Once we have chosen such a frame, the temporal separation of events is definable relative to that frame, and the definition of the magnitude of acceleration can be carried out as was done in the case of neo-Newtonian spacetime. It is interesting to note that even within special relativity a new invariant magnitude, one that serves the purpose of the magnitude of absolute acceleration, is still well defined.

But if we refuse to adopt a particular inertial frame of reference, the only invariantly characterizable magnitude of acceleration is the zero magnitude. It is meaningful to ask, "Is this particle suffering no absolute acceleration at all?" It is not meaningful to ask, "What is the (nonzero) magnitude of the absolute acceleration of this particle?" So we are left with a residual part of the notion of absolute acceleration.

For the Newtonian theory, the absolute position, velocity, and acceleration of a particle all exist. Only the absolute acceleration has empirically noticeable consequences. For the neo-Newtonian theory, absolute acceleration is well defined, but there is no such thing as absolute velocity. In the special relativistic spacetime, whether a particle has a nonzero absolute acceleration is a meaningful question, but there is no such thing as the absolute velocity or the absolute magnitude of nonzero accelerations. And, once again, all of these theories, by postulating an independently existing but "unobservable directly" spacetime arena for the interaction of material things or events, violate the canons of scientific legitimacy of the extreme verificationist.

Only a theory satisfying Mach's constraints of postulating no spacetime whatever, over and above the spatiotemporal relations of things or events to one another, would satisfy the "pure" relationist. Einstein at one time thought that his general relativity theory was the real scientific theory that fulfilled Mach's speculative hopes. I will turn now to a development of some aspects of this theory and to a critical examination of the question as to whether general relativity is really the theory for which Mach longed.

# GENERAL RELATIVITY AND THE PROBLEM OF SUBSTANTIVAL SPACETIME

## 1. *Einstein's Machian Motivations*

The notion of inertial framework plays a crucial role in Newtonian mechanics. If we wish to use the Newtonian laws in their usual form, say $F = ma$, to calculate the behavior of an isolated mechanical system as it evolves from an initial state, and if we wish to invoke only the usual forces of interaction generated by the components of the system on one another, like the electromagnetic and gravitational forces, then we will obtain correct results only if we determine the positions of the components of the system in some inertial frame. Any inertial frame will do, but if we wish to describe the evolution of the system in, say, a rotating coordinate frame, we must invoke centrifugal and Coriolis forces which act on the components of the system but are not generated by the "real" interactions of these components among themselves.

The inertial frames play, as we shall see in Chapter IV, if anything an even more important role in the special theory of relativity. Once again, the usual laws correctly describe the evolutions of isolated systems when we restrict the "forces" to the usual interactions of the components of the system on each other, only if we describe the system from the point of view of an inertial framework.

One of Einstein's motivations in formulating the general theory of relativity was to find a set of laws that could accurately describe the evolution of systems from the viewpoint of any reference system whatever, inertial or not and equipped with the usual Cartesian coordinate system or not. Hence the name "general" relativity. As a matter of fact, as Kretchmann pointed out quite early, one can rewrite the laws of Newtonian mechanics and of special relativity so as to make them uniformly applicable in general reference frames. Of course, doing so is rather a mathematical trick, and the physical distinction between inertial and noninertial frames still exists for these two theories. But as we shall see, the real change involved in going from special to general relativity doesn't obviate this real physical distinction either.

To see why Einstein had hopes that the general theory of relativity would be the theory that finally satisfied the Machian constraints of a good scientific theory, we need only reflect upon some of the considerations Einstein used in connecting gravitation to inertial forces, connections we have seen once before in (II,D,2). The basic chain of reasoning went like this: As far as *local* observation can tell, the mechanical effects of gravitation are indiscriminable from the inertial effects resulting from using an accelerated system in which to perform one's measurements in the absence of a gravitational field. Let us conjecture that this local equivalence of gravity with accelerated coordinate systems holds for all phenomena, not just mechanical phenomena. This will give us, as a result, a theory of the action of gravitation on light, and further, since special relativity convinces us that putting a coordinate system into motion changes the spatiotemporal features of the world as determined by an observer fixed in that system, this will also convince us that gravitation affects the spacetime structure of the world as well.

This conceptual equating of gravity with acceleration of the reference frame, only a *local* equivalence it is important to remember, provides Einstein with his key to a relativistic theory of gravitation, as we saw in (II,D,2). But it also suggested to Einstein that the theory he was constructing would finally satisfy Mach's hopes. Remember that the greatest difficulty with the Machian theory was that there seemed to be no such long-range acceleration-dependent forces that could explain the origin of inertial effects in terms of an interaction between the test system and the rest of the mass of the universe generated because of their relative acceleration. But if gravitation is so intimately connected with the inertial behavior to be expected when we observe a test system from the point of view of an accelerated reference frame as general relativity proposes, perhaps we can account for all "inertial" effects not in terms of some absolute acceleration of the test system with respect to "space itself" as in Newton, nor with respect to "the system of inertial frames of spacetime" as in the neo-Newtonian theory, but rather as the result of the *gravitational* interaction of the test system with the rest of the mass of the universe.

The traditional inverse square gravitational interaction with which Mach was familiar would not suit for this purpose, of course. The force is much too short range, diminishing with the square of the separation of interacting objects, and is in addition not properly acceleration-dependent. But, perhaps, the new relativistic theory of gravitation might just serve the Machian purpose.

Given a specific theory, like general relativity, it becomes possible to "conceptually test" the theory to see whether it really is the kind of theory a convinced Machian relationist should accept as satisfactory. Let me list a few features that Einstein and others have suggested might be expected to be possessed by any genuinely Machian theory. Then I shall say a few things about the structure of general relativity, and finally I will see whether general relativity meets these touchstones of Machianism. In anticipation, let me say that we will discover that in general relativity, as in special relativity, a new invariant notion of absolute acceleration is well defined. This alone should lead one to assert that general relativity is certainly not a Machian theory. But it is useful to explore in some detail some of the specific ways general relativity leads to consequences at variance with Mach's expectations.

1. Despite Mach's skepticism about the rationality of even contemplating "possible worlds" as different from the actual world as would be a world consisting of a small test system and no other matter whatever, it seems reasonable to ask what a theory would predict about a test system in such a world, if it predicts anything at all. If the inertial forces are generated in a test system solely because of its acceleration relative to the remaining mass of the universe, one would expect a Machian theory to predict that there could not be two distinct states for a test system in an otherwise empty universe, one in which it felt no inertial forces and one in which inertial forces played a role. As we shall see, general relativity is applicable to the description of the behavior of test systems in worlds devoid of other matter, and we shall see that it is not in the least clear that general relativity gives the expected Machian results in this case.

2. Since inertial forces are, according to Mach, generated by the relative motion of the test object to the remaining matter of the universe, one should expect a Machian theory that the inertial forces suffered by a test

system depend uniquely upon the distribution of matter throughout the universe and the motion of the test system with respect to this distribution of matter. Again, as we shall see, it isn't clear that general relativity is not compatible with different possible worlds in which (1) the distribution of matter is the same on both worlds, (2) the motion of the test system with respect to the remaining matter is the same in both worlds, and yet (3) the test system experiences different inertial forces in the two possible worlds.

3. According to Mach absolute acceleration, rotation say, is really just acceleration relative to the averaged out mass of the universe. So to speak of rotation, say of the whole mass of the material world, should be an absurdity in a Machian theory. But, as we shall see, there seem to be possible worlds compatible with the general theory of relativity which are naturally described as being worlds in which the total mass of the universe is in rotation.

4. According to Mach, and he asserts this explicitly, the universe consisting of a test system rotating relative to a stationary remaining mass and the universe consisting of the total remaining mass in rotation about the test system are the same universe, and should give rise to the same inertial effects on the test system. But, as we shall see, general relativity does not fully conform to Machian expectations in situations of this kind.

5. If inertial effects are due to the relative motion of the test system and the remaining masses, changes in the mass distribution should be reflected in changing inertial behavior of the "accelerated" system. The general theory of relativity does, in fact, predict that the inertial forces will vary depending on the structure of the overall mass distribution of the universe. But, as we shall see, this can be interpreted in a way that makes the variation seem an "addition" to the basic inertial effect which is independent of the mass distribution—again an interpretation at variance with Machian hopes.

## 2. Spacetime and Gravitation in General Relativity

Let me review a few of the basic components of the general theory of relativity at this point. We will need only a short résumé of the basic features of the theory, and most of what I shall say will merely repeat matters already briefly treated in (II,C,2) and (II,D,2,3).

The spacetime postulated by general relativity is a four-dimensional pseudo-Riemannian manifold. It is the generalization of the pseudo-Euclidean Minkowski spacetime of special relativity and stands to this spacetime much as Riemannian spaces stand to Euclidean spaces. The spacetime can be characterized by assigning to each event location in it four coordinates and then mapping the $g_{ik}$-function relative to these co-ordinates. This first fundamental form contains in its specification a full description of the spacetime relative to the coordinatization chosen. The $\Gamma_{jk}{}^i$-function which describes the geodesic structure of the spacetime can be written as a function of the $g$-function and its derivatives, relative to the chosen coordinatization, as can the $R_{jkl}{}^i$ Riemann tensor function which can also be used to fully describe the spacetime.

Is this curved spacetime treated by general relativity as "an entity in itself, over and above the things in it"? A few remarks are in order. First, the theory treats of the spacetime as substantival in its surface presenta-tion, just as do Newtonian, neo-Newtonian, and Minkowski spacetime theories. Any claim that the theory really affirms spacetime to exist solely as a set of relations among ordinary material things requires, as usual, an argument on the part of the reductionist.

In many ways the spacetime of general relativity appears, if anything, more "real" than do the spacetimes of the older theories. For this space-time is "affected" by the distribution of material bodies in it and "affects" their behavior. As I have already noted in (II,D,2) this idea that the matter distribution "causes the spacetime to have the form it does" must be taken with a grain of salt.

In addition, as I shall discuss in (III,E,4) below, the new dynamical aspect of spacetime in the general-relativistic account, plus the diversity in its structure from point to point, as say the curvature varies, have led some to hope that with general relativity the supersubstantivalist program of not only treating spacetime as having being but of identifying all ma-terial objects with spacetime itself finally may come to fruition. There is nothing in the structure of general relativity as a scientific theory, that is, to give immediate aid and comfort to the pure relationist.

But at the moment we are more concerned with the place of absolute motion in general relativity than with its place in the overall substantival-ist-relationist controversy. To throw light on this we need to remember a few features of general-relativistic field theory and dynamics.

According to general relativity, light rays traveling under no influences but that of gravitation follow null-geodesics in the spacetime. Point masses traveling under no influence but that of gravitation follow time-like geodesics in the spacetime. The elimination of gravitation as a force imposed on otherwise freely-moving particles has come about, remember, by the identification of the gravitational field with the metric structure of spacetime itself. The possibility of this identification was founded on the important fact of the local equivalence of gravitation and accelerated coordinate systems which was Einstein's generalization from the well-known exact proportionality of inertial and passive gravitational mass. These laws of motion may be taken to be an independent element of the theory, or to be themselves derivable from the field equations by the means attempted by Einstein's later work.

To know how particles and light rays will behave, when they are "freely" moving under only the influence of gravitation, we need not only the equations of motion but the specification of the gravitational field, i.e., of the intrinsic geometric structure of the spacetime in the region through which the light rays or particles are traveling. The kind of gravitational field, i.e., spacetime, which exists in a region is intimately related to the mass-energy distribution throughout the region, and this lawlike connection between the two is expressed in the gravitational field equation which is a differential equation relating a function of the $g$-function and its derivatives in a region to the stress-energy tensor describing the distribution of mass-energy and forces throughout the region. Remember that it is rather implausible to view this as expressing a causal relation of mass-energy distribution to spacetime structure, since the stress-energy tensor itself expresses *metric* features of the mass-energy distribution. Rather we should look upon the equation as a kind of consistency constraint upon possible worlds. We describe a world by picking a coordinate system, expressing a $g$-function in terms of that coordinate system which characterizes the spacetime of the world, and then picking a stress-energy tensor function to describe the distribution of the nongravitational mass-energy and forces throughout the world. If the proper function of the $g$-function we have chosen is equal to the stress-energy function, then the world is genuinely possible according to general relativity. Otherwise what we have described is not a possible world at all.

The following question is extremely important in the pursuit of the issue of the place of absolute motion in general relativity and the suitability of the theory as Mach's hoped-for pure-relationist theory of motion: Is the field equation of general relativity such that, in general, a given mass-energy and force distribution (for nongravitational mass-energy and force, of course) is compatible with one and only one spacetime structure?

The answer is negative. Just as a given distribution of charges in classical electrostatics is compatible with any number of electrostatic fields, depending upon the *boundary conditions* imposed upon the fields, so a given stress-energy tensor in general relativity is, in general, compatible with many different spacetime structures. To pick out the right structure for the spacetime to associate with a given mass-energy and force distribution, we need not only the field equation but the imposition of boundary conditions as well. To generate, for example, the famous Schwarzschild solution to the field equations, where the total nongravitational mass-energy of the world consists of a single mass point, we need the additional condition that the spacetime of the world at great distances from the mass point becomes approximately Minkowskian.

As we shall now see, the necessity of imposing boundary conditions in order to uniquely specify a spacetime structure will throw cold water on the hopes that general relativity was the long-sought Machian theory.

### 3. Non-Machian Aspects of General Relativity

In (III,E,1) I gave five conditions that a theory would be expected to meet for it to be called a truly successful Machian theory. Let us see how general relativity fares with each of these conditions.

1. For a theory to be Machian, it should predict that a test system in an otherwise empty universe will feel no inertial forces whatever its state of motion. For, if Mach is correct, there should be no such thing as a state of motion for a test system when there is nothing for the motion to be relative to.

What does general relativity predict about the behavior of test systems in universes devoid of other matter? Well, the stress-energy tensor is surely zero everywhere. But, as I have noted, the form of the stress-energy tensor does not uniquely specify the spacetime structure. If we impose the "natural" boundary conditions that the spacetime be "Minkowskian at infinity," we come to the conclusion that the spacetime of an empty world is just "flat" Minkowski spacetime. And what consequence does this have? The answer is that all the distinctions between the test system being in uniform motion or being "absolutely" accelerated which hold in Newtonian, neo-Newtonian, and Minkowski spacetime hold here as well. In such a world there are two distinct classes of motion for the test system—that in which it feels no inertial forces and that in which it does. Just as Newton claimed, we can still distinguish in such a world between the two weights nonrotating and the weights rotating about their common center of gravity, and the distinction has its observable consequences that in the first case there is no tension in the rope connecting the two weights and in the second there is a tension.

2. According to Mach, the inertial forces suffered by a test system should depend only upon the distribution of matter throughout the universe and the state of motion of the test system relative to this matter. Is this true in general relativity?

The answer is again negative. The inertial forces suffered by a particle will depend upon its state of motion and the structure of spacetime in the region through which it is moving. It is deviation of its trajectories from the timelike geodesics which results in "inertial forces," just as in the Newtonian theory it is acceleration, deviation from a straight-line spacetime trajectory, that induces inertial forces. But since the structure of spacetime is not fully determined by the mass-energy distribution, neither are the inertial forces felt by a particle in motion relative to a given mass-energy distribution.

For example, consider the simplest possible case: a test system in a universe totally devoid of other matter. Even in such an empty universe it has been shown that more than one spacetime structure is compatible with the field equations. Once again, the spacetime structure can be uniquely constrained only by the imposition of boundary conditions. So a given motion of a test system will induce different inertial forces on the particle depending on what the spacetime structure of the region through which it is moving is like, and this spacetime structure cannot be determined solely by the mass-energy distribution.

3. According to Mach it would be absurd to speak of the rotation of the average smeared-out total mass of the universe.

Yet the logician Kurt Gödel has constructed a solution to the field equations whose natural interpretation is just that—the total mass of the universe is in absolute rotation. Now this solution, as we shall see in (IV,D), has other peculiar features as well, but we should not let that fact stand in the way of using Gödel's model of a universe to show us how the notion of absolute rotation of the total matter of the universe is intelligible in the general-relativistic framework.

Why is the natural interpretation of Gödel's solution that of a universe whose total mass is in rotation? The answer is this: Pick a framework at rest in the averaged-out total mass. Examine the trajectory of a light ray or particle sent out from a point and subject to no forces as viewed from the framework fixed in the averaged-out total mass. In Gödel's universe these trajectories would appear to be spirals, just as the trajectory of a bullet fired above a rotating disk would describe a spiral on a coordinate frame fixed in the disk. The universe can be in absolute rotation in general-relativistic worlds, in absolute rotation relative to the "compass of inertia" formed by the paths of freely-moving light rays or material particles (see Fig. 23).

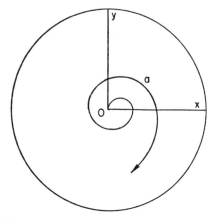

Fig. 23. Free motion in the Gödel solution. The $x$-$y$ coordinate frame is fixed at rest in the average smoothed out mass of the universe. The spiral, $a$, is the path of a "free" particle emitted from $O$. The natural interpretation of the spiral is that the particle is traveling a straight line and the average smoothed out mass is in "absolute rotation," resulting in the spiral path observed from the $x$-$y$ frame.

4. According to Mach, the inertial forces generated on a disk by putting it in rotation relative to the remaining mass of the universe should be the same as would be experienced on the disk if no forces were applied to set it in motion but instead the remaining mass of the universe were set in rotation about the disk.

Now it was first thought that general relativity met this Machian condition. Suppose we imagine a laboratory set on a disk. Around the disk we construct a massive cylinder. We set the cylinder in rotation about the common center of itself and the disk. Will we, according to general relativity, observe centrifugal and Coriolis forces on objects in the disk-laboratory due to the rotation of the shell around the laboratory. Thirring showed that we could expect such forces, as Mach would lead one to expect.

But once again the Machian must be disillusioned. For consider a disk surrounded by two concentric shells of differing radii. Machian considerations would suggest the same inertial effects experienced by an observer on the disk if either (1) the outer shell is put in clockwise rotation and the inner left alone or, (2) instead the inner is put in equal counter-clockwise rotation (the angular velocity being the same in magnitude but opposite in orientation) and the outer shell is left alone. For in both cases the observer on the disk sees two cylinders at the respective radii in the same angular rotation relative to each other. But in general relativity, with the usual boundary condition assumptions, the two situations give rise to differing "inertial forces" on the disk. According to general relativity, spinning the laboratory frame and spinning the world of matter about the laboratory frame are *not* physically equivalent as Mach would lead us to expect.

5. Machian consideration would lead one to suspect that the inertial forces suffered by a test system for different states of its motion should depend upon the particular distribution of masses in the universe.

Now it is true that the distribution of mass-energy in the universe contributes to "fixing" the structure of spacetime in a region, and it is the structure of spacetime in a region that determines what inertial effects an object moving on a trajectory through that region will suffer. But that this should be taken as vindicating Mach, or as indicating the lack of absolute motion in general relativity, is doubtful.

Suppose we adopt the natural boundary conditions that spacetime is Minkowskian in regions far removed from any matter immersed in it. The behavior of a test system in a region affected by the nearby presence of matter will then depend upon (1) the fact that spacetime is "overall" Minkowskian and (2) the fact that the presence of mass-energy nearby has "distorted" the local spacetime from its natural Minkowskian flatness. But Minkowskian spacetime means the ordinary distinction between inertial and noninertial motions familiar from Newtonian and special-relativistic theories. And the distortion from flatness of a region of spacetime due to the presence of nearby matter is naturally interpreted as the imposition of gravitational effects on the system over and above the normal *inertial* effects of the pervasive Minkowski spacetime. In other words, for the version of general relativity which adopts this standard boundary condition, test systems suffer inertial forces because of their absolute accelerations. In addition, they suffer gravitational forces because of the presence of nearby matter. Put this way, the theory looks hardly different from either Newtonian mechanics or special relativity, although, of course, the details of the forces the test system suffers will differ in general relativity from those predicted by the earlier theories. It is certainly not an interpretation of general relativity which would bring joy to the heart of Mach, but it is the natural way of looking at things.

The fact that the full specification of the spacetime "field" in general relativity can be determined only after boundary conditions are imposed, and the fact that it is this element that leads to most of the non-Machian aspects of the theory, may give the non-Machian features the appearance of mere mathematical technicalities. I think that this is false. In classical field theories there is a "natural" field, the zero-field, and the role of charges as sources of fields is to superadd their own contributions over and above the "ground level" zero-field. In general relativity the natural source-free field to adopt is that of Minkowski spacetime, a spacetime in which absolute accelerations are, as we have seen, both well defined and empirically detectable. It is this physical feature, of gravitation being superimposed upon an already present inertial field, that the necessity for the impositions of boundary conditions mathematically represents.

Of course, in classical theories one can sometimes have nonzero fields present, even if they have no origin in charges as sources. For example, radiation can exist in otherwise empty space. Similarly in general relativity, although empty Minkowski spacetime is a natural choice of background spacetime to which to add the contributions of mass-energies to the metric field, it is not the only possible choice of the spacetime field appropriate to empty spacetime, as we have seen. But in any case, the total field can be determined only as a result of the composition (in the case of general relativity, not a simple addition of fields, since the field theory is nonlinear) of the contributions of the distributed mass-energy and of the "background" spacetime itself.

Some contemporary physicists, recognizing the many ways general relativity fails to satisfy Mach's criteria of acceptability for a theory and yet still feeling moved by Machian considerations, have attempted to modify the general theory in various ways to obtain a new theory more suitable from a Machian point of view. One method is to change the law relating the metric field to the matter-energy distribution. Dicke, for example, replaces Einstein's so-called tensor theory of gravitation (the gravitation field is specified by the $g$-function which is a second rank tensor function) by a scalar-tensor theory. There is some reason to believe that in a theory of this sort the spacetime structure may simply disappear in a universe devoid of matter, rather than become Minkowskian as it does in general relativity with the natural **Minkow-**

skian boundary conditions imposed. But the theory remains quite un-developed.

Another approach, suggested by Wheeler, is to insist that the space-time of general-relativistic solutions be spatially closed. Now this topological constraint on the allowable solutions to the field equations turns out to be a little more complicated than one might at first suspect, as we shall see in (IV,D) below. For one thing, it assumes that the spacetime can be split up into spaces-at-a-time, i.e., hypersurfaces of simultaneous event locations all at spacelike separation from one an-other, by choosing a particular state of motion for a reference frame. It is by no means obvious that this can always be done, but still, the motive for imposing this constraint upon solutions is fairly clear. If the spacetime is spatially closed at every time, there is no need to impose the boundary conditions of "Minkowski spacetime at infinity" since there is no boundary to the space at the time and no "points infinitely distant from a given point" or even "points as far from a given point as one likes." Under these circumstances, the connection between mass-energy distributions and spacetime structures imposed by the field equa-tions becomes tighter, so that one may hope for a unique spacetime structure being associated with each stress-energy tensor under these conditions.

But we have achieved this additional conformity of the general theory with Machian expectations only by superadding to the theory an additional independent hypothesis. And, though at least one objec-tion to the claim that general relativity is Machian is obviated—the objection resting on the possibility of many spacetime structures con-sistent with the field equations and a given distribution of mass-energy throughout the universe—it is still far from clear that the resulting modified theory is yet in conformity with Machian expectations in their fullest extent.

## 4. Geometrodynamics and Matter as Spacetime

This seems the appropriate point to make a few brief remarks about the scientific attempts that have been made in recent years to carry out the fulfillment of what I have called the supersubstantivalist pro-gram. From this point of view, not only does spacetime have reality and real structural features, but in addition, the material objects of the world, its totality of ordinary and extraordinary material things, are seen as particular structured *pieces* of the spacetime itself. Curiously enough, the same general relativity, which Einstein saw as establishing Mach's hopes of *eliminating* the postulation of a real spacetime in favor of a pure-relationist account of spacetime as nothing but spatio-temporal relations among material things, has been taken by Wheeler and his collaborators as the ideal theory on which to found a super-substantivalist view.

The reason for general relativity's prima facie suitability for this purpose is clear. Material objects are notable for their variety and their transience. They differ from one another in structure and form, and they change their structure from time to time. Only the curved space-time of general relativity with its variable intrinsic geometric features, differing from point to point and dynamically evolving over time, can provide a sufficiently rich structure to even hope to fulfill the super-substantivalist's aims. The homogeneous, isotropic, and ever-unchanging spacetimes of Newton, the neo-Newtonians, and special relativity are hopeless for this purpose.

The scientific establishment of the supersubstantivalist position would be one case of a scientific reduction. Just as the theory of macroscopic matter is reduced to the theory of matter's microscopic constituents by an *identification* of macroscopic things with arrays of microscopic entities (molecules, or atoms, or elementary particles), and just as physical optics is reduced to electromagnetic theory by an identification of rays of light with a species of electromagnetic radiation, so the scientific supersubstantivalist hopes to achieve a reduction of the theory of matter-energy by showing that this theory can be reduced to a theory of spacetime by means of an identification of each material thing in the world—elementary particle, light ray, or what have you—with a particular segment of structured spacetime.

As noted above (III,B,1), one must distinguish this scientific attempt to reduce the substance of the world to spacetime and nothing but spacetime from a quite different linguistic "trick" espoused by some philosophers. Once we have accepted the reality of spacetime, and accepted the view that all material things are located "in" it, we could trivially adopt a language in which spacetime was the only apparent entity of the world. We do this by replacing expressions that assert the existence of some material thing at a spacetime region ("There is a table in $A$," "There is a light ray throughout region $R$.") by expressions that attribute "material entity features" to the appropriate spacetime region ("Region-$A$ tables," "Region-$B$ light waves," or "Region-$B$ is light-wavish."). But such a replacement of referring expressions ('table,' 'light wave') by attribute expressions ('tables,' 'is light-wavish'), though it might amuse philosophers concerned with the distinction between substances and attributes, would plainly be of no scientific interest whatever.

The scientific reduction is more interesting. It attempts to show that the relation between what we take to be matter and what we take to be spacetime is such that for each distinct kind of matter we can describe, the spacetime region occupied by that matter has its own characteristic spacetime structure—its own characteristic intrinsic geometry, that is. If we knew the full spacetime structure of a region, then, we would already know exactly what kind of matter occupied that region. This suggests that we can do without the matter as a separate individual and simply identify a particular kind of spacetime region as that kind of matter.

Now the logic of such "identificatory reductions" has received some attention in the philosophical literature. Usually standing in the way of such reductions are properties possessed by the entities referred to in the reduced theory, but apparently not possessable by the entities referred to by the theory to which the reduction is to take place. For example, attempts at identifying mental events with brain events are usually objected to on the basis of the claim that the sorts of properties mental events have (being immediately inspectable, for example) can't be possessed by brain events. In the case of reduction we are examining, this objection seems out of place; for the material objects (actually, material event-histories, since we are working in spacetime, not space-through-time) that are to be identified with structured regions of spacetime are already the real material entities of the physicist, devoid of such "secondary qualities" as felt temperature, seen color, etc.

So the prospects of a scientific reduction of all the material stuff of the world to the stuff of spacetime itself seem philosophically no worse than the prospects of a reduction of physical optics to electromagnetic theory. Whether such an identification of matter with spacetime can scientifically be carried out remains to be seen.

What also remains to be seen is whether one can provide a reasonable philosophical answer to a difficult question which arises whenever identificatory reductions are considered. Suppose one physicist claims that the "matter in a region of spacetime" simply *is* the region of spacetime, justifying his claim by means of a scientific reduction of the kind postulated by the geometrodynamic theory we have just been discussing. Suppose another physicist maintains the alternative theory, equally compatible with the data, that the matter and the spacetime region, although lawlike-correlated so that a given kind of matter is always found in a specific kind of spacetime region and a spacetime region of given geometric character is always inhabited by a specific kind of matter, are actually *distinct* entities. How could one decide which theory to adopt—the identificatory or the correlatory? Indeed, are they really distinct theories? I think we can see that a philosophical critique of identificatory reductions will give rise to the debate so familiar from the controversy over the epistemology of geometry I discussed in (II,H,4).

# ABSOLUTE MOTION AND SUBSTANTIVAL SPACETIME—A PHILOSOPHICAL CRITIQUE

If we reflect on the material I have surveyed in Chapter III to this point, I believe that we will be led to two important conclusions:

1. There are important observational facts about the world which cry out for a systematic accounting. These facts clearly reveal an important connection between the forces exerted on a body and its state of motion. What account to give of these forces, what explanation to offer for their presence, and exactly what connection between them and the motion of an object suffering them to postulate, however, depends in important ways not only upon *facts* obtainable by systematic empirical investigation, but upon the adoption of various *philosophical* principles regarding what constitutes sound scientific inference and what constitutes adequate scientific explanation as well.

2. Even given one's basic account of the connection between the forces suffered by an object and its state of motion, the question of the substantival nature of space or spacetime and the absolute nature of time remains open. That is, to go from one's account of absolute motion to the adoption or rejection of either a substantivalist or relationist account of spacetime again requires the invocation of methodological or metaphysical principles whose defense and criticism seem more a matter for philosophical resolution than for scientific decision—

if we can, indeed, distinguish philosophical from scientific reasoning in any interesting way.

I will examine each of these claims in turn.

## 1. Inertial Forces and Absolute Motion

If we swing a rock on a string in a circle, there is a tension generated in the string. And if we are seated in an airplane with its jets firing, we are pressed back into our seats. But how are we to account for these phenomena?

We do notice the important observational fact that there is the intimate connection between the generation of these inertial forces and states of motion of objects which is summed up in the assertion of Galilean invariance. If a system is moving uniformly with respect to a system in which no inertial forces are experienced, no inertial forces are experienced in the other system either. And systems in uniform motion with respect to each other are such that objects at rest in them experience equal inertial forces, even when these forces are nonzero.

To generate a theory accounting for these facts requires many steps of inference. Each brings along its own interesting philosophical issues. First, there is the step of "mere inductive generalization." We observe all of these inertial phenomena, the existence of the forces and their connection with states of motion summed up in the Galilean invariance of mechanics, in our local region of the universe. Even if astronomical observation makes this local region rather large, by human standards, both in space and time, it is surely a bold leap to postulate the same phenomena everywhere and everywhen in the universe. As I have noted, general relativity makes one suspicious indeed of the boldness of this generalization.

Even brasher is the step, taken by Newton and recognized as an independent inference by him, of assuming that the phenomena we observe in the universe constituted as it is would be observed to occur in possible worlds quite different from the world we know. The claim that different states of motion of a test system could be distinguished observationally, by means of varying inertial forces exerted on the components of the system, in a world different from ours in that it is devoid of any of the material universe whose existence constitutes such an important feature of the real world, is a claim that outruns any mere inductive generalization. The inductively-supported claim of the ubiquity of the connection of inertial forces with states of motion in the actual world seems at least "testable in principle," say by getting into communication with experimenters who happen to inhabit extremely remote parts of our universe. But how could we test the claim that even in a world in which the fixed stars are annihilated, spinning buckets could be distinguished from those not spinning by the fact that the water surface would be concave in the former case but not in the latter? To generalize from what is the case in the actual world, even "universally" the case, to what *would* be the case in worlds quite different from the world we inhabit is to make a scientific inference that outruns "intra-actual-world inductive generalization." Such inferences require philosophic inquiry.

I am not claiming, mind you, that such inferences are illegitimate or even senseless, as Mach suggests. Scientists frequently describe the behavior of ideal systems that never were and never will be, and theoreticians of general relativity spend their lives describing what the dynamics of test particles would be in worlds possessing mass-energy distributions grossly at variance with that in the world in which we live. I only want to emphasize the existence of this "inferential leap," its importance for establishing a doctrine of absolute motion, and the way theories established on the basis of the results of thought experiments that take place in possible worlds other than the actual world have their support in considerations that outrun both empirical observations and simple inductive generalization from them.

Finally, and as Mach saw this step is intimately connected with that just noted, there is the inference to the claim that whatever motion of a test system accounts for the appearance of inertial forces in it, it cannot be construed as motion relative to any ordinary, or even extraordinary, material objects. It is this step that Mach denied, maintaining that while the inertial forces did have their origin in accelerations, these were perfectly ordinary accelerations relative to the other masses of the universe. As we have seen, this view is incompatible with that of Newton, who took the relevant accelerations to be with respect to substantival space itself; with that of the neo-Newtonian and special-relativistic theory, which took the accelerations to be relative to "the inertial frames of spacetime"; and with that of general relativity, which takes the source of the inertial forces to be the deviation of the trajectory of the system from the geodesics of spacetime itself in the region through which the system is traveling.

Of course, the Newtonian, neo-Newtonian, special-relativistic, and general-relativistic theories of absolute acceleration differ from one another, although they share a common incompatibility with Mach's thesis. For Newton, absolute acceleration is acceleration relative to some *thing*, substantival space, and its existence proves the reality of absolute position and absolute velocity as well, despite the lack of observational consequences in the variation of the absolute position or velocity of an object. For the neo-Newtonians, absolute acceleration does not lead to the postulation of real but unobservable absolute velocities.

In both the Newtonian and neo-Newtonian accounts, not only is the existence of an absolute acceleration a real fact about the system, but the magnitude of this absolute acceleration is also an invariant quantity. It exists, it can be observationally determined, and it is the same for observers in all states of motion. In special relativity, on the other hand, the distinction between absolutely accelerated and absolutely unaccelerated systems remains intact, but the magnitude of the absolute acceleration, so long as it is not zero, is determined only relative to a reference frame fixed in a particular state of motion.

In general relativity the discussion of absolute acceleration becomes much more complicated. According to this theory, the inertial forces have their source in the motion of the test system relative to the structured spacetime of the region through which it travels. As in the Newtonian, neo-Newtonian, and special-relativistic accounts, this spacetime is taken to be a real constituent of the world with real features in its own right, and not, at least prima facie, a convenient fiction for the set of spatiotemporal relations among material objects. As in the neo-Newtonian and special-relativistic theories, and as opposed to the Newtonian account, this "entity of reference" for absolute accelerations is not a *thing*, an object in space and persisting through time, but rather a spacetime collection of event locations.

But in general relativity, as opposed to the neo-Newtonian and special-relativistic accounts, this spacetime itself is not an unchanging, fixed arena of material happenings. It is an entity that itself can vary in its structure with variations in the distribution of matter and energy within it, and it can undergo dynamic evolution, i.e., it can vary from place-time to place-time unlike neo-Newtonian and special-relativistic spacetimes which are homogeneous and have the same structural features at every one of their constituent event locations.

All of these theories, even the Machian, infer a connection between the existence of inertial forces and an acceleration of some kind. All make use of the kinds of inference, mere inductive generalization and generalization from the actual world to what would be the case in other possible worlds, made use of by Newton. Even Mach makes an inference of this latter kind, despite his disparaging remarks. For it is just as much an assumption that no inertial forces would exist on a test system in an otherwise empty universe, no matter what was done to the test system, as the inference to the effect that some changes in the system would induce differing inertial forces even in the empty universe.

Note that all of these theories blame the inertial forces on an acceleration, and all share Newton's important *relationist* assumption that 'is accelerated' is a two-place relational term expressing a relational feature of the world. To Newton the second place in the predicate must be filled with a term denoting substantival space, if the acceleration is to be counted absolute. To Mach, it must be filled by a term referring to the averaged-out mass of the universe, the "fixed stars." To the neo-Newtonian, special-relativist, and general-relativist, the entity toward which accelerations are absolute accelerations is the geodesic structure of spacetime. Even Leibniz and Berkeley make an assumption of something like this kind when they take absolute accelerations to be species of the genus "relative acceleration." In the next section I shall have need to comment on this assumption, for, in a sense, it underlies the whole of the inference from the existence of absolute motions to a substantival spacetime, if one once rejects Mach's solution as scientifically unacceptable.

## 2. *From Absolute Acceleration to Substantival Spacetime*

Suppose we accept the existence of absolute motions in the following sense: (1) We admit that systems in relative motion with respect to one another can differ in that in these different systems varying inertial forces will be experienced. (2) We agree that we cannot account for these forces in terms of the differing relative motions of the systems with regard to some material entities in the universe, for, in disagreement with Mach and in agreement with Newton, we accept the conclusion that test systems in motion with respect to one another would feel such differing inertial forces even in the possible world in which the remaining mass of the universe were annihilated, and, in this imagined world, we agree that the varying forces cannot be accounted for in terms of the relative motion of the small, widely separated test systems with respect to one another. Having made these assumptions, are we then committed to the acceptance of a "real" spacetime existing over and above the admittedly existing material objects and their admitted spatiotemporal relations to one another?

The answer is negative. We can countenance absolute motions without countenancing substantival spacetime. We can, in fact, maintain a consistent theory that is pure relationist with regard to spacetime, relationist enough to fit the most hard-nosed verificationist, and that yet postulates absolute motions.

How on earth can this be done? The answer is somewhat surprising. To maintain the relationist doctrine of space and time in the face of the acceptance of absolute motions, what we must do is deny that the predicate 'is absolutely accelerated' is a relational term! The expression 'A is accelerated' is incomplete. To complete it we must answer the question, "Relative to what is A accelerated?" But the expression 'A is absolutely accelerated' is a complete assertion, as is, for example, 'A is red,' or 'A is bored,' and unlike 'A is north of.'

There are varying systems, systems in motion with respect to one another. Some of these suffer inertial forces. Call these systems absolutely accelerated systems. But don't confuse 'is absolutely accelerated' with 'is accelerated.' Acceleration is a relation that one material object has relative to some other material object. Absolute acceleration is a property that a system has or does not have, *independently of the existence or state of anything else in the world.* Absolute acceleration is not a relation of a thing to some other material object, even the "averaged-out mass of the universe." It isn't a relation an object has to substantival space or spacetime itself, either. For these latter "reference objects" don't exist. It simply isn't a relation at all.

Mach simply didn't carry his positivist view of laws as summaries and extrapolations of observed experience far enough. All we observe is that (1) objects in relative motion vary in the inertial forces they suffer and that (2) objects in uniform motion with respect to one another suffer similar inertial forces. There is nothing "in the data" which forces us to look "behind" these effects for some motion of a system which *explains* the inertial forces. We simply admit that they are there and that their presence and magnitude varies from system to system depending upon the relative motions of the systems.

But why do some systems suffer no inertial forces, whereas others do? I offer no explanation. This is just a brute, inexplicable fact about the world. It seems clear that if one adopts this stance there is nothing incompatible with accepting "inertial forces," i.e., accepting that some systems are acted upon by forces that others do not experience, accepting that the *variation* in the force suffered is a function of the *relative* motion of the varying systems, and maintaining a pure-relationist account of space, time, or spacetime. What we give up is the hope of "explaining" the occurrence of the forces in terms of some state of motion of the system suffering the forces. We can still explain the *variation* of forces from system to system in terms of the lawlike correlations holding between relative motion and varying force among material systems. It is most curious that Mach, along with the Newtonians, accepted the most crucial premise necessary to argue from absolute motion to substantival space. Picturesquely we can characterize this assumption as the assumption that absolute motion is a *kind* of motion. But why assume this at all? What we observe is that some systems suffer inertial forces and others do not, and that the variation in the forces suffered is a function of the relative motion of the systems. While Mach criticizes Newton's metaphysical leaps beyond the data, he shares Newton's most crucial assumption, that some *explanation* of the inertial forces in terms of the motion of the system suffering the forces must be forthcoming. If we reject this assumption, we can have relationism with absolute motion without falling into Machian speculative physics at all. And from Mach's point of view, isn't it clear that the assumption should be rejected? For, after all, what kind of explanation does a Newtonian account give us in any case? It does not associate the appearance of inertial forces with any other independently observable phenomenon, and from a Machian positivist point of view, isn't this what *real* explanation consists in? In other words, isn't it better to deny the possibility of explanation instead of disguising the absence of explanation by the proposal of a "pseudo-explanation"?

We see then that the inference from the existence of absolute motion to the existence of substantival space or spacetime requires adopting a principle whose defense can only be philosophically justified, and hardly empirically confirmed or unconfirmed. It looks like this: If a phenomenon is found to vary in a lawlike way with the relative motion among material systems, then the existence of the phenomenon, and its "magnitude," must itself be accounted for or explained in terms of some motion of each system in which the phenomenon is observed. But why should anyone adopt such a principle? It is here that Mach and his positivist successors should have launched their severest attack on Newton and those others arguing from inertial forces to substantival space or spacetime. Curiously, none of the critics ever seem to have even noticed that the principle was assumed, much less to have criticized it.

Suppose we reject the principle, for it seems to have little or no a priori warrant and certainly can't be empirically established. How then can we decide between the exponents of substantival spacetime theories and exponents of pure relationism? I am afraid that we are driven back to the purely philosophical disputes of (III,A) and (III,B), disputes that take their form and must be resolved, if they are resolvable at all, quite independently of any scientific theorizing founded on the existence of inertial forces.

What are the philosophical grounds on which the dispute can be resolved? The best case for the pure relationist, as I noted in (III,B,2), is the verificationist or epistemological version of Leibniz's argument against the substantivalist, for, as we have seen, the metaphysical and metaphysical-causal arguments he proposes are much less persuasive. The fundamental argument against substantival doctrines is that these doctrines postulate the existence and structure of entities that are in principle immune to empirical observation. As we have seen in (II,H, 4), such postulation seems all too easily to lead one to skepticism, conventionalism, or unmotivated apriorism.

And what are the substantivalists' best replies to the pure relationist? They are the old familiar philosophical arguments. First, the relationist seems to limit the existing world to the world knowable to human beings through sense perception. Isn't this too hubristic on their part? Of course, the relationist reply will be that although there may be entities of the world and features of them forever unknowable to us, it is at least illegitimate to believe in them and, even stronger, mistaken to think that we can even assert anything intelligible about them.

The second reply to the relationist is also familiar. It is that he is unable to account in his theory for expressions commonly taken as intelligible in ordinary and scientific discourse. The relationist cannot account for the notion of an unoccupied spatiotemporal location or make sense of the notion of a universe totally devoid of material things, it is said. Of course, here the relationist has several "outs": (1) He can invoke possibilia, taking unoccupied locations in a world containing matter as possible but nonactualized spatiotemporal relations to material things and taking talk about the structure of totally empty spacetime as *façon de parler* for talk about the lawlike structure that would govern entities in the world if, contrary to assumption, there were any. As I have noted, the substantivalist believes this invocation of possibility-talk to be illegitimate, since, he says, all intelligible possibility-talk is founded upon the assumption of an actual structure "grounding" the possibilities. But we have already rehearsed this argument. (2) The other alternative for the pure relationist is to simply deny the meaningfulness of talk about empty spacetime locations in the actual world or about empty spacetime altogether. The fact that we talk about these things is no more persuasive to him of their intelligibility than is the persistent occurrence of pure speculative metaphysical talk evidence for him of the intelligibility of metaphysics as a discipline.

Once again I leave the issue unresolved. What we have seen, though, is just how careful one must be in trying to invoke observational facts and the theories invented to explain them to resolve persistent philosophical disputes. We can *make* the observational data informing us of the existence of inertial forces and the theoretical apparatus postulated to explain them relevant to the question of the substantivality of spacetime, but only by choosing to accept the philosophical theses that the existence of inertial forces requires an explanation, that the explanation is to be in terms of the "real motion" of the system suffering the forces, and that such real motion must, like ordinary motion, be motion relative to some entity. Without these unempirical postulates the existence of inertial forces is irrelevant to the question of the substantivality of spacetime, and if we reject these "philosophical" premises we are thrown back on the earlier, purely philosophical arguments Newton hoped to supplant with a proof of substantivality from physics.

Before leaving this topic, there is one more observation it is important to make. In (II,H,4) we saw how certain views about the epistemology of geometry had clear metaphysical or ontological consequences. An exponent of the reductionist position, such as Reichenbach, is clearly committed not only to the view that the "alternative theories saving the phenomena" are really the same theory, but also to the view that, properly speaking, there is no such thing as space or spacetime itself, but only material objects or events and the spatial or spatiotemporal relations among them.

Does the implication go the other way, however? I think it is clear that it does not. One could clearly be an antisubstantivalist, believing that space or spacetime was nothing but the set of spatial or spatiotemporal relations among material things or events, and yet also believe that the metric geometry of the world was in no way a matter of convention. One might, for example, believe that space was nothing but the set of spatial relations, actual and possible, among point material objects, and yet believe that the evidence made it clear that space was Euclidean. And, one could even believe that there were good aprioristic grounds for rejecting such alternative accounts involving "universal forces," etc., which postulated a non-Euclidean space.

Nothing in the antisubstantivalist position automatically commits one, for example, to believing that the congruence of spatially separated rods is not directly available to empirical examination, is a matter requiring a coordinative definition, or is a matter of convention in any way. On the other hand, the very verificationist principles that ordinarily lead one to an antisubstantivalist view are the same principles that frequently lead one to a Reichenbachian position on conventionalism. So I think that we can say that it is not unexpected that many antisubstantivalists can be led to accept reductionist views of the kind I described in (II,H,4), with their so-called conventionalistic consequences, as well. So antisubstantivalism doesn't imply the full reductionist view but is likely to be associated with it.

Another way of looking at this matter which is illuminating is this: The reductionist position rests upon the assumption that the observable is the local and material, global and purely spatiotemporal features being in the realm of theory. Antisubstantivalist positions usually rest upon the epistemological assumption that only the material is the observable, but an antisubstantivalist might well deny the restriction of the observable to the local. Should he do so, he will end up with the nonconventionalist antisubstantivalism I have noted above. But should he accept the claim that global material features are as immune to direct inspection as purely spatiotemporal features, he will accompany his antisubstantivalism with a Reichenbachian view about the conventionality of the metric as well. For just as an alleged epistemic inaccessibility of space or spacetime itself leads to antisubstantivalism, so it is the alleged inaccessibility of such nonlocal features as the congruence of noncoincident rigid rods which leads to the doctrines of conventionality of the metric.

# BIBLIOGRAPHY FOR CHAPTER III

## SECTIONS A AND B

### PART 1

An historical survey of some views on space sympathetic to the substantivalist position is:

Graves, J., *The Conceptual Foundations of Contemporary Relativity Theory*, chap. 2, "History."

### PART 2

For a study of relationism, one cannot do better than start with Leibniz's original arguments. They are most easily accessible in:

Alexander, H., ed., *The Leibniz-Clarke Correspondence*.

A clear recent survey of arguments for and against relationist accounts is:

Hooker, C., "The Relational Doctrines of Space and Time," *British Journal for the Philosophy of Science* 22 (May 1971).

The issues are also discussed, primarily with respect to a relational account of time rather than space, in:

Van Fraassen, B., *An Introduction to the Philosophy of Time and Space*, chaps. 1 and 2.

For a discussion of the problem of identifying individuals across possible worlds more or less in conformity with the view I have adopted, see:

Lewis, D., "Counterpart Theory and Quantified Modal Logic," *Journal of Philosophy* 65 (March 7, 1968).

For the contrary view, in which identity of individuals does not rest upon commonality of essential properties, see:

Kripke, S., *Naming and Necessity* (mimeographed).

## SECTION C

### PART 1

Newton's arguments for substantival space can be found in:

Newton, I., *Principia Mathematica,* the "Scholium to the Definitions."

Also, see the Leibniz-Clarke correspondence cited above.

An alternative account to my reconstruction of Newton's "hidden premises" can be found in.

Lacey, H., "The Scientific Intelligibility of Absolute Space," *British Journal for the Philosophy of Science* 21 (November 1970).

### PART 2

For a review of early replies to Newton's arguments which is sympathetic to the relationist standpoint, see:

Reichenbach, H., "The Theory of Motion According to Newton, Leibniz and Huyghens," in his *Modern Philosophy of Science.*

For a critical survey, arguing that Newton has much the best of the argument, see:

Earman, J., "Who's Afraid of Absolute Space?" *Australasian Journal of Philosophy* 48 (1970).

## SECTION D

### PARTS 1 AND 2

For Mach's theory, see:

Mach, E., *The Science of Mechanics.*

The reader might note that in some versions of his theory, Mach makes the force he postulates totally independent of the separation of the material objects concerned, making a "test" of Mach against Newton by the use of local variations in the surrounding mass distribution impossible.

### PARTS 3 AND 4

For a lucid, nontechnical, discussion of neo-Newtonian spacetime, see:

Stein, H., "Newtonian Space-Time," *Texas Quarterly* 10 (1967).

Also:

Earman, J., "Who's Afraid of Absolute Space?"

A very thorough technical development of the theory is available in:

Havas, P., "Four Dimensional Formulations of Newtonian Mechanics and Their Relation to the Special and General Theory of Relativity," *Reviews of Modern Physics* 36, no. 4.**
This last item also contains full citations of the original mathematical work of P. Frank, E. Cartan, and K. Friedrichs.

## SECTION E

### PARTS 1–3

On absolute motion in general relativity, nontechnical references are:

Grünbaum, A., "The Philosophical Retention of Absolute Space in Einstein's General Theory of Relativity," reprinted in Smart J., *Problems of Space and Time*;
and:

Earman, J., "Who's Afraid of Absolute Space?"
See also:

Goenner, H., "Mach's Principle and Einstein's Theory of Gravitation," in Cohen, R., and Seeger, R., eds., *Ernst Mach: Physicist and Philosopher*;

Dicke, R. "The Many Faces of Mach," in Chiu, H., and Hoffman, W., eds., *Gravitation and Relativity*;

Wheeler, J., "Mach's Principle as Boundary Condition for Einstein's Equation," also in Chiu and Hoffman;
and:

Sklar, L., "Absolute Space and the Metaphysics of Theories," *Nous*, VI (Nov., 1972).

For additional material, including a discussion of non-Machian aspects of Thirring-type experiments in general relativity see:

Dicke, R., "Mach's Principle and Equivalence," in *Proceedings of the International School of Physics "Enrico Fermi,"* Course XX, Evidence for Gravitational Theories.

### PART 4

An intensive study of geometrodynamics from the philosophical point of view is contained in:

Graves, J., *Conceptual Foundations*, chap. 4.
A technical treatise on the theory is:

Wheeler, J., *Geometrodynamics*.**

## SECTION F

### PARTS 1 AND 2

For further arguments in the vein of this section, see:

Sklar, L., "Absolute Space."

CHAPTER IV.

## CAUSAL ORDER AND TEMPORAL ORDER

# PRELIMINARY REMARKS

In this chapter I will pursue a number of important philosophical questions about time: What *is* the temporal structure of the world, that is, what is the ontological or metaphysical status of temporal relations? What is the particular temporal structure that characterizes the world? Can we consider alternative temporal structures, or is there only one possibility for the temporal structure of the world a reasonable man could consider? How do we know what the temporal structure of the world is? Do we know this a priori, or by means of empirical investigation? Or is there, once more, an issue of conventionality to be considered? Finally, is there any sense in which the temporal structure of the world can be considered reducible to some other structure? In particular, can we "eliminate" temporal relations among events entirely, putting causal relations in their place?

Now philosophers have devoted a great deal of attention to the nature of time, and I could, perhaps not unprofitably, pursue their investigations in some detail, outlining the historical evolution of the debates over some major issues and trying to come to my own position on some of the issues as decisively as I can. But this is not the line of attack I will take.

Instead, in keeping with the overall structure of this book, I will pursue in some detail the following question: To what extent can changes in accepted physical theory affect changes in our philosophical conception of time? We shall see very shortly that this approach will not lead us to neglect the older philosophical issues entirely, for one claim that is frequently made is that changes in contemporary physics finally lead to a scientific resolution of traditionally philosophical disputes. When we try to see whether this is correct, we will, of necessity, plunge into some of the older disputes at least briefly. But the primary concern will be with the exposition of the structure of time according to some contemporary physical theories and with the particular philosophical issues that examination of these theories brings to the fore.

The scientific theories whose impact I will examine are special relativity and general relativity, although there will be at least one occasion on which I will need to refer, however superficially, to quantum mechanics as well. The *scientific* role of time in special relativity is quite clear. That is, very few scientists would hesitate to say that they knew just what special relativity told us about time. Occasionally scientists do maintain novel assertions about time in special relativity, but they are generally considered confused by their colleagues.

Special relativity has also received quite a lot of attention from philosophers attempting to give a critique of its conceptual foundations. Here we will find that although a great amount of clarification has been obtained by the philosophers there are still philosophical confusions rampant and unresolved philosophical issues to consider. As usual, my primary goal is not to provide ultimate solutions to the most basic philosophical questions, but rather to sufficiently disentangle the knot of issues so that we can finally become clear about just what the basic philosophical questions really are and what *possible* answers to them can be proposed. As we shall see, even this limited task is one of some complexity.

With regard to general relativity, the "scientific completeness" of the theory comes into question as well. The temporal structures of the world allowed as possibilities by general relativity have recently come in for some intensive physical and mathematical scrutiny. It turns out that once we have adopted general relativity, the variety of temporal orders the world can possess as a whole seems to expand greatly, opening up important questions about the extent to which we must allow these possibilities as genuine possible realities of the world, or on the contrary, the extent to which we must modify or expand the theory in order to eliminate possibilities it allows which we wish to reject on a priori grounds.

In addition, the possibilities allowed by general relativity bring about philosophical questions, questions concerned with the conventionality of the topological structure of the world, with the identity or diversity of events in the world, and with the possible causal structures we will "permit" in the world, which outrun the questions suggested by the special theory alone. On top of this, the consideration of the worlds described by the general theory will throw additional light on the philosophical issues the special theory gave rise to, putting the search for answers to the questions thrown up by special relativity in a slightly different light.

The primary philosophical issue we will be concerned with is the question of the reducibility of temporal structure to causal structure, the so-called question of the causal theory of time, although other more traditional philosophical issues will be touched on briefly. The question of the reducibility of temporal relations to causal relations has many philosophical aspects that will not be unfamiliar to the reader who has reached this point, for it shares issues in common with the questions of the conventionality of geometry and the reducibility of spacetime to the set of spatiotemporal relations among material things. But, as we shall see, this particular philosophical question will bring in novel issues of principle as well. Once again, the main concern is not to resolve fundamental issues of metaphysics and epistemology, but to see how questions raised by particular scientific theories serve both as paradigm examples of deep philosophical disputes and, at the same time, as additional resources that can be brought to bear on resolving the fundamental philosophical problems.

## ORIGINS AND FEATURES OF THE
## SPECIAL THEORY OF RELATIVITY

### 1. *Prerelativistic Spacetime and the "Round-Trip" Experiments*

The scientific revolution of special relativity, like some other scientific revolutions, was initiated by the results of a "crucial experiment." Nineteenth-century physics, with its prerelativistic spacetime assumptions, led to the performance of a series of experiments whose outcomes were of the highest importance. Initially, the experiments were designed to establish an important fact, the state of motion of the reference frame at rest in the medium of transmission of light and other electromagnetic radiations, i.e., at rest in the aether. The results of these experiments were quite different from what had been anticipated; so different, in fact, that it was immediately realized that very severe changes would have to be made in accepted physical theory to account for the observations. Few physicists realized initially, however, just *how* severe the revisions of classical concepts would turn out to be, and it was Einstein's great insight to see that the "best" account of the phenomena was obtained by an extremely radical change in the orthodox ideas.

The basic idea behind the experiments is quite simple. Take a laboratory frame. Transmit light about on this frame following various paths, but such that each light beam emitted is eventually received at its point of origin in the frame. Now the time taken for the light to complete its round-trip journey from point of origin to identical point of reception will certainly depend on the route taken. But it will also, as I have already remarked in (III,D,1), depend upon the state of motion of the laboratory frame relative to the medium of transmission of the light, the aether. If we consider the relative time taken by the light to complete its round-trip journey along two different paths, this ratio will be seen, in general, to depend upon the "state of motion of the laboratory with respect to the aether." I will give an example of this situation below.

It is important to note that while the aether theory was assumed by the experimenters, the predicted results of the experiments do not depend upon the aether theory of transmission of light. Rather, the predictions depend simply upon (1) the standard notions of the structure of spacetime and (2) the assumption that the velocity of light is independent of the velocity of its source. Now (2) was a well-established empirical fact. The way that (1) functions is clear: Consider light traveling through space in time. Suppose that the velocity of light is the same in every direction relative to some one specified frame of reference in a given state of motion. Now any observer in motion with respect to that given frame must, according to the most basic notions of space and time underlying the prerelativistic theories, see the velocity of light as different in different directions. He will see the velocity of light *reduced* in the direction that is his direction of motion relative to the frame, because he is "catching up" to the light in that direction. And he will see the velocity of light *increased* in the direction opposite to the direction of his motion relative to the original frame, for he is "running away" from the light moving opposite to him. The velocity in other directions which he observes will depend upon the angle these directions make with his direction of motion.

The simplest example to illustrate this is the Michelson-Morley experiment. Here light is sent out from a point and the beam at some other point is split into two parts. These are transmitted equal distances at right angles to one another, reflected back along their path to the splitting point, and then directed back along their path to the point of origin. When one of the directions of the beam after splitting is along the direction of motion of the frame with respect to the aether (or the frame in which the velocity of light is isotropic), then that beam will take a longer time for its full round trip than will the beam that spends its time after splitting traveling back and forth perpendicular to the direction of motion of the frame in the aether. Proving this requires nothing more than elementary plane geometry and some trivial vector algebra. It is shown in every elementary textbook on relativity, and I will not rehearse the calculations here. The time difference for the round trip can be determined by letting the returning beams interfere with one another at the observing point (the point of origin and reception) and can be measured to an extremely high degree of accuracy.

According to the theory, we should expect the relative times taken by the two beams, and hence the time difference in their total round-trip time, to vary with the state of motion of the laboratory in the aether (or isotropic frame). We don't know which state of motion is the isotropic state, but we can surely determine the changes in time difference between the two beams' passage times as the state of motion of the laboratory changes, say as the earth's rotation changes the laboratory's direction of motion. So we should see a varying time difference between the beams as the laboratory changes its state of motion. What we actually observe, instead, is that the time taken by each beam is exactly the same no matter what the inertial state of motion of the laboratory frame.

The Michelson-Morley experiment lacks generality in that the distances traveled by both beams are the same. The Kennedy-Thorndike experiment is somewhat more general. Here the beams, once split off from the original beam, are sent on round trips away from and back to the point of splitting along paths of differing length. Now, naturally, the total time taken by each beam will be different, since they travel different distances. What we are led to expect by the orthodox views of space and time is that the time *difference* between the passage times taken by the two beams will again vary with the state of motion of the laboratory frame with respect to the frame in which the velocity of light is the same in every direction. What we observe instead is that this time difference remains constant, irrespective of the motion of the laboratory frame.

One note of caution. All that the experiments show is that the time difference is independent of the state of motion of the frame so long as that *motion is inertial*. Of course frames fixed on the earth suffer small accelerations because of the earth's rotation, but that can be ignored. There is no claim being made that the round-trip time differences would remain invariant if, for example, we set the laboratory in rapid rotation. They will not, as a matter of fact.

A little thought shows that we can construct an idealized thought experiment which both sums up the results of the Michelson-Morley and Kennedy-Thorndike experiments and "inductively generalizes" from them. Consider any experiment that goes as follows: Arrange a set of mirrors so that light sent from a point of origin in a given direction, or in the opposite direction, by being reflected from the mirrors, returns after a time to its point of origin. The polygonal path can have any shape we like, and, in general, the mirrors will not be at equal distances from one another. What we should expect on the basis of the ordinary notions of time and space underlying prerelativistic physics is that the two beams sent in opposite directions will in general take different times to traverse the closed path from origin to origin, and that the time difference of traversal will depend on the relative motion of the laboratory frame to the frame in which the velocity of light is the same in every direction. What we observe, however, is that both beams take the same time to traverse the closed path, so long as the motion of the laboratory frame is inertial motion. This happens wherever and whenever the experiment is performed, so long, of course, as the light is being propagated through a vacuum. This inductive generalization from the observed data is the "fact" for which our revised theory must account.

I might note here, also, that the emphasis upon *light* should not mislead one. I assume that this same truth about "equal round-trip propagation times around closed paths" holds not only for light but for the propagation of other "basic" signals through a vacuum as well—say for the propagation of gravitational influence, or any other fundamental field whose quanta have rest mass zero (see Figs. 24, 25, 26).

How can we account for these astonishing experimental results or for the general law about round-trip propagation times we have "induced" from them? A little reflection on the part of the reader will show that any mere "patching up" of the details of classical prerelativistic physics won't do, for the prediction of varying time difference in propagation time which the standard theory gives us is based upon the most deep-seated considerations of space and time. How could light appear to propagate in all directions with the same velocity to two observers who are themselves in relative motion with respect to one another?

To see how surprising these experimental results are, consider the following: Let an observer at rest in a frame in which the velocity of light is the same in all directions send out light waves in every direction. What he sees at any time is an expanding spherical shell constituting

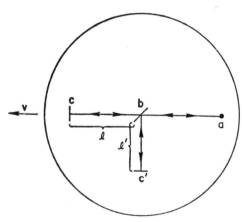

Fig. 24. The Michelson-Morley experiment. The difference in time taken by light beams to traverse the two paths *abcba* and *abc'ba* is measured. If the aether theory is correct, simple calculation shows that when the motion of the apparatus with respect to the aether is in the direction shown by the arrow *v*, the light should take longer to traverse *abcba* than to traverse *abc'ba*. Actually, no difference in time is recorded, irrespective of the state of inertial motion of the apparatus.

the wave front of the emitted light. Now consider an observer in uniform motion with respect to the first system, moving in any direction. The experiments seem to tell us that he will also see the rays of light as having the same velocity in every direction. So at a given time after the light has been emitted, he too should see around him a spherical shell constituting the wave front of the light at that time. But in the time elapsed he has moved relative to the first observer, even if initially they were both at the point of emission of the light. How can there be a spherical shell about the center consisting of the position of the first observer and at the same time a shell that is spherical about the second observer now located at a different point? How can a sphere have two centers at a distance from one another? To resolve this paradox is to arrive at the relativistic account of space and time.

## 2. Prerelativistic Attempts to Account for the Null Results

Let me call the independence of the round-trip time of light around a closed path with respect to the direction of propagation of the light and the inertial state of motion of the laboratory the *null result*. How could

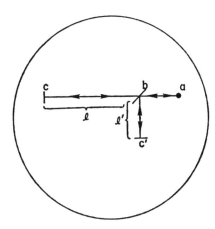

Fig. 25. The Kennedy-Thorndike experiment. The null results of the Michelson-Morley experiment can be accounted for in terms of a shrinking of the apparatus in
the direction of its motion with respect to the aether.

The Kennedy-Thorndike experiment is more general in
that the length of the paths *abcba* and *abc'ba* are no longer
equal, i.e., $l \neq l'$. What one determines is the variation in
the time difference for the light traversing the alternative
paths. The aether theory predicts that this difference will
vary with the state of motion of the apparatus with respect
to the aether, but one observes that the time difference is
independent of the inertial state of motion of the apparatus.

This result cannot be explained away by "length contraction" alone, but requires the "slowing down of clocks
when in motion with respect to the aether" as well.

one account for these surprising observational results on the basis of a
theory that postulated an aether for the medium of transmission of light
or, in any case, postulated a unique state of motion relative to which the
velocity of light really was the same in all directions?

It was seen by Lorentz and Fitzgerald that the null results of the
Michelson-Morley experiment could be accounted for if one simply
assumed that the material objects constituting the laboratory apparatus
shrank when placed in motion with respect to the aether. If the speed
(magnitude of velocity) of the laboratory with respect to the aether is
$v$, call $\sqrt{1 - (v^2/c^2)}$ the shrinking factor, $R$, where $c$ is the constant
velocity of light in all directions in the aether framework. If we assume
that a rod whose length when it is at rest in the aether is $l$ takes on length
$Rl$ when it has speed $v$ with respect to the aether, then the shrinking of

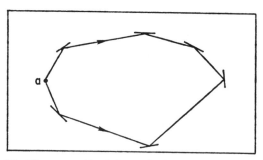

Fig. 26. The generalized "round-trip" experiment. Aether theory predicts that the time taken by the light to traverse the closed path from *a* to *a* will differ for the beams sent around the path in opposite directions, the difference in time depending upon the state of motion of the apparatus with respect to the aether.

Both the compensatory and relativistic theories predict, correctly, that the two beams will traverse the closed path in the same time, irrespective of the detailed shape of the path or the inertial state of motion of the apparatus.

the apparatus in the direction of motion with respect to the aether will be just sufficient to compensate for the "extra time" the light should take on the path parallel to the direction of motion over that on the path perpendicular to the motion, and the results of the Michelson-Morley experiment will always be an equal time taken by both beams, no matter what the inertial motion of the laboratory.

This shrinking won't, however, be sufficient to compensate for a complete independence of time difference for all states of motion of the laboratory for the Kennedy-Thorndike experiment, not for the general round-trip thought experiment. We can guarantee that the time taken by light to make the round-trip circuit will be independent of the direction in which it starts its trip and the inertial state of motion of the laboratory for all such round-trip experiments if we assume, in addition to the "contraction" of rods when they are in motion with respect to the aether, that all physical processes change their rate in time when the processes are undergone by moving objects. In particular, we must assume that a clock which has time $t$ between its ticks when it is at rest in the aether takes time $t/R$ from tick to tick when it is moving with speed $v$ with respect to the aether. So we add to the "real" length contraction of otherwise rigid rods, induced by their velocity with respect to aether, a "time dilation" which causes clocks (and all other physical processes!) to slow down when the apparatus "processing" is in motion with respect to the aether.

A little algebra and geometry and one can easily convince oneself that with these "physical changes" induced in distance-measuring rods and time-interval-measuring clocks, independent mind you of the particular constitution of the rods and the clocks, an observer in motion with respect to the aether, if he performs measurements on the velocity of light using rods at rest in his system to measure distances and clocks at rest in his system to measure time intervals, will come to the conclusion that (1) the velocity of light is the same in every direction and (2) it has magnitude $c$, where $c$ is the magnitude of the velocity of light determined by an observer at rest in the aether frame using his "true" rods and clocks. No wonder the round-trip experiments all give null results!

It is of historical interest to note that Lorentz tried to "explain" the shrinking of otherwise rigid rods when they are in motion with respect to the aether in terms of their atomic constitution and the fact that the distances between atoms are the result of the electromagnetic forces acting between them. His explanation was quite unpersuasive and became of historical interest only since the whole theory, which accounted for the null results in terms of compensatory changes induced in rods and clocks by their motion with respect to the aether, soon gave way before the far more appealing theory of special relativity of Einstein's.

### 3. Einstein's Critique of Simultaneity and the Construction of Minkowski Spacetime

One can approach Einstein's novel account of the null experiments, with its consequent rejection of prerelativistic space and time in favor of Minkowski spacetime, by applying a verificationist critique to the "compensatory" theory I have just outlined. According to the compensatory theory there is one inertial framework distinguished in reality from all the others. It is the framework at rest in the system in which the velocity of light is *really* the same in all directions. Only rods at rest in this frame truly measure distances, for rods in motion with respect to it all shrink; and only clocks at rest in this frame truly measure lapses of time, since clocks in motion with respect to this frame are all "slowed down" by their motion.

But which inertial frame is the frame truly at rest in the aether? The classical way of determining this was to discover the frame that was such that round-trip experiments performed in laboratories at rest in the frame gave null results. But the whole point of the compensatory theory is to guarantee that round-trip experiments will give null results in any inertial frame whatever, including those in uniform motion with respect to the aether.

Well, the true rest frame is the one in which the velocity of light is the same in all directions. Perhaps we can determine this frame, then, by measuring the velocity of light in a laboratory in various directions and seeing if it is truly independent of direction in a frame fixed in this laboratory. But to measure the velocity of light between two points we need (1) the spatial separation between the emission and reception event for the light and (2) the temporal separation between these events.

Let us focus on the latter. We could determine the temporal separation between events at a distance from one another if we could place identical clocks at the two points (i.e., at rest in the laboratory relative to which we are fixing the points and spatially coincident with them) and find some way of synchronizing the clocks at a given time. Now we might think that we could synchronize the clocks by synchronizing a clock with the clock at one point, moving it to the second point, and then synchronizing the second clock with the moved clock. But if we assume that the compensatory theory is correct this method won't work, for it will fail to provide us with a unique synchronization, i.e., a unique event at the second location to call simultaneous with a given event at the first location. For when we move the clock from one point to the other, its rate will vary with its velocity relative to the aether. So different "paths" carrying the clock from the one place to the other, for example, moving it quickly the first two-thirds of the distance and slowly for the remaining third as opposed to moving it with a uniform velocity from one point to the other, will establish incompatible synchronizations. And we cannot compensate for the slowing of the moved clock relative to those at rest in the aether, because we don't know what the rest frame in the aether is. That is what we are trying to discover in the first place.

We can't synchronize the clocks by sending a light signal from one clock to the other, either. For we must allow for the time of transmission of the light from one clock to the other. That will depend upon the velocity of light from point to point, relative to the laboratory frame, and that is just what we are trying to synchronize the clocks to measure in the first place.

Now if we had causal signals of unlimitedly high velocity relative to the inertial laboratory frame, we could synchronize the clocks by sending faster and faster signals from one clock to the other, taking the event at the second location to be simultaneous with a given event at the first location if the second location event is the latest event at the second location not reachable by any causal signal from one point to the other. But, as a matter of fact, *there are no signals that travel faster than the speed of light relative to an inertial framework.* This is an empirical fact, which we can discover by searching for a signal that traverses a round-trip course fixed in an inertial frame faster than a light signal sent in a vacuum around that course. We simply find no such signals. We shall later see that the absence of signals of arbitrarily high velocity—that is, the impossibility of finding for any specified time interval measured at the origin, no matter how small, some signal that traverses the course in that time interval or less—does more than block an attempted coordinative definition for 'events at a distance from one another and simultaneous.' The absence of signals faster than light also will block a possible objection to the definition of distant simultaneity Einstein himself chooses.

So we can't determine the "aether frame" by round-trip measurements on light, and we can't determine it by measuring the velocity of light between two separated points either. The former approach fails since the round-trip experiments give null results in *all* inertial frames and the latter approach fails because the very compensatory theory it is supposed to be used to support, in combination with an empirical fact about light being the "limiting velocity" for causal signals, makes the measurement of the velocity of light, prior to knowing already which frame is the aether frame, impossible (see Fig. 27).

The situation should seem familiar to the reader. From the point of view of dynamical measurements alone, we cannot determine which inertial frame is the frame at rest in substantival space, for all inertial frames are mechanically equivalent. This is the principle of Galilean equivalence. The nineteenth-century electromagnetic theory suggests that we can at least determine which inertial frame is at rest with respect to the aether, for electromagnetic theory is not Galilean-invariant from the point of view of prerelativistic spacetime. This is because it predicts a velocity of light the same in all directions, and from the point of view of prerelativistic spacetime there can be only one such frame. But now we see that we can no more determine the aether frame by optical experiments than we could determine the frame at rest in substantival space by means of mechanical experiments. For both mechanical and optical experiments, one inertial frame is equivalent to any other.

Now we have seen that the three assumptions, (1) there is a velocity

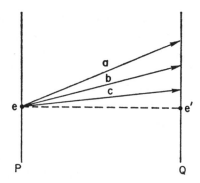

Fig. 27. A coordinative definition for simultaneity for events spatially separated, using "infinitely fast" causal signals. If there were causal signals of arbitrarily high velocity, as determined by round-trip experiments, we could find the event, $e'$, at $Q$ simultaneous with a given event, $e$, at $P$ by sending causal signals from $e$ at $P$ to $Q$ and finding the latest event at $Q$ not reachable by any causal signal sent from $e$. But, of course, there are no such causal signals as $a$, $b$, $c$ of ever greater velocity, for no causal signals travel faster than light which has a definite finite velocity of propagation.

of light constant in all directions, (2) spacetime is as prerelativistically described, and (3) electromagnetic theory is Galilean-invariant, are incompatible. The compensatory theory opts for the solution of saying that the electromagnetic theory isn't really Galilean-invariant; it only appears to be so because of the distortions induced in rigid rods and clocks by their motion with respect to the aether. But this leads us into the verificationistically unhappy situation of asserting that one and only one inertial frame is at rest in the aether, and no observation could ever tell us which one. Einstein saves the verificationist principle by making a different choice: Electromagnetism is a Galilean-invariant phenomenon and light does have a definite velocity the same in every direction relative to *every* inertial frame. It is the orthodox view of space and time which we must reject.

Let us construct a new spacetime that will reconcile the principle that laws of nature obey Galilean equivalence with the characteristic of electromagnetic theory that light is predicted to have a definite velocity the same in every direction. When we have done this we will discover that from the point of view of the new spacetime, Newtonian mechanics is no longer a theory that "takes the same form in every inertial coordinate frame," i.e., that it is no longer Galilean-invariant. The resolution of this will be to drop Newtonian mechanics, replacing it with the dynamics of special relativity. The historical evolution is curious. We are led to believe inertial coordinate systems equivalent by Newtonian mechanics. Taking the principle of the equivalence of all inertial frames and generalizing it to electromagnetic theory, we must modify our spacetime theory. When we have done this, to retain the principle of the equivalence of all inertial frames for mechanical phenomena, we must drop the mechanics that led us to that principle in the first place, replacing it with a new dynamics which is Galilean-invariant relative to the new spacetime structure. This is what Einstein does and it works; for the experimental data, when we check it, confirms the new mechanics against the old, now rejected, Newtonian theory.

Let me use the remainder of this section to "construct" the new spacetime. It turns out to be, naturally, the Minkowski spacetime of (II,C,1), and we are finally seeing the rationale for adopting this new spacetime theory which I promised earlier. At this point I will pretty much just propound the new doctrine. In the next section, I will examine some of its striking conceptual and observational consequences. In (IV,C) and (IV,E) below, I will return to this construction and offer some philosophical reflections on it.

I will follow Einstein's construction of the spacetime from the direction of developing a new concept of simultaneity for events at a distance from one another; the full development of this spacetime picture is Minkowski's. I start off with a few primitive notions. First of all, the basic constituents of the spacetime are event locations. Once again I call them *events* remembering that no concrete "happening" is necessary at an event location for it to still be a location. I assume that we know what it is for event location $a$ and event location $b$ to be one and the same event location, or if we think of happenings at the locations, I assume that we know what it is for one happening to be at the same placetime as another happening.

What is it for two distinct events to be *simultaneous?* Prerelativistically, this is an assumed notion. Simultaneity is assumed to be an invariant notion, two events simultaneous for one observer are simultaneous for any other observer, and it is also assumed to be an equivalence relation, i.e., to be a symmetric, reflexive, and transitive relation among events. I assume none of this. The notion of simultaneity we end up with is not invariant, it is only defined relative to an observer in a given inertial state of motion. For any such observer, simultaneity, so relativized, is an equivalence relation, but if $a$ is simultaneous with $b$ relative to one observer, and $b$ simultaneous with $c$ for another observer, there is no reason to believe in general that $a$ is simultaneous with $c$ relative to either observer.

To give a Reichenbachian coordinative definition for simultaneity for distinct events, pick a particular inertial frame $S$. One says that $e$ and $e'$, two distinct events, are simultaneous relative to $S$ if and only if: (1) a clock at rest in $S$ at the place (relative to $S$) of $e'$ records the time of $e'$ as $(t_1 + t_2)/2$, and (2) a light signal sent from the place of $e'$ (relative to $S$, i.e., from the spatial point said by an observer at rest in $S$ to be the place where $e'$ occurs) at $t_1$, whose reflection at the place of $e$ (relative to $S$) is an event coincident with $e$, arrives back at the place of $e'$ (relative to $S$) at time $t_2$. The diagram should make this clear (Fig. 28). Of course, events can be simultaneous even if no actual

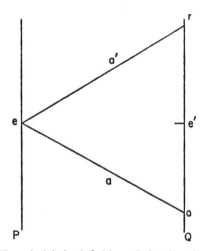

Fig. 28. The relativistic definition of simultaneity for separated events. Relativistically, we take as the event at $Q$ simultaneous with a given event $e$ at $P$, that event at $Q$, $e'$, midway in time between the events $o$ and $r$, where $o$ is the event that is the origination at $Q$ of a light signal which reaches $P$ just at event $e$, and $r$ is the event at $Q$ which is the reception at $Q$ of a light beam sent from $P$ at event $e$.

light ray of the kind described exists. The simultaneity it defined, as usual, in terms of a *possible* operation one could carry out to establish the simultaneity.

Now I will look at some of the consequences of this definition in detail later. For the moment we should note that according to this definition, it is plain that an event simultaneous with a given event for one inertial observer will *not* be simultaneous with that given event for another observer in motion relative to the first (Fig. 29). The accompanying diagram will make this clear. So whatever simultaneity is in the

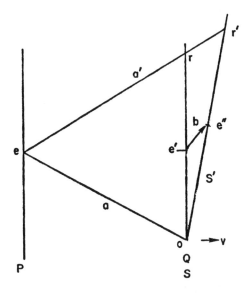

Fig. 29. The relativization of simultaneity to the state of motion of the observer. $S$ and $S'$, who is moving with respect to $S$ as indicated by the arrow $v$, determine which event at their location is simultaneous with $e$ at $P$. They coincide at $t = 0$, and send a light signal from this event, $o$, which reaches $P$ at event $e$. The return signal gets to $S$ at $r$, but by this time $S'$ has moved to the right. The return signal reaches $S'$ only at event $r'$.

    $S$ takes $e'$, midway between $o$ and $r$ and at his location to be simultaneous with $e$. $S'$ takes $e''$, midway between $o$ and $r'$ at his location to be simultaneous with $e$. Since there is a causal signal, $b$, which can leave $e'$ and arrive at $e''$, both $S$ and $S'$ agree, however, that $e'$ and $e''$ are *not* simultaneous.

    So: (1) $e'$ is simultaneous with $e$ for $S$;
         (2) $e''$ *is* simultaneous with $e$ for $S'$;
         (3) but $e'$ is *not* simultaneous with $e''$ for either $S$ or $S'$.

new theory, it is quite unlike the relation of simultaneity which is assumed to hold or not hold among events independently of the state of
motion of the observer in the prerelativistic account.

To complete the construction we need to define three more relations
holding between events: their spatial separation, relative to a specific
inertial frame; their temporal separation, relative to an inertial frame;
and their interval, which is an *invariant* relation and need not be relativized to a state of motion of an observer.

Let us first dispose of the case where $e$ and $e'$ are coincident. Then
the spatial separation of $e$ from $e'$ is zero, and invariant, and the temporal separation of $e$ from $e'$ is zero and also invariant.

For $e$ and $e'$ as distinct events we proceed as follows: Relative to
an inertial frame, we determine the temporal separation of two events
by (1) synchronizing clocks at the places the events occur, that is, at
their places as determined by the observer in the specified inertial
frame, (2) reading off the times of the events on the synchronized
clocks, taking the time of an event as the time recorded on the clock
at the location of the event coincident with the event in question, and
(3) subtracting the two time readings to get the temporal interval between the events.

To get the spatial separation of two events, relative to a specific
inertial frame, let a light ray proceed from an event at the place of the
one event, again as determined by an observer at rest in the frame, to
the place the second event occurs, as determined by the second observer.
Find out how long the light takes to get from one place to the other
by means of sychronized clocks at the two locations. The spatial
separation of the events is this time interval multiplied by the velocity
of light, which is taken to be a constant, independent of the particular
frame chosen and of the direction from one place to another.

What is the interval between two events? If the events have spatial
separation $Dx$ and temporal separation $Dt$, the interval between them
is defined as $\sqrt{Dx^2 - c^2 Dt^2}$, where $c$ is the constant velocity of light.
Now I have claimed that the interval between two events is an invariant,
despite the fact that it is defined in terms of the spatial and temporal
separations, neither of which is itself an invariant quantity. How can
one show that the interval is indeed invariant?

This follows from two assumptions: (1) The assumption that the
velocity of light is a constant and is the same for every observer in any
inertial frame. (2) An assumption about the *linearity* of the relation
of spatial and temporal separations for one observer to those for some

other observer. Let me make a few remarks about these two assumptions in turn.

1. What led us to the new view of spacetime in the first place was the *apparent* constancy of the velocity of light for all inertial observers. The compensatory theory handled this by saying that the constancy was merely apparent, and the appearance was due to the variation in our rods and clocks due to their motion relative to the aether frame. We wish all inertial frames to be truly equivalent. And so we take the apparent constancy of the velocity of light as indicating the *true* equality of the velocity of light for all inertial observers. Let us forgo, for the moment, worrying about whether the assertion of constantcy of the velocity of light states a *fact* or rather provides us with a new *definition*. I will return to this issue in (IV,C) below.

2. The linearity assumption says this: Suppose we have the temporal and spatial separations of a pair of events relative to inertial frame $S$ and relative to inertial frame $S'$. In general these will be different for the two frames. But, surely, the spatial separation of the events relative to $S'$ should depend only upon the spatial and temporal *separations* of the events for $S$, and not upon the particular times and places, relative to $S$, at which the events occur. Otherwise, the individual locations and instants would seem not to be on a par with one another. That is, while the spatial separation of $e$ and $e'$ will be different for $S'$ than it is for $S$, it shouldn't depend upon what particular time $e$ or $e'$ occur or where they occur, but only upon how much spatial *separation* and temporal *separation* they have relative to $S$ (Fig. 30). This amounts to a "homogeneity" assumption for the spacetime, and I will have something to say about it later.

Having made these two assumptions we proceed as follows: Suppose $e$ and $e'$ are events such that a light ray starting at event $e$ goes through event $e'$. If this is true for $S$, it is true for $S'$ as well. Now, whatever $Dx$ and $Dt$ are relative to $S$, $Dx/Dt = c$, since $c$ is the constant velocity of light and light gets from $e$ to $e'$. So for $e$ and $e'$, $Dx = c \cdot Dt$, which implies that $Di = \sqrt{Dx^2 - c^2 Dt^2} = 0$. But looking at $e$ and $e'$ from the point of view of $S'$, again, if $Dx'$ and $Dt'$ are their spatial and temporal separations according to $S'$, $Di' = \sqrt{Dx'^2 - c^2 Dt'^2} = 0$. Notice that $c$ has the same value in each case, for this is the assumption about the invariance of the velocity of light for all inertial observers. So if two events have a zero interval between them for one inertial observer, they have a zero interval between them for all inertial observers, i.e., null intervals are invariant.

What about non-null intervals? Suppose $e$ and $e'$ are such that $Di = \sqrt{Dx^2 - c^2Dt^2} = r$, a nonzero number. From the linearity assumption and from the invariance of null intervals, we can prove, once again, by some elementary mathematics, that $Di' = \sqrt{Dx'^2 - c^2Dt'^2} = r$ equals the same nonzero number. From linearity and the invariance of null intervals, we can prove that all intervals, null or non-null, are invariant functions on the pairs of events.

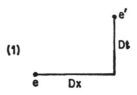

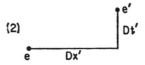

Fig. 30. The linearity assumption. Let $e$ and $e'$ be two non-coincident events. For one observer, $S$, $e$ will have a certain spatial separation from $e'$, $Dx$, and a certain temporal separation, $Dt$, as shown in (1). For $S'$, $e$ and $e'$ will have different spatial and temporal separations, $Dx'$ and $Dt'$.

If we wish to be sure that $Dx'$ and $Dt'$ will be functions only of $Dx$, $Dt$, and the relative velocity of $S'$ with respect to $S$, and not of the particular place or time of occurrence of $e$ and $e'$, then we must invoke the linearity assumption.

So spacetime consists of events that have invariant intervals between them. If the square of the interval is null, the events are connectible by a possible light ray. If it is positive, then no causal signal can connect them, i.e., they have spacelike separation, for no causal signal can outrun light. If the square of their invariant interval is negative, they are connectible by some causal signal slower than light, i.e., they have timelike separation. And all pairs of events have spatial and temporal separations that are not invariant, but such that the spatial separation for a given inertial frame depends only upon the spatial and temporal separation for any other inertial frame and not upon the specific locations of the events in the other frame. That is, I have just constructed the Minkowski spacetime of (II,C,1).

We do need one additional assumption to complete the construction. The set of all events which constitutes spacetime is not "decomposable" in any invariant way into spaces-at-a-time and instants of time. But, relative to a given inertial frame, there is such a decomposition. Choose a particular inertial frame. There is an equivalence relation among events, simultaneity with respect to the chosen frame. Use this to partition the set of all events into equivalence classes. Each equivalence class contains all the events simultaneous, relative to the chosen inertial frame, with each other. All of these events are at a spacelike separation from one another. I call these classes spacelike hypersurfaces and, relative to a given inertial frame, we can view spacetime as an infinite set of spacelike hypersurfaces, one for each "instant of time." Now the important assumption is made that each spacelike hypersurface has the structure of Euclidean three-space.

## 4. Some Important Features of Minkowski Spacetime

The accompanying diagram summarizes many important features of Minkowski spacetime (Fig. 31). The point $O$ is a specific event. The $x$-axis represents all the events simultaneous with $O$ relative to a particular inertial frame. Actually, the single line stands for a whole Euclidean three-space of events, the spacelike hypersurface relative to the chosen frame containing $O$. The $t$-axis contains all the events said by the chosen observer to be at the same place as event $O$. For another observer, in relative motion with respect to the original choice, one could represent his view of Minkowski spacetime on this same diagram by (1) drawing a new $x'$-axis at a slope to the original $x$-axis to represent his spacelike hypersurface containing $O$, (2) drawing a new $t'$-axis sloping with respect to the $t$-axis to represent the points at the same place as $O$ as viewed by the new observer, and (3) recalibrating the $x'$ and $t'$ axes in such a way that spatial and temporal separations for the new observer are correctly represented on the diagram. This is easily done by utilizing the fact that all intervals between events are the same for both observers, allowing one to use "calibration hyperbolae" to transfer the metric along the $x$ and $t$ axes to a new metric

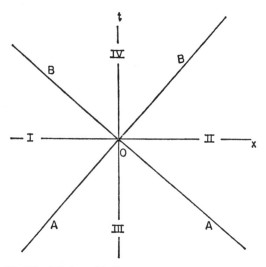

Fig. 31. The Minkowski diagram for an observer. The line
*x* represents all events simultaneous with *O* for the observer. The line *t* represents all events at the same place as
*O*. *A* is the "collapsing light shell" on *O*, and *B* the "expanding light shell." I and II are regions of events with
spacelike separation from *O*. III is the set of events with
timelike separation from *O* and in *O*'s past. IV is the set
of events with timelike separation from *O* and in *O*'s
future.

along the *x′* and *t′* axes. This is illustrated in the second diagram
(Fig. 32).

The lines *A* and *B* in Fig. 31 represent light shells "collapsing" in
the event *O* and "expanding" from event *O*. They represent, in other
words, events at a null separation from event *O*. The lines represent
spheres in both the past and future time directions. The regions I and
II represent all the events at a spacelike separation from *O*, i.e., all
events that cannot be "connected" to *O* by any causal signal whatever.
One can easily prove that for any event in I or II, there is an observer
who sees that event as simultaneous with *O*. Region III represents all
the events from which a causal signal slower than light can be sent to
reach *O*, i.e., all the events prior to *O* in time and with timelike separation from *O*. Region IV represents the events future to *O* and timelike
separated from it, i.e., all the events such that a causal signal traveling
slower than light can leave *O* and reach the event in question. No
observer traveling with respect to the original observer will see events

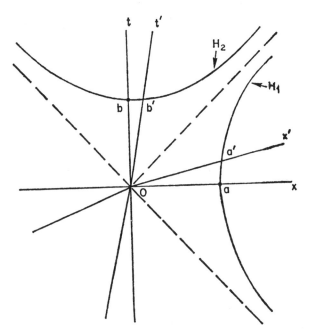

Fig. 32. Coordinatizing events for two observers in relative motion in the same Minkowski diagram. The line x is the set of events simultaneous with $O$ for $S$, and $t$ the set of events at the same place as $O$ for $S$. The line $t'$ represents the motion of $S'$ to the right according to $S$.

The line $t'$ also represents all the events at the same place as $O$ according to $S'$ (note I have assumed that $S$ and $S'$ were coincident with event $O$). The line $x'$ represents all the events simultaneous with $O$ according to $S'$.

The hyperbolas $H_1$ and $H_2$ have equations $x^2 - t^2 =$ constant. So they are all events with a given specific interval separation from $O$ according to $S$ (note that I take the velocity of light as unity for convenience). Since intervals are invariant, the equations of the hyperbolas for $S'$ are $x'^2 - t'^2 =$ constant, since an event with separation of interval $r$ from $O$ according to $S$ is also an event with interval separation $r$ from $O$ according to $S'$, and since $S$'s intervals $x'^2 - t'^2$.

Event $a$ is simultaneous with $O$ for $S$ and a specific distance from $O$. Since $a'$ is (1) on the same hyperbola as $a$ and (2) on the $x'$ axis, $a'$ is the event that $S'$ says is simultaneous with $O$ and just as far from it as $S$ says $a$ is from $O$.

Event $b$ is at the same place as $O$ for $S$, and a certain time from $O$. Since $b'$ is (1) on the same hyperbola as $b$ and (2) on the $t'$ axis, $b'$ is the event that $S'$ says occurred where $O$ occurred and just as much after $O$ as $S$ said $b$ occurred after $O$.

in Regions III or IV as simultaneous with $O$. Events in Regions I and II are sometimes spoken of as being *topologically simultaneous* with $O$, and those in Regions III and IV, as well as those on the light lines, as being *topologically nonsimultaneous*.

The adoption of Minkowski spacetime has many consequences. One of the most important is the necessity of adopting a new mechanics to replace the Newtonian theory which is no longer Galilean-relative in the light of this spacetime. I won't say anything about these important issues here but instead review three famous consequences of Minkowski spacetime: (1) the "shrinking" of rigid rods placed in uniform motion with respect to an observer, (2) the "slowing down" of clocks placed in uniform motion with respect to an observer, and (3) the so-called "clock paradox." I might note at this point that in the construction of Minkowski spacetime I used only the notions of ideal clocks and light rays. As noted in (II,D,3), this would not be the only possible choice for setting up the fundamental coordinative definitions. I could have chosen instead ideal rigid rods and ideal clocks, or ideal clocks and freely-moving particles, or even freely-moving particles and light rays alone. I shall examine the consequences of the various choices in (IV,C,2,3) below.

1. *Length contraction:*  Suppose we have a rigid rod at rest in inertial frame $S'$ and of length $l$. And suppose that $S'$ is moving with uniform speed $v$ in the direction of the length of the rod with respect to frame $S$. What is the length of the rod according to an observer fixed in $S$? It seems natural to call the length of the rod according to $S$, the distance (spatial separation) relative to $S$ of two events that constitute simultaneous (relative to $S$) events of the endpoints of the rod being somewhere. If we do adopt this "definition" for the length of a rod in motion with respect to a given inertial frame, then a little simple calculation will show us that a rod whose length is $l$ relative to a frame in which it is at rest, has length $Rl$ relative to a frame with respect to which it is moving with speed $v$ along the direction of its length. Notice that the shrinking is symmetrical, i.e., $S$ attributes a length $Rl$ to a rod that $S'$ says has length $l$ if the rod is at rest in $S'$. Similarly, $S'$ attributes a length $Rl$ to a rod at rest in $S$ and said by $S$ to have length $l$. The diagrams below illustrate this "symmetric shrinking" although they do not, of course, prove that the common shrinking factor is $R$ as we have maintained (Figs. 33 and 34).

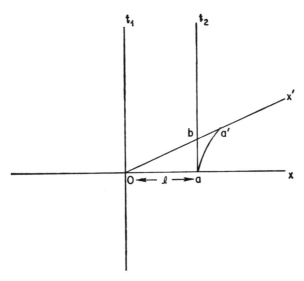

Fig. 33. A rod at rest in $S$ appears shortened to observer
$S'$. The lines $t_1$ and $t_2$ represent the endpoints of a rod of
length $l$ (according to $S$) at rest in $S$ through time. Event
$a$ is distance $l$ from $O$ according to to $S$, since according to
$S$, $a$ is simultaneous with $O$ and the rod has length $l$.

The event simultaneous with $O$ and a distance $l$ from it
according to $S'$ is $a'$, since it is on the $x'$ axis and on the
same hyperbola as $a$. But $b$ is the event locating the end of
the rod at a time simultaneous with its other end being at
$O$, according to $S'$. Since the distance $Ob$ is less than $Oa'$,
and since $Oa' = l$, $S'$ says the rod is shorter than $l$.

How real is the shrinking predicted by special relativity? One might
be tempted to think that since the adoption of special relativity is
founded on a new definition of simultaneity, and since our definition for
the length of a rod in motion requires the notion of simultaneity, the
change from prerelativistic to relativistic theory is merely one of how
the facts are described. But we can see that this isn't so. Imagine two
rods, each of length $l$ when at rest in $S$. The rods are charged at the
endpoints such that when the left ends meet a spark is emitted, and
similarly when the right ends meet. The rods are put in uniform
motion with respect to $S$ parallel to the $x$-axis, one going in the positive
$x$-direction and the other in the negative $x$-direction. A strip of film
is pasted, at rest with respect to $S$, along the $x$-axis. The rods pass by
one another, the sparks are emitted, and they leave marks on the film.

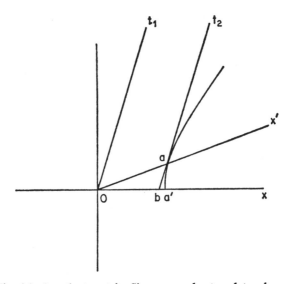

Fig. 34. A rod at rest in $S'$ appears shortened to observer $S$. The lines $t_1$ and $t_2$ represent the endpoints of a rod at rest in $S'$ through time. Since the rod has length $l$ according to $S'$, and since $x'$ is the set of events simultaneous with $O$ according to $S'$, $a$ is an event whose interval from $O$ is $l$. Since $a'$ is the event simultaneous with $O$ according to $S$, and since $a'$ is on the same hyperbola as $a$, $a'$ is the event whose distance from $O$ is $l$ according to $S$. But $Ob$ is less than $Oa'$, and $b$ is the event that, according to $S$, is the end of the rod being in a place at the moment its other end is at $O$. So, according to $S$, the rod has length less than $l$.

How far apart will the marks be, marks, mind you, on an object at rest in $S$? According to the prerelativistic theory they will be $l$ meters apart. According to special relativity they will be $Rl$ meters apart, and it is the special-relativistic prediction that is confirmed by experience. Now there may be ways of "interpreting" this result which are at variance with the usual "language" of special relativity, as we shall see, but here is certainly a hard fact differentiating the length predictions of special relativity from those of prerelativistic theories (Fig. 35).

2. *Time dilation:*   Suppose we have a device at rest in $S'$ which suffers two events that are, according to $S'$, at a separation of $t$ seconds. How far apart in time are these events as determined by an observer at

rest in $S$? The answer is that he calculates the events to be $t/R$ seconds apart. Since $R$ is always a number less than one, $S$ says that the events are "further apart in time" than $S'$ does, i.e., $S$ says that clocks at rest in the $S'$ system run slow as determined by clocks at rest in his system. Again the effect is symmetrical, for $S'$ will determine, according to clocks at rest in his system, that the clocks at rest in the $S$ system have their period between ticks dilated by the factor $l/R$. The diagrams below illustrate the existence of the symmetrical time dilation effect, although, once again, they do not show that the dilation factor is equal in the two cases or equal to $l/R$ (Figs. 36 and 37). Once again one could, if one wished, show that this effect was sufficiently "real" to guarantee that special relativity had quite different empirical predictions from the prerelativistic spacetime theory, but I shall leave this for the reader to carry out should he wish.

3. *The clock paradox:* To set up the discussion of the famous "clock paradox" in special relativity—not a paradox at all, incidentally, but simply a surprising result of the theory—we need one more definition. This is the *proper time between two events along a curve.* The "co-ordinate time" between two events is relative to a given inertial frame, as we have seen. It is just the time separation of the events according to an observer at rest in the frame. It is relative to the chosen inertial frame, but does not require the specification of any particular curve

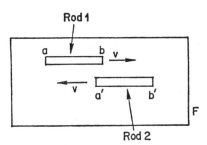

Fig. 35. A prediction of relativity at variance with pre-relativistic theories. The two rods that have length $l$ when at rest in $S$ are put in motion with respect to $S$ with the same speed in opposite directions. When $a$ passes $a'$, a spark jumps to the film $F$ which is at rest in $S$. Similarly when $b$ passes $b'$

What is the distance between the two marks on the film? According to prerelativistic theory, it is $l$. According to relativity it will be $Rl$.

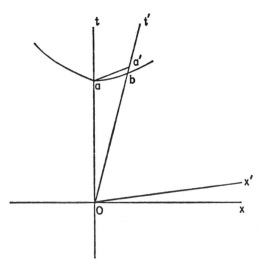

Fig. 36. A clock at rest in S appears slow to observer S'. a is the event that is the reading "one hour" on the clock that follows t, i.e., which is at rest in S. S' says that a is simultaneous with a'. But b is the event that S' says is one hour after O, since b is on the same hyperbola as a and on t'. So S' says, "The clock in S reads 'one hour' at a time after one hour has passed, so the clock at rest in S is slow."

connecting the two events. Proper time is defined only for events at timelike separations and only relative to a particular spacetime curve between the events. On the other hand, it is an invariant notion. The proper time between a and b along curve C is the same for all observers. Suppose we set an ideal clock at t = 0 making this reading of the clock an event coincident (i.e., at the same place and time) with event a. We "move" the clock to event b along curve C, i.e., we let the clock move through space in time such that the record of its trajectory is curve C. We then read the clock time at the event in the clock's history coincident with event b. This is the proper time between a and b along curve C. It obviously *must* be an invariant notion, as a little reflection on the part of the reader will convince him.

Suppose we take two clocks and set them at t = 0 when they are spatially together, that is, clock One's reading t = 0 is an event coincident with clock Two's reading t = 0. We then leave one clock at rest in an inertial frame, any inertial frame, and move the other clock along a spacetime curve which characterizes accelerated motion. We choose a trajectory for the second clock which takes it away from the first clock, but which eventually brings it back into spatial coincidence

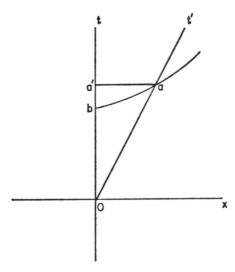

Fig. 37. A clock at rest in $S'$ appears slow to observer $S$. $a$ is the event one hour after $O$ according to the clock whose path is $t'$, i.e., which is at rest in $S'$. $a'$ is the event that is simultaneous with $a$ according to $S$. $b$ is the event that, according to $S$, is one hour after $O$, since it is on the same hyperbola as $a$ and on $t$. Therefore $S$ says, "The clock at rest in $S'$ read 'one hour' only after more than one hour has passed, so the clock at rest in $S'$ is slow."

with the first clock. At the end of this trip we compare clock readings. We can prove from special relativity that the total elapsed time recorded on clock Two will always be less than that recorded on clock One. This holds independently of the particular constitution of the clock, so that biological processes will do the same. If we start with two identical twins, leave one at rest in an inertial frame and accelerate the other away from and back to his stay-at-home sibling, the accelerated twin will be found to have aged less than the stay-at-home twin over the elapsed period between the two events.

Since this phenomenon has been treated at length and with great clarity in the literature, I need not pursue it in detail. I should note a few things, however: (1) Nothing here violates the principles of "symmetrical time dilation" discussed above. The symmetrical situation was two inertial observers discussing the coordinate time between events in the life-history of objects at rest in the other inertial observer's system. Here, however, we are comparing elapsed proper times of objects, one of which remains inertial while the other is accelerated, and this is not the same thing at all. (2) The two clocks have not been treated symmetrically, so there is no reason to be too surprised that they have behaved so differently. One clock has been accelerated, and, as pointed out at length in (III,D,4), special relativity retains the Newtonian and neo-Newtonian distinction between systems that are absolutely accelerated and those that are not. (3) If we idealize the accelerated clock's history into two inertial motions with a "jump" between inerital frames at the turn-around point, as illustrated in the diagram, we can seen that the situation looks to the accelerated observer like this (Fig. 38): (a) from $O$ to $e$, I calculated the stay-at-home

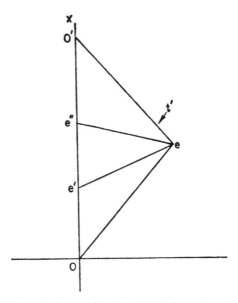

Fig. 38. The clock paradox. The total time elapsed on the clock carried on the broken path $t'$ from $O$ to $O'$ is less than that elapsed on the inertial clock which gets from $O$ to $O'$ along the path $t$. This is because the clock following $t'$ has been accelerated at the break in the path.

The accelerated observer calculates that the inertial clock runs slow according to his from from $O$ to $e'$, and also from $e''$ to $O'$; but he sees himself as moving instantaneously from one inertial path to the other (at event $e$, his acceleration being "almost instantaneous"), yet $e$ is simultaneous with $e'$ in his first inertial frame, and with $e''$ in his second. It is the life of the inertial clock from $e'$ to $e''$ which makes the inertial clock read a greater time interval from $O$ to $O'$ than does the accelerated clock.

clock as running slow from event $O$ to $e'$, simultaneous with $e$ in my "outgoing" inertial frame; ($b$) from $e$ to $O'$, I calculated the stay-at-home clock as running slow from $e''$ to $O'$, since $e''$ is the event simultaneous with $e$ in my "homecoming" inertial frame; ($c$) although it is true that I calculate the stay-at-home clock as running slow from $O$ to $e'$ and from $e''$ to $O'$, it is no wonder that its total elapsed proper time is greater than mine, for it also ticked off the seconds from $e'$ to $e''$ while my clock saw no elapsed proper time between those events; (4) does this mean that the inertial-frame-jumping observer "sees" the stay-at-home clock as having "gaps" in its life history? No, for what he sees is illustrated in the next diagram (Fig. 39). It is clear

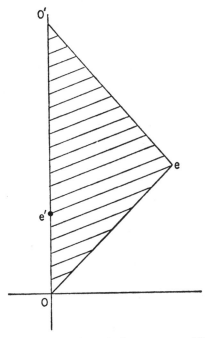

Fig. 39. What the accelerated observer *sees*. The parallel lines represent light signals sent out from the inertial clock at equal intervals. From $O$ to $e$, the accelerated observer sees a "slow motion" picture of the inertial clock. From $e$ to $O'$, he sees a "fast motion" picture. Since the interval from $e'$ to $O'$ is greater than that from $O$ to $e$, the accelerated observer's net film is of the inertial clock marking off more total time intervals than the clock the accelerated observer carries with him. Once again, the proper time elapsed on the inertial clock is greater than that elapsed on the accelerated clock.

that what he *sees* is an *image* of the stay-at-home clock running slowly through part of its history, while the traveler is outgoing, and then running fast through the rest of its history during the homecoming. But since there are more "fast movies" of the history than "slow movies" the total image is a "film" of a clock elapsing a greater proper time than that elapsed on the traveler's clock. One must always be careful to distinguish in special relativity what an observer *sees*, literally, from what he computes to be the case. After all, just because the speed of light is invariant in the theory, that does not mean that it isn't finite. And one will always want to take it into account when one wishes to infer from what one *saw* to what actually occurred.

## 5. *Special Relativity and Some Traditional Philosophical Disputes about Time*

I have deliberately eschewed up to this point discussion of the debates of traditional philosophy about time. It might be profitable now, however, to comment briefly on a few of these traditional issues; for the advent of special relativity, with its radical changes in the notion of simultaneity and hence of time, has been claimed by some to provide a scientific resolution of what had been taken to be metaphysical issues. What we will see, and this should no longer surprise the reader, is that special relativity throws novel light on the philosophical questions, but it is unable by itself to resolve fully the long-standing philosophical issues.

Here are two traditional philosophical claims: (1) Only those things exist which exist at the present moment. Things whose existence is in the past, and things whose existence is yet to be, cannot properly be said to exist at all. (2) Whereas the present and past, having occurred or currently occurring, can be said to have determinate reality, the events of the future, whatever they *will* be, cannot be said to have any kind of "determinate" reality at all. They are at best mere possibilia without the same "full reality" possessed by the happenings of the present and the past.

Trying to make clear sense of these claims is not an easy task, if indeed it is a possible one at all. Usually critics of the claims argue somewhat like this: Insofar as we really need a notion of "existence" or "determinate existence" we can make do perfectly well with a completely *timeless* sense of the notions. Past and future things have just as much reality as present things, and future happenings as much "fully real determinate existence" as happenings of the present and past. The present moment is simply the moment at which an assertion is being made. To deny existence to entities because they fail to exist at the moment of assertion is as silly as to deny objects existence because they fail to exist at the *place* at which the assertion is being made. We should no more deny *reality* to the "then but not now" than we would to the "there but not here." And similarly for the future happenings. They are as determinately real as present and past happenings, and to deny this would be as pointless as to deny determinate reality to events on our left, but admit it for events where we are and on our right. This is not, of course, to deny the importance of the question of whether the universe is *deterministic,* but that issue is an issue about the *relations* future events bear to present and past events, not about the determinate reality of the future events.

The apparent plausibility of the claims being denied by these philosophers is accounted for by them in terms of the espousers of the claims being misled by the surface form of ordinary language. In ordinary discourse we do treat space and time asymmetrically. We have a full tense structure for making clear different temporal references in our assertions: objects were, are, or will be; events happened, are happening, or will happen. But our spatial reference is handled differently, by using relational terms relating things and events to places: the object is here or there, the event happens here or elsewhere. But, the argument goes on, we could easily replace ordinary discourse with a language symmetrical in its temporal and spatial reference, by simply handling all temporal and spatial locating of objects and events by relating objects or happenings to spatial locations, temporal instants, or, from a relativistic point of view, to spacetime locations, i.e., to events in spacetime.

I don't wish to pursue this debate, but instead to focus on an argument of the following sort: The philosophers who maintain that past and future objects are not real existents, or that future events do not have determinate reality, are refuted out of hand by special relativity. Imagine a given observer at a particular spacetime location. Consider an event future to him, or an object that has existence only in his future. According to the philosophers whose claims we will rebut, the future events have no determinate reality and the wholly future objects no true existence. But suppose the event future to our observer, or all the future events relative to him which make up the future object, are at spacelike separation from the observer's spacetime location. From special relativity we know the following: There will be another observer, in relative motion with respect to the first, whose location is coincident with the location of the first observer and such that the events future to the first observer about which we have been concerned will be simultaneous with the common spacetime location of the two observers relative to the frame of motion in which the second observer is at rest. Now the second observer certainly has reality and determinate reality relative to the first, since they are coincident, let alone contemporaneous. And the events in question are certainly both real and determinately real relative to the second observer, since they are in his present. But how could an event be real and determinately real to an observer who is himself real and determinately real relative to the initial observer and yet not be real and determinately real relative to our original observer? This is surely an absurdity, and we see, therefore, that one cannot accept special relativity without rejecting the traditional philosophical positions (Fig. 40).

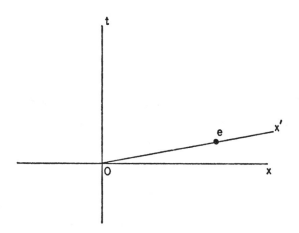

Fig. 40. The argument that relativity refutes the "not fully-determinate reality of the future." Suppose an observer is at event O, and that e is an event in his future but space-like separated from him. Then there will be another observer, also at O, such that e is in his present. So, the argument goes, how can e not have fully-determinate reality for the first observer if it has fully-determinate reality for another observer at the same place and time.

But when we put the argument this baldly, the inability of the physical theory to resolve the philosophical dispute becomes obvious.

In special relativity the notion of which events are simultaneous with one another becomes a relative notion, relative to an inertial frame. And, as I have noted, even if $a$ is simultaneous with $b$ relative to $A$, and $b$ simultaneous with $c$ relative to $B$, $a$ may not be simultaneous with $c$ relative to either $A$ or $B$. So if we wish to tie up reality or determinate reality of events and objects with their temporal relation to an assertor, we must admit these notions to be just as relative to an inertial state of motion of the assertor and just as "nontransitive across observers in different states of motion" as we have made the simultaneity relation. Now the traditional philosophical exponent of these claims may indeed be reluctant to take this "relativization" step. If so, he must either drop his claims or reject special relativity. But he need not do either, for he can simply accept the consequence, surprising no doubt, but by no means inconsistent or patently absurd, of relativizing his notions of reality and determinate reality at the same time that he relativizes his concepts of simultaneity. The science can change the philosophy and put the dispute in a new perspective, but it cannot resolve the dispute in any ultimate sense.

# CAUSAL ORDER AND TEMPORAL ORDER
# IN SPECIAL RELATIVITY

In this section I will begin a philosophical critique of some of the fundamental features of the relativistic theory of time. The critique will not be complete, for I shall focus on issues relating to the conventionality of the special-relativistic theory of time. I will not pursue at length the question with which we will ultimately be concerned: To what extent does the adoption of a relativistic theory of time support the reductionist claim that the temporal relations among events are reducible to causal relations? To fully tackle this question, it will be extremely useful to have before us some important aspects of the nature of temporal relations in general relativity. So I will break off the philosophical critique in midstream to bring out some of these important features of general relativity in (IV,D). In (IV,E) I will return to the philosophical discussion and continue where I leave off at the end of this section.

## 1. *Einstein's Redefinition of Simultaneity*

As Einstein himself motivated the change from prerelativistic to relativistic spacetime through a critical analysis of the notion of simultaneity, it will be reasonable for me to begin the philosophical critique of this "scientific revolution" by making a careful investigation of just what is going on in the shift from the old concept of simultaneity to the new.

To begin with, let me rehearse the "facts" the new theory must account for. These facts are actually rather bold generalizations drawn from the specific scientific observations made by the experimentalists. But this, of course, is the way the theoretician always works, constructing theoretical explanations not of particular experiments but rather of general classes of possible observations whose nature is inductively inferred from the actual experiments:

FACT ONE: *All round-trip light propagation experiments give equal times for the propagation of the light in either direction around the closed course, so long as the laboratory is at rest in some inertial frame.*

FACT TWO: *There are no round-trip signals of arbitrarily high velocity, that is, it is not the case that for each round-trip course and for each time interval, no matter how small, a signal can be found which traverses the course in less than the specified time interval.*

FACT THREE: *If two distinct events cannot be connected by a causal signal that travels no faster than a light signal, then they cannot be connected by a causal signal at all.*

What I will argue is this:

1. The whole Einstein program is motivated by a fundamental epistemological assumption: Although *coincidence* of events is a relation among events directly open to observational inspection, simultaneity between noncoincident events is a relation not open to direct inspection. To have an intelligible notion of simultaneity for distinct events in one's theory, therefore, this relation between events must be tied down to some directly observational features of the world. One can view the ties as coordinative definitions, if one likes, or perhaps instead as the most basic *laws* connecting simultaneity to other features of the world, just as one can view the connection of geodesics with paths of free particles as either a coordinative definition or as the fundamental dynamical law. But some such connection there must be.

2. The facts summed up in the "three basic facts" make the prerelativistic connection of simultaneity to observable facts impossible, at least if one rejects the compensatory-type theories discussed in (IV,B,2). That is, faced with these facts one must either adopt some

version of the compensatory theory or drop the prerelativistic concept of simultaneity for noncoincident events.

3. Having disposed of the prerelativistic concept of distant simultaneity, one must put a new notion in its place. The existence of the round-trip experimental results suggests a natural replacement for the prerelativistic notion. One can generate this notion by reflecting on the basic facts we have reviewed, and then choosing a definition as close to the prerelativistic as possible.

4. The new concept of simultaneity will lead to some assertions that are at variance with the prerelativistic theory. The new theory does have different experimental consequences from the old "uncompensated" theory. This is as it should be, since some of the facts on which we are basing the theory were unexpected on the basis of the older theory. Further, the new theory of simultaneity will lead us to say different things about simultaneous events than we used to say. This is no surprise, for simultaneity is a different relation among events than was the old simultaneity relation. There is no relation in nature fitting the requirements of the old notion, at least if we reject the compensatory theory, and the new relation that takes its place can't be expected to be identical with the old. Some things true about the old relation are, however, true about the new. In particular, both relations are equivalence relations relative to a particular inertial frame and both relations are compatible, given the facts about the world, with the assertion that causal signals never proceed from one event to an earlier event.

Let us examine each of these features of the conceptual revolution in turn.

1. I am not going to comment on the justification of the epistemological assumption but only point out how crucial it is to Einstein's argument, to the development of Minkowski spacetime in general, and to the philosophical controversy that has swirled around such issues as whether special relativity rests upon a "convention." It is clearly the "inaccessibility to direct observation of simultaneity for noncoincident events" that forms the whole framework in which Einstein argues for the *possibility* of accounting for the unexpected facts by some alternative to the compensatory theory. This assumption also provides the whole framework in which issues as to the conventionality of the special-relativistic theory as opposed to alternative accounts arise.

If we could "directly" tell when events are simultaneous with one another, even when they are noncoincident, it would not take an Einstein to reject the compensatory theory, if that theory is wrong, or to establish it if it is correct. If we could tell when events at a distance from one another are simultaneous by direct inspection, we could simply *see,* for example, in which states of motion clocks really slowed down. For we could check the period of a clock in motion against the events themselves and ask ourselves: Do clocks at rest in this particular inertial frame record events at a distance as simultaneous when they really are, or do they misindicate simultaneity for events at a distance? Without an assumption of the direct inaccessibility of simultaneity for noncoincident events, the whole program of seeking for possible empirical *tests* of distant simultaneity and of examining the adequacy of the proposed tests "in the light of the facts" would never get underway.

On the other hand, the assumption that the coincidence of events, or if we wish the simultaneity for events at the same place, *is* a matter for direct observational inspection, underlies the whole of Einstein's program. Only this allows us to take the results of the round-trip experiments as giving us the "hard facts," for example. The whole virtue of the round-trip experiments is that they rely upon the observation of simultaneity or nonsimultaneity of events *at a single point* for determining their outcomes. It is this that allows us to take the results of the round-trip experiments as "givens," observationally established prior to the job of fixing tests for simultaneity for noncoincident events.

2. What is wrong with the prerelativistic notion of simultaneity for noncoincident events? This turns out to be a matter requiring some thought.

First, we should be clear that if we will tolerate one of the compensatory theories then there is nothing wrong, relative to this toleration, with the prerelativistic notion of simultaneity. Events at a distance are either simultaneous or they are not. And if they are simultaneous, they are simultaneous with respect to every observer in every state of motion. Of course, we have no way of finding out just which events really are simultaneous with one another. We can't determine this by direct inspection, as we saw in (1) above, and we can't determine it by using clocks or light rays either, since the velocity of light is not the same in every direction except in the frame at rest in the aether and we have no way of telling which frame is at rest in the aether. Clocks properly record time intervals only when they are at rest in the aether, and we have no way, again, of knowing when a clock is so at rest or when it is really in uniform motion with respect to the aether and hence running slow and misleading us about real time intervals.

Now these last remarks show us what is objectionable about the compensatory theories. This can be put in two ways: (1) Compensatory theories violate the canons of verifiability. (2) There are simply too many such theories and we can't find out which one is right. They violate the canons of the verificationist since the compensatory theories all postulate the existence of the aether frame, relative to which light rays have their velocity the same in all directions, rods at rest properly record length intervals, clocks at rest properly record time intervals, etc. But by their very assertions, the compensatory theories deny us the possibility of ever finding out just which inertial frame really is the frame distinguished by being the rest frame in the aether. There are too many compensatory theories in the sense that one can construct a new compensatory theory for each choice of inertial frame as rest frame in the aether, and there is no way whatever of experimentally selecting one of this infinite class of compensatory theories as the correct one. As von Neumann has remarked, the problem with a nonrelativistic explanation of the facts is not that one can't be given but that too many can be given, and no reason can be given for selecting one rather than the other. In a clear sense, the motivation behind special relativity is the elimination of arbitrary choice from physics.

So suppose, on good verificationist grounds, we reject the compensatory theories. Why should we now object to the prerelativistic notion of simultaneity as an unacceptable concept? Once again the argument is not as simple as might first appear. I think it has two stages: (1) One natural empirical test for simultaneity for events at a distance is made impossible by the nonexistence of arbitrarily fast signals relative to an inertial frame. (2) The other natural correlate for simultaneity for distant events is shown to be objectionable since it violates the canons of "good definition." This is established by the results of the round-trip experiments and by the nature of the compensatory theories that would have to be imposed to explain them prerelativistically.

First the role of the "nonexistence of arbitrarily fast round-trip signals," Fact Two. A natural way to test empirically for simultaneity for events at a distance, prerelativistically, would be like this: We are at point $P$, and we select event $e$ at point $P$. We wish to know which event at point $Q$ is simultaneous with event $e$ at $P$. We sent a round-trip causal signal that leaves $P$ at an event earlier than $e$ at $P$, is reflected at an event $h$ at $Q$, and returns to $P$ at event $e$. We find the earliest event at $Q$, $f$, such that no causal signal, no matter how rapid its round trip, can be launched from $P$ before $e$, be reflected coincident with $f$ at $Q$, and return to $P$ at $e$. This is the event at $Q$ simultaneous with $e$ at $P$.

Now if we had causal signals whose round trips were as far as we liked over a specified course, this technique would give us the pre-relativistic notion of simultaneity complete with an empirical test condition. But there are no such signals, so we can no more use this "method" to establish an empirical test for simultaneity for events at a distance than we could use the method whose instructions were, "Use the simultane-ouscope to directly observe whether $f$ at $Q$ is simultaneous with $e$ at $P$." There are no "simultaneouscopes," and there aren't any arbitrarily fast round-trip causal signals either.

But the absence of such arbitrarily fast round-trip causal signals is not by itself enough to lead us to reject the prerelativistic notion. The null results of the round-trip experiments are crucial as well. To see this imagine the following possible world: There is an aether. The round-trip experiments give null results only when at rest in the aether frame. When in uniform motion with respect to the aether they give the expected non-null results. It just happens to be the case, though, that material reference frames never move at more than a certain uniform velocity with respect to the aether, and hence never more than a certain specified velocity with respect to each other. In such a world there would be maximally fast causal signals, presumably the velocity of light in a given direction plus the maximum velocity of a material frame moving in the opposite direc-tion with respect to the aether would be the "fastest observed causal signal between events." Yet we would not drop the prerelativistic notion of simultaneity in such a world, I believe.

Why should we? We know what the aether frame is in this world; we know, therefore, the frame in which the velocity of light is the same in all directions, the one and only frame in which this is so. Two events at a distance are *really* simultaneous when they appear to be simultaneous on the basis of optical experiments carried out in a laboratory frame at rest in the aether. In other words, to find the event at $Q$ simultaneous with event $e$ at $P$: Choose a laboratory frame at rest in the aether; send out light rays from the place $Q$, relative to this frame, which are reflected at the place $P$, at event $e$, and returned to place $Q$; find the event $f$ at $Q$ which is equidistant in time from the emission and reception events at $Q$ of the light beam reflected at $e$: $f$ at $Q$ is simultaneous with event $e$ at $P$. This notion of simultaneity is clearly invariant, since the experiment is carried out in a specific reference frame, the frame agreed by all observers to be at rest in the aether.

So if the round-trip experiments gave their expected non-null results, results that varied as the relative motions of the laboratories in which the experiment was carried out varied with one another, we would have no reason not to retain the prerelativistic concept of "absolute" simultaneity.

To see how it is the null results of the round-trip experiments which really forces upon us the rejection of the prerelativistic notion in favor of its relativistic successor, we must see how the null results make the alternative "natural" empirical test for simultaneity for events at a distance an inadequate test.

Definitions are sometimes said to be arbitrary, but this is only partially true. We can label a feature of the world with any linguistic term we like, but whether this feature exists is hardly arbitrary. For example, we can see how putting into our language a name for an object that doesn't exist or for an entity when there is actually more than one entity having the defining feature, can lead to *contradictions* in ordinary arithmetic.

Define $n$ as 'the largest integer.' Now $n$ is certainly greater than 1. Hence, $n^2$ is greater than $n$. But $n^2$ is an integer, and $n$ is the largest integer. Hence $n$ is greater than $n^2$. Contradiction. What went wrong? There is no largest integer, and to define $n$ as the largest integer is illegitimate. Define $m$ as 'the odd number between 2 and 6.' Now 3 is odd and 3 is between 2 and 6. So $m = 3$. But 5 is odd, and 5 is between 2 and 6. Hence, $m = 5$. So $3 = m = 5$, and $3 = 5$. Absurdity. What went wrong? You cannot define $x$ as 'the thing that is $F$' if more than one thing is $F$ without getting a contradiction.

A "natural" prerelativistic definition for simultaneity for events at a distance is, as we saw, this: We wish to find event $f$ at $Q$ simultaneous with event $e$ at $P$; set up a clock at rest at $P$; synchronize a clock at $P$ with the rest clock; move the second clock to $Q$, and synchronize a rest clock at $Q$ with it; count the event $f$ at $Q$ which has clock reading $t$ coincident with it with the event $e$ at $P$ which has clock reading $t$ on the clock at $P$ coincident with it.

But this definition is no good, for both in the compensatory theory and in the relativistic theory, that is, in any theory adequate to account for the null results of the round-trip experiments, the synchronization of a clock at $Q$ will depend on *how* the moving clock is moved from $P$ to $Q$. In both accounts, the time the moving clock attributes to event at $Q$ will depend on how the clock was moved from $P$ to $Q$ and not just upon its original synchronization with the clock left behind at $P$. So more than one event at $Q$ will be called simultaneous with the event $e$ at $P$, and we cannot call each of these distinct events *the* event at $Q$ simultaneous with $e$ at $P$.

As we see, the rejection of the prerelativistic notion of absolute simultaneity depends not just upon the absence of arbitrarily fast causal signals, but, crucially, upon the unexpected null results of the round-trip experiments.

It should be clear at this point that another alternative suggestion for establishing a notion of absolute simultaneity for distant events which is plausible prerelativistically also must be rejected because of the null results of the round-trip experiments. If we were able to determine the frame at rest in the aether, we could synchronize clocks at a distance from one another by means of light signals transmitted from clock to clock, allowing in the synchronization for the time of transmission of the light by choosing as the event at $P$ simultaneous with a given event at $Q$, that event at $P$ midway in time between the event at $P$ which corresponds to the transmission of a light signal to $Q$ which reaches $Q$ just at the event in question, and a second event at $P$ which corresponds to the reception at $P$ of the light signal reflected at the event at $Q$ and transmitted back to $P$.

Prerelativistically this "radar" method for establishing simultaneity would be adequate only if carried out in a frame at rest in the aether. For in any other frame the velocity of light would differ in going from $P$ to $Q$ and then in returning from $Q$ to $P$, and the choice of the midtime event at $P$ as that simultaneous with the reflection at $Q$ would be incorrect.

But then, of course, the impossibility of finding out empirically just which frame is genuinely at rest in the aether means that the supposed coordinative definition for absolute simultaneity fails to establish for an event at $P$ a unique event at $Q$ simultaneous with it. And this is what would be required of any adequate definition of absolute simultaneity.

As we have seen, the solution of special relativity is to retain this radar method for establishing simultaneity, vitiating the prerelativistic inadequacy of its nonunique specification of simultaneity for distant events by simply denying that there is a single unique event at $Q$ simultaneous with a specified event at $P$.

3. So there is no such relation as "event $e$ at $P$ simultaneous with event $f$ at $Q$" in the prerelativistic sense, for to assume there is such a relation violates the canons of verifiability. If the relation exists, we can never know when it holds between two separated events and when it doesn't. On verificationist grounds we reject this relation out of hand.

What do we put in its place? The answer is the relativistic relation of "simultaneity for events at a distance from one another relative to a specific inertial frame." The empirical correlate of two events being simultaneous relative to inertial frame $A$ is clear: Choose your inertial reference frame, and work in a laboratory at rest in the frame; set up a clock at point $P$, at rest relative to the chosen laboratory frame; find the time at $P$, on this clock, of the two events that are such that a light ray emitted from the first event and reflected at event $f$ at $Q$ is received at the second event at $P$; find the event, $e$, at $P$ which is equidistant in time from the emission and reception events at $P$ on the clock at rest at $P$; call $e$ the event at $P$ simultaneous with $f$ at $Q$ *relative to inertial frame* A. This looks just like the "optical determination of simultaneity" in the pre-relativistic theory, but it is different. Which events will be determined to be simultaneous with one another will depend upon the inertial frame in which the experiment is performed, as we saw in (IV,B,3) above. But this is no longer an objection to the definition, for simultaneity is now defined only relative to a particular inertial frame of reference. That $e$ is simultaneous with $f$ relative to $A$, but not to $B$ in motion with respect to $A$, is no more destructive to our definition than would be the fact that my brother is *my* brother, and not yours, shows anything wrong with the standard definition of 'brother.'

4. So we have a new simultaneity relation for events at a distance from one another. It is obviously quite different in essential ways from the prerelativistic notion. Most essentially, 'is simultaneous' is now a three-place relation for events at a distance, not a two-place relation. To make an assertion we need reference to two events and to a particular inertial frame of reference.

But the new relation does have some features in common with the old one; here are three of these features:

A. In prerelativistic physics simultaneity is an equivalence relation. In relativistic physics, once one has selected a particular inertial frame, simultaneity with respect to that frame is again an equivalence relation.

B. In prerelativistic physics it is a "truth," I won't say now "analytic" or "synthetic" in nature, that causal signals never propagate from a later event to an earlier event, nor from one event to another event simultaneous with the "emission" event but at a distance from it. These truths remain true in relativistic physics.

To demonstrate this we need our third basic fact, the fact that no causal signal can connect events that are not connectible by a signal that is no faster than light. If there were causal signals that traveled faster than light, the relativistic definition of simultaneity would lead to the situation in which there were two events simultaneous according to at least one observer, at a distance from one another but connectible by a causal signal, and, worse, to the situation of there being two events, $e'''$

and $e$, such that at least one observer says $e'''$ precedes $e$ in time, yet a causal signal can be propagated from $e$ to $e'''$. The proof of this is most clearly brought out by means of a diagram; the illustration shows how the relativistic definition would lead to "causal paradox" were light not the "fastest signal" in nature (Fig. 41). The null results of the round-trip experiments and the nonexistence of arbitrarily fast round-trip sig-

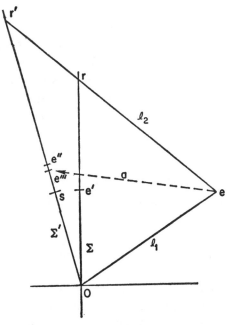

Fig. 41. If there were arbitrarily-fast causal signals, the relativistic definition would lead to causal anomaly. $\Sigma$ and $\Sigma'$ are in relative motion as shown. $\Sigma$ declares $e'$ to be simultaneous with $e$ using the relativistic definition and the light signals $l_1$ and $l_2$, picking $e'$ halfway in time between $O$ and $r$. $\Sigma'$ declares $e''$ simultaneous with $e$ using the same method, but picking $e''$ halfway in time between $O$ and $r'$.

Now $e'''$ is later than $s$ which is simultaneous with $e'$ for $\Sigma$. Since $e'$ is simultaneous with $e$ for $\Sigma$, then $e'''$ is later than $e$ for $\Sigma$. Since there are arbitrarily-fast causal signals, there should be a signal from $e$ to $e'''$. But since $e''$ is before $e''$ in time, and $e''$ is simultaneous with $e$ according to $\Sigma'$, a causal signal has been sent, according to $\Sigma'$ from $e$ to an event, $e'''$, earlier than $e$ in time.

As a matter of fact, since no causal signal is faster than light, no causal signal from $e$ can reach $\Sigma'$ before $r'$. And all observers agree that $r'$ is later in time than $e$.

nals show the inadequacy of the prerelativistic concept of simultaneity for events at a distance. The fact that light is the fastest causal signal has its role in showing that whatever one thinks of the new relativistic notion of simultaneity relative to a given inertial frame, it is at least blessed with the virtue that we are never led by it to attribute infinite speed to a causal signal (a signal propagating between events simultaneous to a given observer) or to maintain that there are causal signals that propagate from later to earlier events from the point of view of at least one inertial observer.

Some recent authors have proposed that it is compatible with special relativity that there be causal signals that propagate faster than the velocity of light. I shall forego pursuing this question here, but only remark that to date such theories have failed, in this author's opinion, to adequately treat the apparent problem of causal paradox introduced by the existence of such signals and the relativistic notion of simultaneity.

C. Prerelativistically, one gave an "empirical test" for distant simultaneity which utilized light rays and another that utilized transported clocks. It was always assumed prerelativistically that the simultaneity assignments arrived at by these two methods would be in agreement. We saw that the former method fails because it requires the establishment of the aether rest frame and the latter since it assumes, incorrectly, the synchronization of clocks initially synchronized and then transported to a new location by different courses of motion.

Is there a notion of "simultaneity established by clock transport" definable in special relativity, and will it give agreement with the assignments of simultaneity established by the light-ray method?

The answer is yes. To prove this we need the additional assumption made by Einstein about the linearity of the relation between assignments of position and time made by two inertial observers, i.e., the assumption that whatever the spatial and temporal separation assigned two events by observer B will be, they depend only upon the spatial and temporal separations assigned by observer A and the relative motion between A and B, not on the specific locations of the events assigned by the two observers. When we have added this assumption, we can describe the full structure of spacetime and the full relative behavior of rods and clocks at rest in one inertial frame to those at rest in another inertial frame.

With these results we can show that there is a notion of simultaneity establishable by clock transport and that it is in agreement with the simultaneity established by the light-ray method. Naturally, this new notion of simultaneity is only a notion of simultaneity relative to a particular inertial frame. How could it otherwise be in agreement with the light-ray definition, since that clearly establishes only a relativized concept of simultaneity?

The clock-transport method of establishing simultaneity relative to a reference frame works like this: (1) Pick a particular inertial frame. (2) At a place at rest in the frame set up a clock. (3) Synchronize another clock with the first and move it to a new location *infinitely slowly relative to the chosen inertial frame.* Of course we can't transport objects infinitely slowly, but we can transport them from *P* to *Q* slower and slower and see what happens "in the limit." This is what is meant by "infinitely slowly transporting a clock." (4) At the new location, synchronize a rest clock with the transported clock. (5) Call an event at the second location simultaneous with an event at the first if they have the same time-reading relative to their respective clocks at rest at their locations in the inertial frame chosen. One can *prove* in special relativity that the simultaneities established by this method will be in complete agreement with those established by the light-ray method. Why does this method only fix a relativized notion of simultaneity? The answer is obvious: It is because what constitutes "infinitely slow transport" of a clock is defined only relative to a particular inertial state of motion.

Let me end the critique of the relativistic notion by remarking on an apparently paradoxical result in relativity, that a spherical wave front of light can be a sphere about two distinct points at a distance from one another. The solution is now obvious. According to one observer, the simultaneous locations of the wave fronts of the rays of light emitted by him simultaneously are at equal distances from him. The other observer asserts the same thing. But the set of events which is "the simultaneous locations of the wave fronts of the various light rays" for the first observer is not the same set of events which constitutes a light shell for the other observer at a given time. For the events that are simultaneous for the first observer simply aren't simultaneous for the second observer at all!

## 2. How Conventional Is Einstein's Definition of Simultaneity?

The issue of conventional elements in Einstein's definition of simultaneity, and in the structure of the special theory of relativity in general, has its origin in the common source of all conventionality theses, i.e., in the possibility of the construction of alternative theoretical accounts, all apparently incompatible with one another, yet all having the same total set of observational consequences.

The problems one encounters in trying to unpack conventionality theses, elucidate possible solutions to crucial questions raised by conventionality theses, and come to some resolution of these questions have, of course, been expounded at length (II,H,4); we can survey the issues somewhat more briefly here. As the reader will remember, the basic assumption on which all of these debates rest is that one can, in fact, make a coherent and intelligible distinction between the observational consequences of a theory and those consequences of it not purely observational in nature. Without this basic assumption, the whole intelligibility of alternatives equally capable of saving the phenomena goes by the board.

Let us assume that in the case of special relativity and its "alternatives" such a distinction can be made. As usual, "observational" is usually identified with "material and local." Remember the assumption, necessary to get the whole Einsteinian program under way, that it is the coincidence and noncoincidence of events which is assumed to be the raw observable data upon which the theory is to be constructed. Local simultaneity and coincidence in space for simultaneous events, i.e., coincidence of events, is taken as a given notion by the construction. With it we can establish synchronization of two clocks "at a point," congruence of rigid rods at rest with respect to one another and with respect to the laboratory frame of the observer, and coincidence of a given event with the event consisting of the reflection of a light ray—these being the basic ingredients in the construction of the notions of simultaneity for events at a distance, of spatial and temporal separation of distinct events, etc.

As we shall see in (IV,E) below, there may be another "basic observable phenomenon" posited by the theory as well. We may wish to maintain that it is simply a basic observational fact whether a given set of events constitutes a continuous history in the life of a single object. This is because we assume, throughout the construction of the theory of special relativity, that we can tell of any two events, coincident or not, whether they constitute events that are part of a causal chain. We assume this, for example, when we use the light-ray definition for establishing simultaneity for events at a distance, for we assume that we know that the emission event, reflection event, and reception event are three successive events in the history of a chain of events which constitutes the history of a single light ray through an interval of time.

Further, we assume in constructing the theory that the priority of events at a place, i.e., which events at a place are before or after which other events, is simply an observationally determinable fact. I will reflect some on this matter in (IV,E) as well, and throughout Chapter V.

What we don't assume to be observational in nature are such things as the structure of spacetime itself or the relations even among concrete "happenings" which hold for distinct and noncoincident happenings at a distance from one another or (aside from temporal priority for events at a place) at a separation from one another in time. It is these relations, and the underlying spacetime responsible for their structural interrelationships, that we "construct" in the theory.

How can we characterize the alternative theories that will save the data? One thing we know is this: All of them will have observational consequences that will make them incompatible *at the observational level* with the total prerelativistic theory. For the only reason for exploring these novel theories in the first place was the appearance of observational facts incompatible with the prerelativistic total theory, for example, and most crucially, the null results of the round-trip light experiments.

It is worthwhile to explore some of these alternative theories in a little detail. A great deal of attention has been paid in the philosophical literature on this subject to the existence of alternative accounts of the data that Reichenbach would describe as "differing merely with regard to descriptive simplicity," but insufficient attention has been paid to characterizing the general framework of the philosophical dispute, as treated in (II,H,4) for example. Nor has enough been done to characterize the full range of descriptive alternatives that there really are. Instead, much attention has been focused on a few alternatives out of the total number and on a few of the features characterizing them.

A general remark on likeness and unlikeness among alternatives, and among the various alternatives and other theories incompatible observationally with all of them, is in order here. The situation looks something like this: One can construct the various alternatives by making "choices" at various points in the construction. As we have seen, in going from the data, or better, from the inductive generalizations built upon the data —like the general thesis of equal round-trip times for light around closed courses independent of their direction around the course and the inertial frame in which the course is located—we need to make certain assumptions to arrive at a full theory. Einstein, for example, makes the assumption that the apparent constancy of the velocity of light in all directions and for all inertial observers is to be taken as a *real* constancy. He also assumes that spatial and temporal locations attributed to events by two inertial observers are to be linearly related to one another, in order to guarantee the dependence of spatial and temporal separations for one observer on only the spatial and temporal separations for the other observer and their relative motion with respect to one another. In this section and the next I will have something to say about whether these assumptions should be construed as physical assumptions or coordinative definitions.

Whenever such an assumptive element is added to the inductive generalizations on which the theory is founded the possibility of alternative assumptions suggests itself; it is from the alternative choices at these points where one "leaps beyond the data to the full theory" that the possibilities of "alternative leaps" and hence "alternative theories" arise.

Let us look at three such assumptions and the alternatives that arise when one makes assumptions alternative to Einstein's. What one discovers is that the various choices, presumably all agreeing on the observational consequences, give rise to theories that *look* different. That is, some of the resulting alternatives contain propositions that seem to appear in some other alternatives as well, even in the prerelativistic theories and other propositions that are "denied" in the alternatives. Of course the basic philosophical question, essentially the question whose answer differentiates reductionists from antireductionists as I described them in (II,H,4), is whether we should take commonality of assertion to show that the theories really are saying the same thing, for at least this particular feature of the world, and difference of assertion to indicate that the theories are incompatible, or whether, as the reductionist would assert, commonality and difference among theories should be taken as wholly a matter of observational consequences. To the reductionist, assertions in two theories that look alike is no guarantee at all of commonality of asserted content, for the meanings of the terms in the two theories may very well differ. And difference in meaning can also make theories that look different actually "say the same thing," for the theories may have the same total set of observational consequences despite the apparent differences in their theoretical superstructure.

I will return to this issue shortly, but let us first look at the three assumptions made by Einstein and what results when one makes an alternative choice at each possible opportunity:

Assumption One: *The spatial and temporal measurements made by the use of measuring devices at rest in any inertial frame truly indicate the real spatial and temporal relations among events.*

If we, like Einstein, make this assumption, we end up with both the equivalence of all inertial observers in general (i.e., with the principle of Galilean relativity extended beyond mechanics to physical theory in general) and the necessity for relativizing the notions of spatial and temporal separation to a given observer. For although the spacetime world has the same structure according to all inertial observers, the particular spatial and temporal separations among events as determined by devices at rest in the differing inertial frames differ from frame to frame. If these are the "real" spatial and temporal separations among the events, then spatial and temporal separations must, in fact, *exist* merely relative to a particular inertial frame of reference.

If we make an alternative assumption, one possibility we can come up with is just one of the compensatory theories discussed in (IV,B,2); for it is characteristic of these theories that they deny Galilean relativity. Only in the inertial frame at rest in the aether do the laws of electromagnetism take their standard form; they deny the veracity of spatial and temporal separation measurements made with instruments in any inertial frame but the preferred frame. As we have seen, the compensatory and relativistic theories do have the same observational consequences. Of course the compensatory theories suffer from some alleged difficulties, all associated with their postulation of an aether frame that we can never determine, but that isn't what is at issue here.

ASSUMPTION TWO: *Spatial and temporal separations, as they exist relative to one inertial frame, are functions only of the spatial and temporal separations among the same events as they exist relative to a different inertial frame and of the relative motions between the frames.*

Einstein clearly assumes this and the assumption is very important, indeed. It is essentially this assumption that forces the spacetime we end up with to be homogeneous, i.e., to have the same intrinsic geometric structure at each of its points. It is this assumption that tells us that spatial and temporal separations relative to one observer are independent of the particular place-time of the events in question, but instead depend only upon their separations relartive to some other observer.

Now this is not true in general if the spacetime is not homogeneous, if, for example, it has the variable instrinsic geometry of the spacetime of general relativity. What we can see here is this: General relativity accounts for the observational facts in terms of a curved spacetime. We can construct an alternative flat spacetime theory that will have the same observational consequences as any particular curved spacetime theory, as shown in (II,H). We now see that we can also account for the observational facts usually accounted for by special relativity with its postulated flat spacetime, by means of a "conventionally alternative" curved spacetime theory.

ASSUMPTION THREE: *The velocity of light is the same in all directions for all inertial observers.*

This assumption is "built in" to special relativity by means of the definition of simultaneity for events at a distance by the fact that we take as the event at $P$ simultaneous with event $f$ at $Q$ that event *midway* in time between the time of an event at $P$, which is the emission of the light ray reflected at $f$, and the time of the event at $P$, which is the reception of the reflected light ray. If $t_1$ and $t_2$ are the times of the emission and reception events at $P$ respectively, then we take as the event at $P$ simultaneous with the reflection of light at $Q$ that event at $P$ which occurs at $t = (t_1 + t_2)/2 = t_1 + \frac{1}{2}(t_2 - t_1)$. This assumes that the light takes the same time going from $P$ to $Q$ as it does going from $Q$ to $P$, or, in general, that the velocity of light is the same in all directions relative to this arbitrarily selected inertial frame.

Now suppose, instead, we assumed that the event at $P$ simultaneous with the reflection of the light at $Q$ was some other event than the one at $P$ which Einstein "chooses." What alternatives do we have? If we wish to preserve the principle that no causal signals connect simultaneous events at a distance and that causal signals are never propagated from a later to an earlier time, then we must choose some event at $P$ after $t_1$ and before $t_2$. If we wish to preserve Assumption Two, the linearity assumption, it is easy to show that the time of the event at $P$ simultaneous with the reflection of the light at $Q$ must be taken to be $t = t_1 + \epsilon(t_2 - t_1)$ where $\epsilon$ is some constant independent of $t_1$ and $t_2$ and the point $P$.

Suppose we chose $\epsilon$ other than $\frac{1}{2}$. How would the theory differ in appearance from the normal theory of special relativity? This question has received extensive, and I believe, inordinate attention in the literature; it has been dwelt upon to the neglect of the more general and more important issues we have been considering here.

But it might be illuminating to spend a few moments considering the consequences of adopting an $\epsilon$ not equal to $\frac{1}{2}$. In such a theory, the velocity of light would not be the same in all directions. Another result of choosing $\epsilon$ not equal to $\frac{1}{2}$ is that the following principle is violated: For all inertial frames, if the velocity of one with respect to the other is $v$, then the velocity of the second with respect to the first is $-v$. And so on. In addition, the characteristic of special relativity that simultaneity for events at a distance would remain an equivalence relation, at least relative to specific inertial frames, would break down. But all this requires only a lot of algebra. What we see is that if we modify one feature of a

theory, but wish to maintain its compatibility with the observational predictions of the unmodified theory, we may have to change other assertions of the theory as well. Not a surprising result, surely. If we are to
reject such "$\epsilon \neq \frac{1}{2}$" theories, it can only be on such grounds as the fact
that they "unnaturally" single out certain directions in space for no good
reason, much as we might reject the compensatory theories for giving preferred status to some arbitrarily selected inertial frame. But we can't reject the theories because they are incompatible with the facts, if we
restrict the notion of fact to "observational fact" as we have been doing.

How should we view this situation in philosophical perspective? We
require no new insights here at all, for the situation is identical in every
detail with that rehearsed at such great length in (II,H,4). We may reject the notion of "observational consequence," as distinguished from
consequence in general, out of hand. To do this is to obviate the reduc-
tionist-antireductionist debate, to be sure. But it is also to make a mock-
ery of the whole of Einstein's reasoning that led to the formulation of
special relativity in the first place; for, as we have seen, it is just the
denial of observational status to, for example, simultaneity for events at
a distance which motivates Einstein's critique of simultaneity in the first
place.

If we accept the observational-nonobservational distinction, then we
must decide whether to opt for reductionism or antireductionism. If we
accept the former then we must view all the alternative theories, including the prerelativistic compensatory theories it is important to note, as
just "special relativity rewritten." Or, contrariwise, we must view special
relativity as nothing more than a different expression of, for example,
the Lorentz-Fitzgerald compensatory aether theory.

If we opt for antireductionism then we must choose among the alternative positions compatible with it: Must we forever remain ignorant of
the true nature of the world, even where we can meaningfully assert just
what the alternatives are, as the skeptic claims? Should we take the choice
of, say, special relativity over the compensatory aether theory to be
merely a matter of a conventional choice, as the conventionalist would
argue? Or should we rather, as the *apriorist* alleges, select, say, special
relativity on the grounds that the a priori principles of scientific methodology should make us reject as "scientifically unacceptable in principle" theories with as much *arbitrariness* built into them as compensatory
theories, with their arbitrarily selected aether frame, or "$\epsilon \neq \frac{1}{2}$" theories, with their arbitrarily selected directions in which, for example, light
has its maximum velocity of propagation?

Once again, the reader will have to decide for himself. I might note this, however: The *apriorist* position, seemingly philosophically dubious at first, certainly appears in a better light when one realizes that, in the case of special relativity and its alternative competitors, a rejection of *apriorism* leads one to saying either that Einstein's theory doesn't differ at all from the compensatory theories or that one must choose relativity over the compensatory alternatives as a matter of "free choice" if one is to make any decision between them at all. Doesn't everybody know that relativity is the *correct* theory, and that the older compensatory theories are totally unacceptable scientifically, despite their obvious equi-compatibility-with-the-data with special relativity?

### 3. *Axiomatizations of Special Relativity*

There have been numerous programs of axiomatizing special relativity. It might be worthwhile to say a few things about these, not because the summarizing of a theory into a few basic assertions from which all its consequences follow by logic alone is of particular interest philosophically, but because the examination of these axiomatic programs will throw a little light upon the question of the analytic or synthetic status of the various assertions of the theory.

The axiomatizations work something like this: A number of basic concepts are introduced as primitive. Normally these will consist of idealized "basic observables" like ideal freely-moving particles, light rays in a vacuum, ideal rigid rods, ideal clocks, etc. One makes various assumptions about the observable features of the observables by postulating basic axioms. One introduces new concepts as needed by explicit definition in terms of the basic concepts, introducing the existence and uniqueness axioms needed to "legitimatize" the definitions as necessary. For example, one might take a particle's being at rest in a reference frame as basic, being a free particle as basic, and define an inertial frame as any frame in which a free particle is at rest. Or, having asserted in one's axioms, or proved from them, the existence and uniqueness of an event specifying certain characteristics, $F$, one might define 'the $F$ event' as 'the event that is $F$.'

Now different axiomatizations of a theory are possible, in the sense that different such systems will have total sets of consequences of the axioms which are identical. This is familiar to us, for it is just like the possibility of mapping the curved spacetime of general relativity on the basis of alternative sets of "basic measuring devices," say rods and clocks, or light rays and clocks, or light rays and free particles. There is nothing in the theory to indicate the preferability of one axiomatization to another.

Each axiomatization, however, suggests its own partitioning of the classes of assertion of the theory into analytic and synthetic. There is nothing wrong with this, so long as we don't let it mislead us. For example, if we start with light rays and clocks, as in (IV,B,3), we will define distances in terms of relations among events characterized by the passage of light rays and the temporal relations at a point characterized by a clock at that point. It will then become "synthetic" that ideal rigid rods at rest in an inertial system measure spatial distances between events relative to that inertial frame. On the other hand, if we start with rigid rods and define spatial separations in terms of them, then the light-ray/ temporal features of events which we used to define their spatial separation in the original axiomatization will become synthetically associated with spatial separation only by means of an axiom. So assertions of a theory appear as "analytic," i.e., based on definition alone, or "synthetic," i.e., postulated by an axiom or deducible as a consequence of the axioms, depending on how we build the system in the first place. But the only real constraints upon the adequacy of the axiomatization are the demand that its set of total consequences exhaustively capture the total consequences of the original theory, and nothing more, and "aesthetic" demands on the elegance, simplicity, etc., of the axiomatization.

But theories don't abide forever. We change our science. And just what changes we will make, in the light of new data or in the light of new theoretical insights, is unpredictable in advance. We may have a theory in which we assert $S$. In a particular axiomatization of the theory we may take $S$ to be analytic. But we may, as scientists, at some point adopt a new theory in which we assert not-$S$. How can we deny an analytic proposition? Aren't they true by definition? The only reasonable thing to say seems to be this: Within the theories in which they appear, propositions do not come with "analytic" or "synthetic" markers attached. If a particular axiomatization of a theory takes a proposition as analytic, so be it. But this should not be construed as any constraint on the scientific community to cleave to the truth of the proposition come what may. If the definition that made the proposition analytic rests upon existence and uniqueness axioms that turn out to be false, we will certainly have to drop the proposition. Even if it does not become false, there is nothing to prevent us from adopting a new theory in which the proposition is denied, its analytic status in *some* axiomatization of the rejected theory notwithstanding. When we drop the old theory, we drop its axiomatizations as well. And we also drop the "analytic" and "synthetic" statuses given to propositions of the old theory which were, after all, only relative to a particular axiomatic reconstruction of the theory anyway.

# CAUSAL ORDER AND TEMPORAL ORDER
# IN GENERAL RELATIVITY

## 1. *Global Aspects of Curved Spacetime*

The special theory of relativity forced upon us deep revisions in our philosophical views about time. In particular, the inextricability of spatial and temporal relations, at least in any *invariant* way, is a consequence of the theory totally unexpected from the prerelativistic point of view. The realization that there could be such spacetimes appropriate to prerelativistic physics, such as neo-Newtonian spacetime, came only after relativity first forced upon us a fully spacetime theory. Along with this "intertwining" of time and space, the theory forces us to move from the older conception of space-through-time to the more profound notion of spacetime, and from places and instants as the basic spatiotemporal constituents of the world to event locations as the irreducible basic entities of spacetime.

The move from special relativity to general relativity also forces us to take stock of many of our most commonly held and deeply entrenched beliefs about time and to reexamine them with regard to their inevitability and, indeed, with regard to their truth.

The most profound effects of the adoption of the general theory of relativity, or even of the contemplation of it as a *possibly* true theory, are due to the fact that once we have countenanced the possibility of spacetime being curved, the possibility arises of it deviating from pseudo-

Euclidean spacetime in global topological features as well as in metric features. As we saw in (II,B,6), spaces can differ from one another not merely with respect to their intrinsic metrical features but with regard to topological aspects as well. Even spaces that are locally topologically similar can differ from one another radically in the large, with regard to such features as closure, orientability, completeness, etc. Further, the topological features are not totally independent of the metric features. To be sure, spaces can be metrically different and topologically the same, like Minkowski and Euclidean two-space. And they can be metrically alike and topologically different, like the surfaces of the Euclidean plane and the cylinder. But *some* metric constraints can "force" the topology. For example, any two-space whose curvature is positive and constant at every point and which satisfies some additional topological constraints, such as geodesic completeness, must be, topologically, a two-sphere.

In considering possible spacetime worlds of different topological form, we need not consider every possible topological space. We are confining ourselves to the spacetimes compatible with the general theory of relativity. All of these are four-dimensional differential manifolds whose metric (with regard, remember, to the *interval* between events) has "signature two" (i.e., the three spatial parameters are taken as positive and the time parameter negative, or vice versa). It is a pseudo-Riemannian manifold in the sense that it can be approximated at every point (with the possible exceptions of some singular points) by a pseudo-Euclidean four-dimensional spacetime, that is, by Minkowski spacetime.

In addition, the spacetimes with which we deal are constrained by the field equation of general relativity, connecting the local features of the metric structure of the spacetime with the distribution of mass-energy in the spacetime. But many different spacetimes, of extraordinarily varied topologies, are consistent with the field equation, just as in the classical theory of electromagnetism fields as varied as the electrostatic field about a charged pith ball, the magnetic field about a dipole permanent magnet, and the radiation field in a wave guide are all describable by the same set of differential equations: Maxwell's equations.

We have already seen (III,E) that the connection of mass-energy distributions with the structure of the spacetime field must be considered with care. Even a specific mass-energy distribution is compatible with many different spacetime worlds, since the spacetime structure is not, in general, uniquely determined by the mass-energy distribution. When we consider the variety of mass-energy distributions we can imagine in different possible worlds, it is no wonder that an enormous diversity of spacetimes, of widely varying topological forms, can all be worlds consistent with the general theory of relativity. And if this is true of *possible* worlds, can we be sure, in advance of experience, that the *actual* world does not have an overall topology wildly at variance with our Euclidean or even pseudo-Euclidean prejudices?

In this section I will focus on a discussion of the various ways spacetimes can differ from one another topologically. My primary concern will be with topological features affecting our view of time, but it should be obvious—given the most basic feature of spacetime, its interworking of space and time—that topological features of space, or more correctly of spacetime, will be inextricably entangled with topological features of time. Some of the possible worlds I will describe will be of deep philosophical interest. In particular, these questions will exercise us at length: (1) Can we exclude a spacetime with this topology as an impossible spacetime for the world on a priori grounds? (2) If the world has this topological spacetime feature, what other features must it have to be a consistently-described world? I will reserve most of these considerations to the next section, however, and will reserve to the section after that the question: To what extent can we claim the topology of spacetime to be a matter of convention?

The topological aspects of spacetime I will treat are its (1) partitionability into space and time relative to an observer, (2) orientability, (3) geodesic completeness, (4) singular structure in placetimes, (5) closure, (6) connectedness, and (7) its suitability vis-à-vis classical ideas of determinism.

One preliminary note: In (II,B,4) I introduced the notion of a general coordinate system in a curved space. One constraint on such coordinatizations of the points of a space was that in an $n$-dimensional space each point receives $n$ coordinates, one and only one from each of $n$ families. Now once we consider spaces of general topology, it may become impossible to find a single group of $n$ families of coordinate curves that coordinatize the whole space, despite the fact that the space is perfectly "smooth" in an intuitive sense. For example, the coordinatization of the two-sphere by parallels of latitude and meridians of longitude breaks down at the poles, since the poles have no single longitude. This is not a feature of the "real nature" of the poles, obviously, since any pair of antipodal points will do as poles. The solution is to cover the sphere not with one coordinatization, but with "patches." We have a collection of coordinatizations. Each covers a sufficiently small region of the space so that a one-to-one mapping of points of the space to $n$-tuples of coordinate numbers is possible. Where the coordinatizations "overlap" there is a smooth one-to-one mapping from one coordinate assignment to another (a diffeomorphism). All this can be done rigorously. It is of some importance when we come to talk about singularities in spacetime, for one wants to make sure that what appears to be a singular region of the spacetime is not just a nonsingular smooth region badly represented by a singular coordinate system.

1. *Partitionability:* As we have seen, Newtonian and neo-Newtonian spacetime both have the feature that once we have picked a particular event we can speak in an invariant way about "the space at the time of the event." It consists in the set of all events simultaneous with the given event, and simultaneity is an invariant equivalence relation. In Minkowski spacetime there is no invariant set of "the events simultaneous with a specified event." However, once we have chosen a particular coordinate frame, we can utilize simultaneity relative to that frame to define "the set of events simultaneous with the specified event relative to this inertial frame." So whereas Newtonian spacetime is just one and the same space, enduring through time, and neo-Newtonian spacetime a set of invariant "spaces" glued together in time differently than in Newtonian spacetime, Minkowski spacetime (1) can't be invariantly spoken of as "splittable" into spaces-at-instants at all, but (2) can, relative to a particular inertial observer's frame of reference, be split into *his* spaces (spacelike hypersurfaces) at *his* instants.

In a general-relativistic spacetime we may not be able to split spacetime into spaces-at-a-time, even relative to an observer in a state of motion. That is, we may not be able to partition the class of all events into a set of disjoint classes whose union exhausts the class of all events, such that each of the disjoint classes consists of a spacelike hypersurface and such that the hypersurfaces can be "ordered in time." One way this can occur is if there are "closed timelike lines" in the spacetime, a notion I will examine in (5) below. In this case an event may not have a unique place in a time order relative to *any* observer. Even if the spacetime doesn't have closed timelike lines in it, it may still be impossible to find a function that smoothly attaches to each event a real number, the "time" of the event, such that the number assigned to $x$ is less than that assigned to $y$ whenever there is a causal signal propagable from $x$ to $y$. In such universes, the notion of the time of an event is ill defined, even after we have picked a particular frame of motion for an observer. This will be the case, for example, if the world has what I will call 'almost closed timelike lines.'

2. *Orientability:* In (II,B,6) I briefly characterized the difference between an orientable and a nonorientable space. In the case of space-time, various notions of orientability are important. A spacetime can be spatially orientable or nonorientable, and it can be temporally orientable or nonorientable. The former notion is familiar and only requires some formal modification from the definition of orientability for a space to be applied to spacetime.

But what is a temporally nonorientable spacetime? The notion has no applicability at all for the prerelativistic notion of time but comes to the fore once we move from space-through-time to spacetime and once we allow general curved spacetimes as in general relativity.

First, consider Minkowski spacetime. Relative to each event there are two important classes of events—those in the forward light-cone of the specified event and those in its backward light-cone. The former class is the class of events toward which causal signals can be propagated from the specified event and the latter class is the class of events from which causal signals can be propagated which reach the specified event.

In a temporally nonorientable spacetime, the distinction between the forward and backward light-cone is not drawable on a global level. What happens in such a space is this: At a given event we draw a forward-pointing timelike vector. This characterizes a particular forward motion along a timelike line leading "into the future" from the specified event. But now it turns out, in this spacetime, that we can, by following a closed path through the spacetime and moving the "originating" event along this path, continuously transform the forward-pointing timelike vector attached to the origin event, keeping it always timelike, in such a way that when we arrive back at the origin event, having traversed the closed path, the vector attached to the point which we end up with is a timelike vector attached to the origin event pointing in the backward time direction!

What this means is that we cannot globally characterize the sets of points causally connectible to events as the forward or backward light-cones of those events. For by a continuous transformation, which always keeps timelike intervals timelike, we can transform a forward light-cone at a point into a backward light-cone (Fig. 42). The notion of "forward direction of time at an event" is not well defined in such a spacetime, any more than the notion of "right-hand glove" is well defined in a spatially nonorientable three-space in which we can, by a continuous transformation around a loop, make a right-hand glove come out congruent to a left-hand glove at the point of origin of the motion.

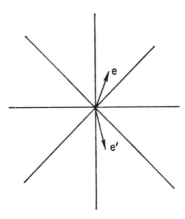

Fig. 42. A temporally nonorientable spacetime. In a temporally nonorientable spacetime it will be possible to convert the timelike vector **Oe,** which points in the future direction from *O*, to the time like vector **Oe′**, pointing in the past time direction from *O*, by a continuous motion that always keeps the vector timelike. In such a world, there is no global distinction between forward and backward lightcones.

3. *Geodesic completeness:* It is possible to describe spacetimes in which spacelike-geodesics or timelike-geodesics or both have boundary points, i.e., points at which they come to an end. If the geodesic is spacelike, space simply "comes to an end" along that straightest curve. If it is a timelike-geodesic, an object freely moving along that curve simply "comes to the end of its history." This may seem physically obnoxious, but there is certainly nothing inconsistent about it. One can generalize these notions to the spacetime being, as a whole, spatially bounded or temporally bounded. For example, a spacetime is future time bounded if from a spacelike surface in it every future directed timelike curve has a boundary point.

Notice that one must not confuse *finiteness* of a curve with its having endpoints. A spacetime can be split, sometimes, into a spacelike hypersurface at a time such that every spacelike-geodesic has less than a maximal length. For example, the space at the time might be the three-sphere whose maximally long geodesics are the "circumferences" of the three-sphere. But these don't have endpoints, although all are finite in length and have, in fact, the same length.

As an example of a timelike curve bounded in the future time direction consider the curve describing a particle falling freely into a "black hole." They are the result of a mass of a sufficient magnitude and density which collapses under its own self-gravitational attraction forever and ever. An observer falling into the hole, and carrying a clock with him, will record a proper time on the clock which approaches a finite limit. He will "get to the center" in a finite time. An outside observer will see him asymptotically approaching the center but never quite getting there, but the falling observer will record his total elapsed time to the center as finite.

4. *Singularities:* Can there be points of the spacetime which are quite irregular, like the cusps on an ordinary curve, for example? Yes, but it is quite hard to characterize them fully. For one thing, one can replace the spacetime with a singular point by a spacetime one gets by simply eliminating the existence of the point and still have a spacetime satisfying general relativity. The modified spacetime will have geodesics that are "incomplete" but will lack the original singularities. The only things I will note about singularities are that (1) some of the nicest spacetimes from a physical point of view can be shown to have singularities in their histories, and (2) one must be very careful to make sure that what appears to be a singularity in a spacetime, which one has described by giving the g-function for the spacetime relative to a particular coordinate system, is not merely an artifact of the coordinate system chosen.

5. *Closure:* If a spacetime is decomposable into spaces-at-a-time relative to some observer, these spaces can be closed and finite. For example, all the spacelike-geodesics in the hypersurface have less than a maximum finite length, and the volume of the hypersurface is finite. Of course this does not mean that the space has a boundary. For example, the universe that consists, relative to a particular observer, of the three-sphere enduring from a past to a future infinite time, the so-called cylindrical universe, has for this observer a finite, closed space at every time.

Spatial closure is one thing, but can a spacetime have closed *timelike* lines? Yes, at least in the sense that the spacetime can be given a description internally consistent and consistent with the field equations of general relativity. Suppose the spacetime is time orientable. Then for events sufficiently close together in time, if a causal signal can be propagated from the first to the second, no causal signal can be propagated from the second to the first along the same spacetime path. But it is still possible that we can pick events such that a causal signal can be propagated from the first to the second, a causal signal can be propagated from the second to the first, and so, a causal signal can be propagated from an event, through distinct events, and back to the original event! The kind of causal "anomaly" introduced by closed timelike lines is not like that introduced by time nonorientability of the spacetime. In this latter case, the very distinction at an event of "forward" and "backward" time direction breaks down. This is not so in the case of closed timelike lines. The direction of time at an event is still clear, it is just that some events can cause chains of events, always in the forward time direction, which eventually "cause" the initiating event itself.

It is easy to construct such a world consistent with general relativity. Take a finite stretch of spacetime which is, to a given observer, ordinary Euclidean three-space through a finite segment of time. Now identify the endpoints of the spacetime in time, i.e., the space at the beginning of the interval is identified with the space at the end. We "wrap the spacetime up in time" just as one "wraps" a strip of the plane infinite in the $y$-direction and finite in the $x$-direction up into a cylinder by identifying the $y$-axes that bound the strip (Fig. 43).

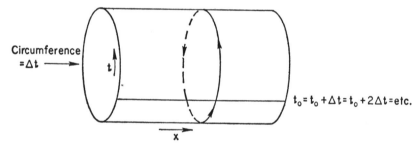

Fig. 43. A closed time world. The $x$ axis represents 3 dimensions of space. The time of the world is closed. Each timelike line from an event gets back to "the same time" by following it out always moving in the forward time direction. For the world to be consistently described, the world at $t_0$ must be identical to the world at $t_0 + \Delta t$, $t_0 + 2\Delta t$, etc.

Now spacetimes with closed timelike lines have interesting physical features. Since an event "causally affects itself" there are severe restrictions on what happens in such a spacetime. I will examine some features of such a world in the next section and, in (IV,D,3) below, ask whether we can always change our picture of the world "by convention" from one with closed timelike lines to one without any.

We have seen so many varieties of spacetime compatible with general relativity already, that one begins to wonder whether the theory imposes any restrictions on the topology of the world at all. It does. There is a theorem to the effect that a spacetime that is (1) simply-connected, i.e., has no holes (see [6] below), and (2) has the topology of the four-sphere, i.e., is closed up in time just the way the two-sphere is in space and closed up spatially like that as well, is incompatible with the condition that the space have a metric of the kind demanded by general relativity, i.e., a metric with Lorentzian signature.

A spacetime almost as disturbing as one with closed timelike lines is one with "almost closed" timelike lines (Fig. 44). What does this mean?

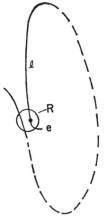

Fig. 44. Almost closed timelike lines. In a world without closed timelike lines but with almost closed timelike lines the following occurs: Pick event *e*. There will be no time-like curve passing through *e* such that if one follows it out in the future time direction one eventually passes through *e* again. Yet for any region about *e*, *R*, no matter how small, there will be a timelike line through *e* which when followed out in the future time direction eventually passes through *R* once again.

Well, there are events such that if we consider a region about them, no matter how small, there is always a timelike line from the event which sooner or later, when we follow it in the forward direction of time, gets back into the region. So even if the event doesn't "cause itself" it causes events arbitrarily close to itself in space and in time, *including events arbitrarily near to events that precede it in time* in the sense that causal signals can be propagated from them to the event in question.

6. *Connectedness:* With regard to connectedness, Minkowski space-time is just like Euclidean four-space. Here are some possible spacetime worlds that are peculiar in their connectedness:

*A.* The spacetime that consists of two sets of event locations $X$ and $Y$ such that no event in $X$ is connected to an event in $Y$ by either a space-like, null, or timelike curve. Such a world consists of two spacetimes totally out of contact with each other, like two two-spheres not touching one another in three-space (Fig. 45). But here we are not thinking of these spacetimes as embedded in any higher-dimensional spacetime.

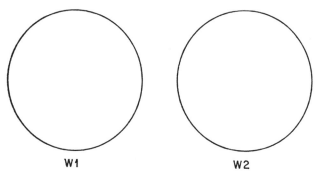

Fig. 45. Totally disconnected worlds. An event in $W_1$ is not connected to any event in $W_2$ by a timelike, lightlike, or spacelike curve.

*B.* Trousers worlds: In these there is a class of events such that every event in it is spacelike, null, or timelike connected to every event in the world. But there are also sets that are such that for every event in them, there is another event, in another set, which is not connected to the given event by any null or timelike curve, or by any spacelike curve except those that go through the first set whose events connect with all events. Such a world is the "trousers world" illustrated below (Fig. 46). It is simply-connected but quite different from the ordinary spacetime we would expect.

*C.* Worlds that are connected—every event in them is joined to every other event by a spacelike, null, or timelike curve—but not simply-con-

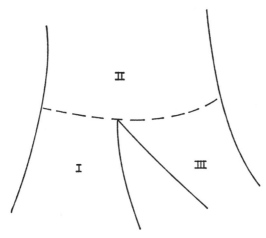

Fig. 46. A trousers world. This world is connected. But it is peculiar in that all curves—timelike, lightlike, or space-like, which connect events in region I with those in region III, pass through region II. In fact, all these curves are such that they cannot describe the passage of a causal signal, so that events in region I and region III are never causally connectible, although events to both I and III can have effects in region II. But events in region I are not connected to any events in region II by spacelike curves either. Hence, it is plausible to assert that events in region I are neither simultaneous with, nor before or after in time with respect to, events in region II.

nected: This world might be like the region contained in a jug and its hollow handle, which opens at two points into the main volume of the jug (Fig. 47). There are closed curves in this world which simply cannot be shrunk, continuously, to a point. They connect event $1$ to event $2$ through one part of the spacetime, and event $1$ to event $2$ through the other part as well. Wheeler has speculated on such "wormhole" worlds as providing geometrodynamic pictures of the elementary particles, the particles being identified, in pairs, with the "openings" connecting the wormholes to the main body of spacetime.

7. Some worlds are peculiar from a deterministic point of view. Suppose the spacetime has a global time function, i.e., there is a function that assigns to each event a real number such that $f(y)$ is greater than $f(x)$ whenever a causal signal can be propagated from $x$ to $y$. The world is said to have a Cauchy surface if and only if there is a spacelike hyper-

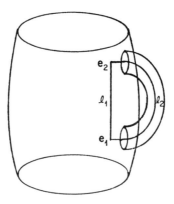

Fig. 47. A world not simply-connected. Two curves connect $e_1$ and $e_2$, $l_1$ and $l_2$. The complete loop consisting of both $l_1$ and $l_2$ cannot be continuously shrunk to a point.

surface such that every timelike curve without an endpoint goes through the hypersurface once and only once. In such worlds, we can specify the "initial state" of the world on this hypersurface, and, if the laws are deterministic, the past and future of the spacetime world will be determined once and for all by the situation on this hypersurface, i.e., at this "instant of time."

Spacetimes with closed timelike lines lack a Cauchy surface. So do some spacetimes without closed timelike lines. In such a world, even if the laws of nature are fully deterministic, one can specify the total state of the world "at a time" and yet have different possible futures compatible with the initial data and the laws of nature. This is because causal effects come into the future along timelike curves that never passed through the spacelike hypersurface on which we specified the initial conditions.

This should be enough to convince the reader that our ideas of time may be shaken up quite violently by considering the possible spacetimes suggested by a theory that, allowing for curved spacetime, allows for spacetimes of varied topology as well. In the next section I will talk about what *physical* sense one such "deviant" world might have, and in the section following I will return to the question of the conventionality of the topology of spacetime, briefly mentioned in (II,G,2), to see what must be done to "conventionally" exchange such a bizarre world for one more pleasing to some—at least as far as descriptive simplicity is concerned.

## 2. *Topological Structure and Causal Anomaly*

I have now described mathematically consistent spacetimes compatible with the field equations of general relativity, but spacetimes that appear, in some cases, quite bizarre. For the reader to be convinced that it is worthwhile to consider such spacetimes from the physical and philosophical point of view, he must be convinced that one cannot reject them out of hand as physically objectionable. In this section I will look at one of the deviations spacetime may have, topologically, from the topologically-familiar Minkowski spacetime. We shall see that a spacetime having this topology is describable, physically, in a perfectly coherent way. We shall also see, however, that if spacetime has such a structure, we must carefully review our philosophical attitudes toward many features of the scientific description of the world.

How would a world with closed timelike lines in it appear to one of its inhabitants? It is easiest to begin with the simplest of all worlds with closed timelike lines, the temporal segment of Minkowski spacetime wrapped up into a cylinder with respect to time. We construct the world as follows: Consider a Minkowski spacetime viewed by a particular inertial observer. At any moment, for him, the world looks like ordinary Euclidean three-space at an instant. Take two separated instants, relative to this observer. Assume that the total state of the world at these instants is qualitatively identical, i.e., if the first instant is $t_0$ and the second $t_1$, we assume that whatever is happening at a spatial point at $t_0$ is happening at that same spatial point at $t_1$. Now identify $t_0$ and $t_1$, that is, simply assert that the state of the world at $t_1$ is *the same state* as the state of the world at $t_0$. Notice that in doing this we are making something of an assumption, at least if we are viewing the construction as beginning with an already-presented world-history in Minkowski spacetime. The assumption is that the evolution of the world following $t_1$ is identical to that following $t_0$, since the states at $t_0$ and $t_1$ are qualitatively identical. In one sense, this isn't an assumption. For the properties of the world at $t_1$ include relational properties of the world at other instants with respect to it. That is, the state at $t_1$ really is identical (qualitatively) with the state at $t_0$ only if it is preceded by and succeeded by states qualitatively identical to those preceding and succeeding $t_0$.

But now we are viewing the spacetime not as a Minkowski spacetime whose temporal segments repeat ad infinitum with strip following qualitatively-identical strip but as a spacetime with the topology of the four-dimensional generalization of the cylinder. Corresponding to the straight lines that form one "dimension" of the two-dimensional cylinder there are the infinite Euclidean three-spaces of the spacetime at any instant, relative to the chosen inertial observer. Corresponding to the circles that characterize the other direction on the two-cylinder are the closed time-like lines, which proceed from a spatial point at a time, through the same point (relative to a particular observer) at later times, and then back to their initial event—and through the same class of events all over again.

An observer at rest in an inertial frame sees a history of events in which, starting from a particular event, there is a sequence of events at the place which consists of events, each later in time than the last until, finally, the original event occurs again and the sequence repeats. Now, of course, an observer need not follow a closed timelike line in this world. He can continually change his spatial location in such a way that his history consists of events that are each never coincident twice with one and the same spacetime location. Yet he will still see the features of the world which will indicate to him its "cylindrical" nature in time. He starts at $t_0$ in a particular inertial frame, seeing the world at $t_0$ as a spacelike hypersurface of simultaneous events, which hypersurface has the topological and metric structure of Euclidean three-space. Let him start examining the world at any time; then the following is true: There is always a later event in his history, such that were he to return to his initial inertial motion at that time, he would see the world at that time consisting of just the same class of simultaneous events as was the class of simultaneous events for him at $t_0$. He may see himself elsewhere in space at that time, but he sees himself at the same time.

Yet we must be careful here. For if he has "changed places at the time," being as it were in two disjoint places at the same time, there must be a double of him at his original location at the time; for, remember, whatever is going on at that place at that time at $t_1$ must be identical to that which occurred there at $t_0$ for the identification of $t_1$ and $t_2$ to go through.

We see that assuming closure in time forces us to adopt other constraints on the world we are imagining, as well. The spatial description of it, for example, must be such (static or periodic in space) that whatever happens, the state at $t_1$ is qualitatively the same as what happened at $t_0$.

There is another constraint on a world like this, one quite familiar from the work of science-fiction writers who treat their possible worlds with a concern for consistency. Consider the usual interpretation of a deterministic universe. The intuitive picture is of a world where the states of the world at various times are so related by laws of nature that we can say that, relative to the laws of nature, only a single state at a given time is compossible with the state at any other time. That is, once we have specified the world at a given time, the specification of all its states at future (or future and past) times is fixed by the laws of nature, given them to be future (future and past) deterministic.

In ordinary deterministic worlds, we imagine that in the specification of a possible world compatible with the given laws, we are fairly free in the initial specification of the world at a given time. We can, for example, in a world of point masses moving under the influence of their mutual gravitational attraction according to the Newtonian laws of motion, "pick" as the state of the world at $t_0$ any distribution of initial positions and velocities for the point masses we choose.

But if the world is to have closed timelike lines, such an arbitrary specification of initial conditions will be impossible. Consider the specification of the events in spacetime at a given instant of time relative to a particular inertial observer. Since the universe must return to exactly this state after a period $\Delta t$, where $\Delta t$ is the time period of the cylindrical world, and since the state of the world at $t_0 + \Delta t$ is lawlike determined by its state at $t_0$, not just any specification of the state of the world at $t_0$ will do. It must have a specification with the right "periodicity" built in, in order that it be, relative to the laws of nature, compossible with itself at the "later time."

As I have noted, this is a favorite concern of good science-fiction writers and they have often pointed out how the possibility of closed timelike lines imposes severe constraints upon the allowable states of the world. If one does go "back in time" one must act in such a way that when "now" comes around again, one does "now" just as one "did now" the "last time." We can indeed imagine worlds with closed timelike lines; we can even give a pretty fair description of what they would look like to an inhabitant of them. What we see is not that such universes are "physically incoherent" but that they are coherent only if the scientific description of the world we intuitively accept is somewhat modified. We can tolerate worlds with closed timelike lines if we will abandon the belief that the arbitrary specification of at least one state of the world at a time is always possible for whatever possible world we construct.

Now the world we have been looking at is quite simple. The topology of the cylinder is not that of the plane, but it is after all rather easy to describe. I will not pursue the detailed investigation of other possible models of worlds containing closed timelike lines, or lines whose topological behavior is otherwise anomalous with regard to temporal features; for example, worlds with "almost closed" timelike lines or worlds that, like the "cylinder" world described above, are such that every event in the world is causally connectible with any other event whatever, not locally but by means of timelike lines which "go around" the universe. We should be aware, however, that the topologies of these worlds can be bizarre indeed and much more complex than that of the "time cylinder."

One last remark is in order here. The time-cylindrical world is a world with closed timelike lines constructed in a very artificial way from the temporally unanomalous world of Minkowski spacetime. Are there any worlds that arise naturally out of attempts to solve the field equations of general relativity for various mass-energy distributions which turn out to have closed timelike lines as an unexpected feature of them? The answer is yes. The Gödel solution of the field equations, briefly discussed in (III,E,3), is such a world. The solutions describe a spacetime uniformly filled with matter. The solution is interesting since its intuitive interpretation is that of a world in which the matter of the world is, as a whole, in absolute rotation. The rotation is "relative" to the "compass of inertia" constructed at any point by following the paths of freely-moving particles and light rays. But the solution is also interesting in that it contains closed timelike lines. A particle in motion under the correct forces will, in this world, go through a history that, after a certain period of time, brings it back to the event from which it started. So closed timelike lines can be "forced upon us" by apparently innocent solutions of the field equations.

## 3. Global Temporal Order and Conventionality

Given a spacetime world with anomalous topological features, especially features like closed timelike lines which introduce deviations in the global temporal structure of the world from that in Minkowski spacetime, it is natural to ask if one can eliminate these bizarre aspects of the world by simply "redescribing" the world, that is, by finding a description that differs from the original account only with regard to "descriptive simplicity," which is yet not possessed of the apparent deviant topology of time.

In the case of the artificially-constructed time-cylindrical world, the "conventional alternative" is plain. As Reichenbach so clearly pointed out, and as I noted briefly in (II,G,2), the descriptions of the world which have the same observational consequences and yet differ in the topology they assign to the spacetime of the world will generally differ in that (1) one description will count as a single event that which is counted as a multiplicity of events in the other description and (2) one description may require "causal anomalies" of various kinds not requisite in the other description.

Suppose we live in a world that appears, initially, to be the cylinderized-in-time-Minkowski-temporal-segment described in the last section. What other world, not possessing closed timelike lines, will have the same observational features? The answer is obvious. We constructed the closed-in-time world by taking a temporally bounded strip of Minkowski spacetime and identifying its temporal boundaries; construct, instead, the spacetime that consists of an infinitely repeated set of such temporal strips, one after the other in time (Fig. 48). That is, replace

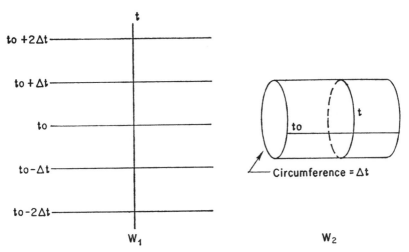

Fig. 48. Topological "conventional" alternatives. $W_1$, which is the Minkowski world of infinitely repeating strips of events that are identical and $\Delta t$ in time-width, appears just the same to an observer as the "time-cylindrical" world $W_2$, where the cylinder has a "time circumference" of $\Delta t$.

the picture of the world as closed in time with a picture of a world open in time but cyclic in its features, with a qualitatively identical history of events being repeated over and over again as one moves in either the future or the past temporal directions from an instant.

Such a picture differs from the original, as we would expect, in its rules for individuating events. What I previously called a single event is now looked upon as a sequence of events of order-type $\omega^* + \omega$, i.e., of the same order-structure as the negative-zero-positive integers. One event becomes a countably infinite number of qualitatively identical events. Of course, in such a world the version of Leibniz's "identity of indiscernibles" principle which says of two individuals in the actual world that they are really but one individual if they are identical in every qualitative feature is false. There are an infinite number of events qualitatively the same in every way, but numerically distinct from one another. But, as we saw in (III,B,2) where Leibniz's arguments were discussed, this is not a very serious objection to this world-picture, since that version of the principle of identity of indiscernibles appears pretty implausible anyway.

Does this world have causal anomalies? Well, it is an unusual world in that its history is cyclic without breakdown, but other than this it is not causally unusual at all. If causation appeared to be continuous in the original description, it remains so in the new description, for all the *local* relations among events, that is, laws that tell us which kind of events have in their vicinity events of which other kind, are preserved in the new description. Each event is "disindividuated" into an infinite sequence of events. But each neighborhood of an event is also disindividuated into an infinite sequence of neighborhoods. In the final description, each event of the world of a particular kind is surrounded by nearby events of just the kind that the event from which our new event came by disindividuation was surrounded by. Causality looks the same in the new picture as it did in the old.

But can we always move from a topologically unusual to a topologically unnoteworthy world at such a small price as the rejection of one form of Leibniz's law? To go further into this question, we will need a technical notion, the concept of a covering space for a given space.

Given a space $S_1$, we call another space, $S_2$ a *covering space* for $S$ if the following conditions hold: (1) There is a function that maps $S_2$ into $S_1$, that is, there is an $f$ such that each point, $x$, of $S_1$ can be viewed as $f(x')$, with $x'$ in $S_2$. Of course, many $(x')$'s in $S_2$ can be mapped by $f$ into the same $x$ in $S_1$. But for each $x'$ in $S_2$ there is a unique $x$ in $S_1$ to which $f$ maps it. (2) Every point $x'$ of $S_2$ has a neighborhood, in the topological sense (II,B,6) which is mapped *one-to-one* into a neighborhood of $f(x')$ in $S_1$. In other words, although many points in $S_2$ are, in general, mapped into the same point of $S_1$, for each point, $x'$, of $S_2$ there is a small enough region about it (i.e., open set containing it) such that there is a small enough region about $x = f(x')$ in $S_1$ so that ($a$) each point of the neighborhood of $x'$ is mapped to a unique point of the neighborhood of $x$ in $S_1$ and ($b$) each point of the neighborhood of $x$ is mapped (by the "local" inverse of $f$, $f^{-1}$) into a unique point of the neighborhood of $x'$ in $S_2$. (3) Each point $x$ of $S_1$ is the image under $f$ of some point $x'$ of $S_2$. That is, $S_1$ is "covered" by $S_2$.

The Minkowski space with repeated strips of qualitatively-identical sets of events is a covering space of the time-cylindrical spacetime. It is condition (2) of the definition of a covering space which is most important, here. It tells us that although one point in $S_1$ corresponds in general to many points of $S_2$, still for any point of $S_2$ there is a small enough region about it so that if we go from points in that region back to the points from which they came in $S_1$, we obtain points ($a$) that are near to the point from which $x'$ came in $S_1$ and ($b$) for each point near to $x$ in $S_1$ we obtain at most one point in $S_2$ near to each of the points in $S_2$ that $x$ is mapped into.

In general, the mapping from $S_1$ to $S_2$ takes a single point in $S_1$ into many points in $S_2$. But if $S_2$ is a covering space of $S_1$, then the mapping does not take a single point in $S_1$ near to $x$ into many points in $S_2$ all of which are near to any one of the images of $x$ in $S_2$. Each "near neighbor" of $x$ in $S_1$ is mapped into many points. Each of these is near *some* image of $x$ in $S_2$, but no two are near the same image point of $x$. What this amounts to is that although going from $S_1$ to $S_2$ will exchange single points for a multiplicity of points, the *local* features of $S_1$ and $S_2$ are topologically the same (Fig. 49). Since this, in the case of spacetime, is the same as saying that the local causal relations are similar in $S_1$ and $S_2$, it is natural to speak of a mapping that changes the topology of the spacetime as being one that introduces no genuine causal anomaly if the mapping is from a space to a covering space of it.

Is it the case that any spacetime with anomalous temporal structure has a covering space without this bizarre temporal aspect? Or in other words, can we always move to a description of a world that ($a$) has the same observational consequences as the original description and ($b$) introduces no "genuine causal anomaly," by simply disindividuating certain events in the original description into a multiplicity of events in the new description?

The answer is negative. It does turn out that every time-nonorientable spacetime has a time-orientable covering space, so that time nonorientability does seem to be an unpleasant feature of the world we can

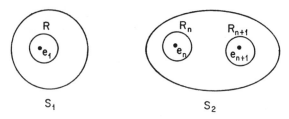

Fig. 49. Covering spaces. Suppose $S_2$ is the covering space
for $S_1$. Then a point in $S_1$, $e$, may have many images in $S_2$,
say $e_n$, $e_{n+1}$, etc. But for a region $R$ about $e$ in $S_1$ the following will be true: About each image of $e$ in $S_2$ there
will be a sufficiently small region in $S_2$, say $R_i$, about $e_i$,
such that the mapping will be one-to-one from $R$ to $R_i$. So
while $e$ may be "duplicated" many times in $S_2$, the local
features of the neighborhood about $e$ in $S_1$ are preserved
in the local regions about the images of $e$ in $S_2$.

dispense with by merely rejecting the version of Leibniz's law which
tells us always to reject a multiplicity of qualitatively identical events
whenever possible. But there are spacetimes with closed timelike lines
which are such that every covering space also has closed timelike lines.
This isn't true in the case of the time-cylindrical world; it has a covering
space with no timelike lines. But it is true of the Gödel universe. Any
covering space of this world still has closed timelike lines. If we can find
an alternative theory with the same observational consequences as the
Gödel theory but a picture with no closed timelike lines, we will have had
to purchase the absence of closed timelike lines by introducing into the
new picture many events where we had one before, and by violating the
principles of the continuity of causality which held in the original picture.

The reader can get some idea of what is going on here by considering
mappings of the cylinder and of the two-sphere into the two-plane. We
can map the two-cylinder into the two-plane with the two-plane as its
covering space by picking the one-to-many map that repeatedly maps the
cylinder into "strips" of the plane. But every map of the sphere into
the plane will "tear" the sphere at some point, thereby making the
neighborhood relation on the image different from that on the original
sphere. That is, there will be points on the sphere which are close, but
whose images will be such that at least one image of one point will have
no image of the nearby point anywhere nearby.

So the problem of "conventional alternatives saving the phenomena"
becomes mathematically much more interesting in the topological case

than in the case of merely making metric changes in the theory designed to save appearances. And yet, philosophically the situation remains much the same. We do have a few new considerations, such as the role played by the version of Leibniz's law which is violated in the move from space to covering space and the role played in choosing a description by the desirability of retaining a view of causality as a continuous phenomenon. In the topological case, unlike the metric, we must pay some disturbing prices for retaining "orthodox" topology. If we wish to hold on to the topology of Newtonian spacetime or Minkowski spacetime, at any cost and irrespective of the observational data, we must tolerate at least the existence of universes in which a qualitatively identical segment of the world repeats itself ad infinitum spatially and temporally, and, perhaps, a world in which causal influences can "jump" regions of spacetime in extraordinarily disturbing ways.

Aside from these considerations, however, all the old problems introduced by considering apparently incompatible descriptions of the world all of which have the same observational consequences remains just as it was in the case of metric alternatives, of the alternatives for absolute acceleration, and of the various alternatives to the normal special-relativistic account of the round-trip experiments.

## THE CAUSAL THEORY OF TIME

### 1. *Some Basic Features of Causal Theories of Time*

The causal theory of time is not a single philosophical theory, for many different theories have been put forward which are labeled as causal theories of time. For the sake of clarity and brevity I will not expound any one philosopher's developed theory and then criticize it but rather put forward a number of claims that, I believe, capture the fundamental features of some of the important positions that have been called causal theories of time. I believe that these positions are sufficiently like those presented by particular thinkers that my critiques will not constitute assaults upon straw men, but whether anyone takes the causal theory to be one of the theories I criticize is less important than the fact that these claims have interest in their own right.

Causal theories of time are reductionist theories that outrun in their content the kind of claims made by the other reductionist accounts I have so far examined. The claim that assertions about the structure of spacetime are reducible to assertions about actual and possible spatiotemporal relations among concrete unextended "happenings" is a typical reductionist position. Some propositions are alleged to be "reducible to" or "eliminable in terms of" propositions of an apparently different kind. For example, to affirm that a set of event locations constitutes a segment of a timelike- or lightlike-geodesic is taken to be "reducible" to the claim that the set of locations constitutes a set of locations of the events that constitute a fragment of the history of a possible freely-moving particle or light ray in a vacuum.

The causal theories of time are especially interesting in that they avow a reduction of assertions about temporal relations among events, either concrete happenings or mere locations of possible happenings, to assertions about relations among these events which are not prima facie spatiotemporal at all. So the claim of these theories is not simply that some spatiotemporal talk can be reduced to a certain fragment of spatio-temporal talk, like the claim that all spatiotemporal assertions can be reduced to assertions about the local spatiotemporal relations among material things, but the stronger claim that some spatiotemporal talk can be reduced to assertions about the existence and nature of relations that are not, on the surface, spatiotemporal at all. Crudely, it isn't merely that some spatiotemporal relations form a foundation upon which all spatio-temporal features can be constructed "by definition," but that all tem-poral relations, at least of a certain kind, can be founded upon a structure that is not spatiotemporal in its intrinsic nature at all. This is quite crude, but a more sophisticated version of this claim will be forthcoming.

Initially, it is very important to distinguish two different types of causal theories of time. The first, let me call it the scientific theory, alleges that the "identity" of temporal relations with causal relations is a scientific discovery, established by empirical investigation and scientific theorizing upon it. From this point of view, the reducibility of temporal to causal relations is much like the reducibility of salt to arrays of sodium and chlorine ions or the reducibility of light to electromagnetic radiation. To say, "This is a piece of salt," doesn't amount, in terms of *meaning,* to asserting, "This is an array of sodium and chlorine ions." But pieces of salt are as a matter of fact such ionic arrays. From this point of view, it is a scientific *discovery* that, as a matter of fact, for events to be temporally related *is* for them to have a specifiable causal relation among themselves.

The other approach is the philosophical theory. From this point of view, it is not merely true that temporal relations are causal relations—one could not, in any possible world, have discovered otherwise. For, according to this claim, an assertion that a temporal relation of a certain kind holds among some events can be *translated* into an assertion about a certain causal relation among these events. And the proof that this translation is correct, that the causal assertion exhausts the meaning of the temporal proposition, rests upon an analysis of the meaning of the original temporal assertion.

As we shall see, the two distinct claims, the scientific and the philosophical reductionist, rest upon different justifications for their support and can have quite different critical attacks launched against them. Naturally, the claims are not completely independent of one another. If the philosophical reductionist is correct, then surely the scientific reductionist is correct insofar as he avows an identity of temporal and causal relations in the actual world, although he is incorrect in believing that empirical support must be forthcoming for the assertion of this identity. If 'brother' really means 'male sibling,' then, surely, all brothers are male siblings; but if the identity of meaning of the terms really exists, it hardly takes a scientist to discover that brothers are male siblings.

Which temporal relations are reducible to causal relations according to causal theories of time? And to which causal relations are the particular temporal relations to be reduced? For the moment, let us work within the context of the Minkowski spacetime of special relativity, although we shall see in the next section that it is dangerous to confine philosophical speculation to the context of a particular scientific theory's being accepted as correct.

Most temporal relations among events are not invariant according to special relativity. For example, for two events, $e_1$ and $e_2$, which are at spacelike separation, the theory tells us that $e_1$ can be prior to $e_2$ in time relative to one inertial frame and yet simultaneous with $e_2$ or temporally subsequent to $e_2$ relative to other inertial frames. But there are some temporal relations that one can specify which are, according to the theory, invariant. If $e_2$ is in the "backward light-cone" of $e_1$ (including the backward light lines that are the backward light-cone's boundaries, but not including $e_1$ itself), then we can call $e_2$ *absolutely temporally prior* to $e_1$. If $e_2$ is in the forward light-cone of $e_1$, we can call it *absolutely temporally subsequent* to $e_1$. And if $e_2$ is distinct from $e_1$ and not in either of its light-cones, we can call $e_2$ *absolutely simultaneous* with $e_1$. Usually it is these notions that are said to be reducible to certain causal relationships by causal theories of time.

Some causal theories don't go this far. They do allow the reducibility of absolute simultaneity to a causal relationship, but they reject the reducibility of absolute pastness or futurity to causal terms. Instead, it is the relationship of one event being *absolutely temporally between* a pair of events which is alleged to receive a causal reduction. From the point of view of absolute temporal priority being given, $e_3$ is absolutely temporally between $e_1$ and $e_2$ if and only if either $e_1$ is absolutely prior to $e_3$ and $e_3$ to $e_2$, or $e_2$ is absolutely prior to $e_3$ and $e_3$ to $e_1$.

Let me start off with a fairly simple version of this theory, reserving the discussion of some complications of it to the two succeeding sections.

The events at a spacelike separation from a given event, $e_1$, are such that no causal signal whatever can be propagated from them to $e_1$ or from $e_1$ to them. Let us introduce the notion of a *genidentical* collection of events. For the idealized material world of point masses and light rays, a genidentical set of events constitutes a set of events each member of which is an event in the life-history of a single mass point or light ray as it moves through space in time. We can think of a pair of events as being causally connected if there is a genidentical set of events containing the pair of events in question. So a first try might be that $e_1$ and $e_2$ are absolutely simultaneous if they belong to no set of genidentical events. This won't do, for a pair of events might be such that a causal signal *could* propagate from one to the other but that no causal signal actually so propagates. This makes it clear that the "definition" of absolute simultaneity in causal terms will have to invoke the notion of possibility in the defining expression. We end up with something like: $e_1$ and $e_2$ are absolutely simultaneous if there is no possible set of genidentical events to which $e_1$ and $e_2$ both belong. As we shall see, and this should no longer surprise us, the invocation of possible events here will lead to important difficulties with the causal theories.

To give "causal definitions" of temporal priority, we need the additional notion of a genidentical set of events having a "direction," that is, constituting a set of events which is a fragment of the life-history of a single particle or light ray *from* one time *to* another. The definition would then look something like: $e_1$ is absolutely temporally prior to $e_2$ if and only if there is a possible set of genidentical events *from* $e_1$ to $e_2$. A similar definition would be given for $e_1$'s being temporally subsequent to $e_2$.

It is sometimes claimed that one can give a causal definition of absolute temporal betweenness even if one cannot give one for absolute temporal priority. Suppose, for example, that we are deprived of the notion of a set of genidentical events being *from* one event in it *to* another, i.e., of the temporal directedness of the set of events. We can still define temporal betweenness if we are given the notion of a set of genidentical events being continuous—intuitively, containing no "gaps," but for the moment a notion taken as primitive. The definition looks like this: $e_2$ is absolutely temporally between $e_1$ and $e_3$ if and only if (1) there is a possible continuous genidentical set of events containing $e_1$, $e_2$, and $e_3$, and (2) if we delete $e_2$ *and all events absolutely simultaneous with it* from our class of possible events, then there will be no continuous genidentical set of events containing $e_1$ and $e_3$ (Fig. 50).

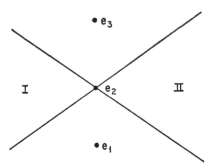

Fig. 50. A "causal definition" of temporal betweenness. In this causal theory of time, $e_2$ is said to be temporally between $e_1$ and $e_3$ if, when we delete $e_2$ and all the events absolutely simultaneous with it from the spacetime, i.e., all the events in regions I and II, then there is no causal path possible from $e_1$ to $e_3$.

Clause (1) guarantees that an event absolutely temporally between two others is not absolutely simultaneous with them. Clause (2) is designed to exclude events absolutely "outside the interval between $e_1$ and $e_3$." We can't do this just by insisting that there is no possible continuous genidentical set containing $e_1$ and $e_3$ when $e_2$ alone is deleted, for even if the deletion of $e_2$ breaks the continuity of one path connecting $e_1$ and $e_3$, there may be other possible paths of genidentical particles and rays which are continuous and connect $e_1$ and $e_3$, i.e., the paths that "go around" $e_2$. Dropping all the events absolutely simultaneous with $e_2$ breaks the continuity of all possible paths connecting $e_1$ and $e_3$ in just the case we want, the case where, intuitively, $e_2$ is absolutely *between* $e_1$ and $e_3$.

These definitions are quite simple. First we must ask how these proposals look as ingredients of a *scientific* reduction of temporal to causal order. We will discover that ending the restriction of attention to special relativity makes the reduction seem quite a bit less plausible. Further, if we accept special relativity as correct, there are still *philosophical* issues remaining to be resolved, even if it is *scientific* reduction we have in mind.

When we change our point of view and look upon these definitions as components of a philosophical reduction, far greater difficulties will be seen to afflict the causal theory. These are difficulties not with the particular form of definition chosen, but with the whole motivation and intent of the reductionist program.

## 2. The Causal Theory of Time as a Scientific Hypothesis

How does the causal theory of time look as a scientific hypothesis? I shall examine just how good the theory I outlined looks in the light of a number of contemporary scientific theories which we might have some reason to believe actually describe the world as it is. The soundness of the causal theory, when it is taken as a scientific hypothesis, will not surprisingly depend upon what theory of the world we take to be correct. After all, we do believe that light *is* electromagnetic radiation; but should we reject all current scientific ideas about light and electromagnetism, we might also reject the identification of light with a kind of electromagnetic radiation which, after all, depends for its plausibility on the acepted theories. Actually, as we shall see, even if we accept as correct the scientific theory in which the causal theory of time as I have presented it looks best, special relativity, there will still be important philosophical questions to ask about just how, and to what extent, this particular theory supports the causal theory of time alleged to "follow" from it.

First some preliminary remarks about the use of the notion of possibility in the definitions. From the point of view of the causal theory of time as a scientific hypothesis, possibility is always to be interpreted as meaning "possibility relative to the truth of the accepted theory." For example, it is not implausible to argue that the theory of the identity of light with a kind of elctromagnetic radiation tells us not simply that all actual light rays are electromagnetic rays, but that any possible light ray would also be a form of electromagnetic radiation. This follows from the fact that the connection (identification) of light and electromagnetic radiation follows from the lawlike sentences of our accepted theory. That there could be a possible world (according to some accounts of possibility) in which there were light rays but no electromagnetic radiation, that is, possible worlds in which the whole of the lawlike structure of the actual world was violated, is irrelevant to the truth of the "counterfactual identification" above, which is supported by the actual lawlike nature of the world. When I come to considering the causal theory of time as a philosophical hypothesis, we shall see that the invocation of possibility in the definitions requires the consideration of a far wider class of possible worlds, including possible worlds wildly at variance with the lawlike structure of the actual world.

For the causal theory of time to be a plausible scientific hypothesis we need to show that there is a close connection, following from the true scientific theory of the world, between the temporal relations among events, as intuitively understood—at least as intuitively understood relative to the acceptance of the theory and the other, *causal,* relation to which the temporal relation is "reduced" by the causal theory of time. Suppose the causal theory of time reduces temporal relation $T$ to causal relation $C$. Suppose, in addition, accepting some particular scientific theory, that the theory itself tells us that there are possible pairs of events which have the $T$-relation to one another, but not the $C$-relation, or vice versa. Then, relative to this theory, the causal reduction of $T$ to $C$ is incorrect. What this entails is the following: It is necessary for the truth of the reduction of $T$ to $C$ that whenever the $T$-relation holds among events the $C$-relation holds as well, and vice versa. But this alone is not sufficient. For it might just be a *contingent accident* that events are $T$-related when and only when they are $C$-related. But there *could* be events with one relation holding but not the other, even given the correctness of the theory. What we demand as a necessary condition of the correctness of the causal theory (again, not necessarily a sufficient condition, as we shall see) is that it follow from the laws of the correct theory that every $T$-related ordered set of events be $C$-related as well, and vice versa. That is, for the causal theory of time to even get off the ground as a scientific hypothesis, relative to the acceptance of a particular theory, it is necessary that the theory entail on the basis of its laws that the two "identified" relations be coextensive. Mere happenstance coextensiveness, i.e., $T$ as a matter of fact holding among events when and only when $C$ does, won't suffice. For in this case it is possible, even relative to the correctness of the accepted theory, that there be events that are $T$-related but not $C$-related, or vice versa, and this is surely enough to force us to reject the *identification* of the $T$-relation with the $C$-relation.

Now the definitions in the last section can be construed, given the "scientific" construal of the causal theory of time, as the assertion that certain temporal relations are identical with certain causal relations. So we require, for the plausibility of the causal theory of time so construed, that the accepted scientific theory at least guarantee a lawlike coextensiveness of the "identified" relations. How does this lawlike coextensiveness look from the point of view of various current scientific hypotheses?

Naturally, the lawlike coextensiveness goes through when the accepted theory is special relativity. Indeed, the spacetime of special relativity, with its invariant notions of absolute simultaneity, absolute temporal priority, and absolute temporal betweenness, was "designed" just so as to make the association of spatiotemporal and causal relations that characterized by my "definitions." The elements required on the right-hand side—genidentical sets of events, directedness of such chains of events in time, and continuity of such "causal chains"—are just the elements presupposed in the arguments used to establish the structure of Minkowski spacetime in the first place.

Does this mean that there is nothing more to say, if we accept the special theory of relativity, about the causal theory of time as a scientific hypothesis? No, for there are two important questions to ask in this situation: (1) Given the lawlike coextensiveness of certain temporal relations and the appropriate causal relations in special relativity, is it then obvious that we should *identify* the temporal with the causal relations? (2) Seeing the manner in which Minkowski spacetime was *constructed,* just so as to make the coextensiveness between certain temporal relations and the appropriate causal relations come out right, should we take the identity of the relations to be a scientific result of adopting the theory of special relativity or should we rather view the identities as *analytic,* generated out of the very meaning of the temporal notions, and view the theory of relativity as the scientific attempt to account for the empirical data in terms of temporal concepts whose meaning is antecedently fixed by the "causal definitions"?

Question (1) I shall discuss at the end of this section, after we have looked at my "definitions" in the light of some theories alternative to special relativity. Question (2) is essentially the proposal that the philosophical version of the causal theory of time is correct, and that the association of certain temporal relations with appropriate causal relations in special relativity is not a consequence of the theory but rather a consequence of the very meaning of the terms designating the temporal relations. I shall return to this in great detail in the next section.

How does the causal theory of time fare in the light of general relativity? It is clear that once one admits into the realm of possible worlds those worlds with novel topologies, in the temporal sense, which are still worlds physically compatible with the general theory of relativity, one will have to modify one's causal theory of time in important ways. For the connection between invariant temporal relationships and causal relationships in general relativity is quite a bit more complex than the relation that holds according to the special theory of relativity if one once admits the possibility, say, of closed timelike lines.

Consider, for example, the simplest model of a world with closed timelike lines—the temporal strip of Minkowski spacetime wrapped up into a cylinder in the temporal direction. It is clear that the association of temporal betweenness and continuous genidentical classes of events proposed in the last section will have to be severely modified. Consider three events, $e_1$, $e_2$, and $e_3$, such that $e_3$ is "between" $e_1$ and $e_2$ in the sense that a possible causal signal propagated from $e_1$ to $e_2$ will go through $e_3$ between them in the intuitive sense. Yet $e_3$ will not be between $e_1$ and $e_2$ according to the definition, for even if we delete $e_3$ and all those events absolutely simultaneous with it there will still be a possible continuous genidentical class of events connecting $e_1$ and $e_2$, that is, the other segment of the closed timelike line which runs from $e_2$ "around time" and back to $e_1$ (Fig. 51). What this amounts

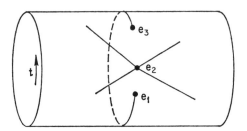

Fig. 51. Temporal betweenness and closed timelike lines. In the universe closed in time, $e_2$ can be intuitively between $e_1$ and $e_3$ in time, yet when we delete $e_2$ and all events absolutely simultaneous with it from the universe, $e_1$ and $e_3$ are still connectible by a timelike, or casual, curve, the one that "goes around the world in time."

to, essentially, is that the deletion of a single point splits a line open in both directions into two disjoint segments, but deletion of a point from a circle does not split the one-dimensional space into disconnected parts.

The response to possible worlds of this kind, compatible with general relativity, is usually to modify the definition connecting the temporal and causal relationships. One proposal has been to study the "connectivity" of the circle and form a new definition connecting absolute temporal betweenness and causal connectivity appropriate for this curve. But this is a wrong-headed response. For, as I noted in (IV,D,1), the time-cylindrical world is only one of the worlds with an anomalous temporal topology allowable by general relativity. These worlds can have timelike lines with a connectivity far more complex than either that of the line open to infinity in both directions or that of the circle. There are, for example, those possible worlds in which every single event is connected to every other event by a timelike line. A new definition of absolute temporal betweenness for each world would indicate that the connection between the temporal relation and the causal relation was not one guaranteed by the lawlike structure of the theory, for each different world with each of the different and incompatible relations of the temporal to the causal relation is compatible with the laws of the theory.

There is a class of possible solutions to the general-relativistic model which allows a modified version of the connection between causal and temporal relations of the sort we have been discussing. Even if a space-time has a global causal topological structure quite at variance with the causal structure of Minkowski spacetime, it will generally still be the case that it is everywhere locally like Minkowski spacetime in its causal features. That is, around any point there will be a small enough region such that within that region causal and temporal relations are as they are stated to be in the causal theory of time based on Minkowski spacetime.

For all the worlds in this class, we can propose a new causal theory of time which associates an appropriate *local* causal relation to each local temporal relation. So, for example, $e_3$ is locally absolutely temporally between $e_1$ and $e_2$ if there is a sufficiently small region containing the three events such that, neglecting all events outside of this region, there is no causal chain in the region joining $e_1$ to $e_2$ when $e_3$ and all events in the region absolutely simultaneous with it are deleted.

So for limited classes of worlds compatible with general relativity we can find appropriate modified causal theories of time. Of course if the world is not of this class, as would be, for example, a world that is general-relativistically possible but not even causally Minkowskian in its local structure at some points, then even the modified causal theory of time will not be correct. So far, however, none of this is disturbing. For if the causal theory of time is just a scientific hypothesis resting upon the assumed correctness of certain theories, then it is no surprise that adopting a novel theory of the actual constitution of the world will force upon one a modification of one's causal theory of time or even its entire rejection.

When we consider the changes in the accepted theory which come about because of the acceptance of the quantum theory, we encounter the adoption of a new theory that seems at variance with the causal theory of time in a very deep-rooted way. It is not like the case of the change from the special to the general theory of relativity, where the basic concepts of a genidentical class of events, direction in time through such a class, or continuity of such a class are retained and only the details of the "definition" connecting temporal to causal relations need be changed. For, it is plausible to claim, the very concept of a genidentical class of events, or of a causal chain connecting two events, disappears in the light of the new theory. And if the basic concepts of the causal definitions of the causal theory of time are now lacking, how could any causal theory of time be scientifically plausible?

To explain *why* it is claimed that adoption of a quantum-theoretic view of the world eliminates the possibility of conceiving of causal chains connecting events, and even the possibility of conceiving of "one and the same particle or light ray passing through space in time," would take us much too far afield. What is interesting for our purposes is this: Despite the rejection in quantum theory of the very notions used in the original justification of the construction of the spacetime of special relativity, it is still possible to formulate quantum theory in terms of the spacetime constructed in special relativity. In the development of the Minkowski spacetime of special relativity one utilized arguments that assumed the existence of causal chains, genidentical classes of events, etc., in the world. Yet now one adopts a novel theory in which the old spacetime is retained as the arena of concrete happenings in the world, but in which the whole scaffolding used in erecting the spacetime picture has been discarded. There is nothing unusual about this in science, and such a retention of the old theory despite the rejection of the original reasons for its proposal is hardly unique.

Something of the old connection between spacetime and causation is retained in the new theory. For example, in relativistic quantum field theory there is usually postulated a principle of "microcausality." Expressed by means of an algebraic relation among the operators that characterize the quantum fields (i.e., by the claim that the operators of a field at two spacetime points will always commute if the points are at spacelike separation), the theory intuitively asserts that happenings at spacelike separations have nothing to do with one another. But the fields of quantum field theory are not like the fields of classical or relativistic physics, and the vanishing of commutators for spacelike separations cannot be given a simple interpretation in terms of the absence of "genidentical causal chains." We retain the spacetime of special relativity, originally justified by the connection between temporal and causal relations, but with the causal relations vanished from the theory, and with it the simple-minded scientific causal theory of time.

Yet we are tempted to be cautious at this point. To be sure, the general theory of relativity forces upon us a more complex relation between invariant temporal relations and causal relations than that postulated by special relativity, and quantum theory forces us to change to a relation that can only be called a causal theory in a Pickwickian sense. Still, in both of these theories elements harking back to the rather neat association of invariant temporal and causal relations guaranteed by special relativity remain. This, in conjunction with the intuition that it is the causal theory that imposes constraints upon the construction of the spacetime of special relativity suitable to explaining the observational data rather than the given spacetime of special relativity having the causal theory as a mere consequence of its structure, suggests that we will not be misguided in pursuing the assertion of the causal theory of time as a *philosophical* thesis at least some distance.

That the philosophical version of the thesis is worth consideration is also suggested by an investigation of the answer to question (1) above: Given that we accept special relativity as the correct picture of spacetime, should we assert that the invariant temporal relations are to be considered *identified* with certain causal relations, or merely that the theory guarantees a lawlike coextensiveness of the associated but distinct relations?

The problem here is not unrelated to one I discussed in (III,E,4). There it was in the context of the geometrodynamical thesis that all of the matter of the world could be identified with regions of spacetime possessing characteristic intrinsic geometric structure. The scientific problem there was this: Can we find enough variety in the characterization of spacetime regions by means of their intrinsic geometry, so that we could establish an appropriate lawlike connection between a material object of a given kind being in a spacetime region and the spacetime region having its characteristic intrinsic geometric structure as determined by the field equations? The metaphysical problem was: Once we have a one-to-one lawlike association between material objects of a given kind being "in" a spacetime region and the region having its characteristic intrinsic geometric structure, how can we then go on to justify the *identification* of the material objects of the specified kind with the spacetime region itself?

As we saw, the possibility of alternative theories, one with the objects and spacetime regions merely lawlike-correlated and one with the objects identified with their appropriate spacetime locations, led to all the typical issues faced by the person confronting "alternative" theories equally compatible with all possible observational data. As I remarked, the situation occurs whenever a "reduction by identification" is proposed, say, for example, the reduction of physical optics to electromagnetic theory carried out by the identification of light rays with a form of electromagnetic radiation.

The issues become somewhat more complex where the question is the choice between the alternatives of maintaining that the absolute temporal relations are either identified with or merely lawlike-correlated with their appropriate causal "correlates." For one thing, reductions by identification usually take place by the identification of objects, or their spacetime surrogates "world-histories," with one another. Thus, pre-relativistically at least, light rays and electromagnetic rays are substantival entities, as are salt crystals and arrays of sodium and chlorine ions. But here it is a question of the identity or mere lawlike coextensiveness of two *relations*. Now we do have cases of reduction in science where, superficially at least, it is a *property* expressed by one predicate which is said to be identical with a property expressed by a different predicate, just as in substantival identifications an object designated by one referring expression is said to be the same object as an object designated by some other referring expression. For example, it is sometimes claimed that the macroscopic theory of matter reduces to the microscopic partly through such "property identifications" as the identification of temperature with mean kinetic energy of the molecules of a substance.

The pursuit of this question would take us much too far afield. Even a cursory reflection on what really goes on in science when thermodynamics and the macroscopic theory of gases are "reduced" to statistical mechanics quickly reveals that whatever it is, it can't be simply described as "identifying one property expressed by a predicate with a property expressed by a different predicate." Furthermore, the problems involved in questions about the existence and identity of properties and relations provide some of the most intractable, difficult problems of contemporary metaphysics. For the moment, the best I can say is that even given a lawlike coextensiveness between an absolute temporal relation and some appropriate causal relation among the same events, the question of *identifying* "scientifically" the temporal relation with the causal relation remains an open philosophical issue fraught with perplexity.

There is another feature of the connection of temporal with causal relations we should note, which gives it a somewhat different appearance from the usual questions of reduction by identification in science. Usually the "identification situation" looks like this: We have a theory that postulates the existence of an entity of a certain kind. The theory contains within it its own specification of what kinds of *empirical tests* are appropriate for establishing or disconfirming the existence of one of the entities postulated by it in any particular situation. Another theory postulates its own characteristic entities, again specifying test conditions for the presence or absence of the entities. The test conditions of the second theory are, in general, quite different from those specified by the first. For example, we test for the presence of a light ray by using our eyes, but for the presence of electromagnetic radiation we test by using field meters to measure the electric and magnetic fields present. The reduction of the one theory to the other comes about by the identification of the postulated entities, so that after the identification it becomes correct to test for the presence or absence of the entity in question by using either test condition, that of the first or that of the second theory. Once we know that light rays *are* electromagnetic rays of a particular kind, we can test for the presence of light rays (= electromagnetic rays of a specified frequency range or wavelength) by *looking* ("optical test" for light rays) or by performing the appropriate measurements by means of field meters ("electromagnetic test" for light rays).

But in the case of the absolute temporal relations and the associated causal relations are there such "independent" tests? Some philosophers have been tempted to say, "No," but for quite different reasons and with quite different consequences for these views about the "identity" of these relations. Many proponents of the causal theory of time would argue as follows: Even within the context of the spacetime theories postulating the existence of invariant temporal relations among events, the only *test* proposed for the existence of a particular absolute temporal relation in a particular case is the existence or nonexistence of a particular kind of causal relationship. What this shows is that the "identity" of the temporal with the causal relation is not an "identification established a posteriori by scientific discovery" but instead "a necessary a priori identification establishable by considering the meaning of assertions attributing a certain temporal relation between the events." In other words, on reflection we see that the thesis of the causal theory of time as a *scientific* hypothesis is too weak, for one can actually establish the reduction of temporal to causal relations by considerations of the *meaning* of assertions about temporal relations. In other words, the claim of those who take the causal theory of time as a *philosophical* thesis is correct.

On the other hand, many who *reject* the causal theory of time would argue like this: The proponents of the causal theory of time speak as though one could establish, independently, the existence of certain temporal relations among events and the existence of certain causal relations, and then go on to show the identity of the relations. As a matter of fact, upon reflection we see that (1) one establishes the existence of the temporal relations independently of establishing the existence of any causal relations, but that (2) to establish the existence of the causal relations requires an *antecedent* knowledge of the existence of the temporal relations. What this shows is that both the "scientific" and the "philosophical" versions of the causal theory of time are untenable.

We are beginning to see that the possibility of discriminating a scientific from a philosophical causal theory of time begins to look dubious. This is even more apparent if we once reflect upon the appearance of "possibility language" in definitions. When it is claimed that science establishes that light rays are electromagnetic rays, what is being asserted is that each *actual* light ray is identical with some *actual* eletromagnetic ray. But all our causal theory identifies with a particular actual absolute temporal relation is some *possible* causal relation among the events. The only way such a definition can be made plausible is to maintain that it is not founded on an empirically established identity, but that the temporal relation must be associated with the possibility of a certain causal relation, since the association is founded upon the very meaning of the expression attributing the temporal relation to the events. So it is incumbent upon us at this point to drop the scientific causal theory and move on to the philosophically-motivated "reductionist" account of absolute temporal relations. For if the proponent of a scientific version of the causal theory of time is satisfied with the claim that certain temporal and causal relations are coextensive or lawlike coextensive, then his position is intelligible if, as we have seen, usually dubious in fact. But if he is making the stronger claim that one could scientifically establish an *identity* of temporal and causal relations, then, as I have just noted, it is hard to see how he can keep his position from rapidly sliding into that of the philosophical reductionist's.

## 3. *The Causal Theory of Time as a Case of Philosophical Reductionism*

The "philosophical version of the causal theory of time differs from the scientific version in that it attributes to the fundamental propositions relating certain temporal relations to their associated causal relations the status of *definitions* or *analytic propositions*. The claim here is not merely that the temporal and causal relations are coextensive, or lawlike coextensive, or even empirically established to be identical, but that they are necessarily coextensive where the necessity rests upon an analysis of the *meaning* of the predicates expressing the temporal relations. For example, according to this view what it *means* to attribute absolute temporal priority of $e_1$ to $e_2$ is that there is a possible "directed" genidentical set of events "from" $e_1$ "to" $e_2$. It isn't simply the case that events related by absolute temporal priority have the causal relation specified, or that this follows from the laws of nature, or even that these relations are identical. Rather, one could know, a priori, that this association of the temporal to the causal relation holds, just as one could know, independently of all experience, that all brothers are male siblings, for 'brother' simple *means* 'male sibling.'

As we shall see, this "reduction by meaning analysis" rests upon *epistemic* claims, just as does, for example, the pure relationist's view of the reduction of all spatiotemporal language to language about the spatiotemporal relations among concrete material happenings, or even the phenomenalist's alleged reduction of material objects to "permanent possibilities of sensation."

I will examine the epistemological rationale for this reductionist account of temporal relations shortly, and I shall criticize it in some detail. First, however, it might be profitable to examine some *consequences* of the philosophical reductionist's position and see just how plausible they are. The most apparent consequence of a philosophical reductionist's position is the promotion of a number of assertions to *analytic* status. If what it means for a pair of events to have the $T$-relation is that they have the $C$-relation, then all pairs of events $T$-related are $C$-related, and this is an analytic truth. This is not a consequence of any reductionist position whatsoever, as we have seen in (II,H,4), except in a weakened sense. As we saw in (II,H,4), the basic claim of any reductionist account is that any two *theories* with the same totality of observational consequences "say the same thing." One could hold to this line without partitioning the particular sentences of the theories into analytic and synthetic classes. But if the reduction proceeds by a term-by-term definition, as does the philosophical version of the causal theory of time, then it does follow that particular sentences of a theory are *a fortiori* assigned analytic status.

Now some philosophers are dubious of the analytic/synthetic distinction altogether. We can offer some critical comments of the alleged analyticities of the proponents of the philosophical version of the causal theory of time, however, without any excursion into the general problem of analyticity.

If a proposition is analytically true, no evidence could stand to refute it. If we reject the proposition, in the light of scientific progress, it can only be because we have changed the meaning of some of the terms of the proposition. But all of the propositions that have analytic status according to the exponent of the philosophical causal theory of time seem to be such that, intuitively, we can indeed reject them in the light of new evidence; it seems, again intuitively, a misconstrual of this rejection to classify it always as "changing the meaning of a term." As I noted in (II,H,4), the difference between "changing the view of the facts" and "changing the meanings of terms" is not all that clear, so intuitive arguments are, perhaps, the best we can rely on here. We should note this, however: The cases where it is most plausible to maintain that a switch from theory 1 to theory 2 is founded merely upon "changing the meanings of some of the terms in theory 1" are those cases where theory 1 and theory 2 have "the same empirical consequences." But the cases we shall be looking at are those where the theoretical revolution brings about new predictions with regard to observations as well. Now one could, of course, try to hold that insofar as theory 2 rejected an analytic proposition of theory 1, that at least some of the terms of theory 2, ones which appear in the rejected analytic proposition of theory 1, have "changed their meaning," but it would be incumbent upon the exponent of the analyticity of the proposition to demonstrate this to us. And he can't do so by simply referring to the analyticity of the proposition in theory 1, for it is just the analytic status of this proposition which is in question.

We have already seen ways theoretical change seems to force us to reject as false propositions that a causal theory of time urges us to accept as analytic, at least in the philosophical versions of the causal theory of time. For example, in going from special to general relativity, the alleged analytic proposition connecting absolute temporal betweenness to its associated causal relation no longer remains true in all possible worlds satisfying the newer theory. In going from prequantum-theoretic physics to the quantum theory, all of the analytic propositions of the causal theory of time seem to fail, since the very causal language used to "define" the temporal relations becomes totally inappropriate.

I might add one more example here. In the definition of absolute temporal priority, it is assumed that causal propagation is always from events to other events absolutely temporally subsequent to them. In the philosophical versions of the causal theory of time, this is analytic and simply specifies what 'absolutely temporally subsequent' *means*. Yet there are physical theories, speculative to be sure but realistic candidates for acceptance nonetheless, in which causal influence can propagate from an event to another event absolutely temporally prior to it. The Wheeler-Feynman action-at-a-distance theory of electromagnetic interaction, a theory that dispenses with fields in favor of "time retarded" and "time advanced" causal interaction among masses at a distance from one another, is just such a theory. Now this theory does have startling consequences. Just as in the case of general-relativistic worlds with closed timelike lines, as we saw in (IV,D,2), the theory violates the traditional notion of the state of the world at at least one instant of time being arbitrarily specifiable. But as we also saw above, a novel theory of this kind is by no means blatantly inconsistent or otherwise scientifically reprehensible in principle. If such a theory is accepted, we will have to tolerate the possibility of event 1 being absolutely temporally subsequent to event 2, but with the possibility of a causal signal propagating *from* event 1 *to* event 2. In this case, an alleged analytic truth of the philosophical version of the causal theory of time becomes simply a falsehood.

We should pause for a reflection here. It seems as though the claim that at least certain propositions relating temporal order to causal order are analytic is dubious, to say the least. But if this is so, how can we account for the important role of the empirical fact that light is the maximally fast causal signal in legitimating the new relativistic definitions of temporal order, a role discussed in (IV,B,3)? Remember the use of this "fundamental fact" in the construction of Minkowski spacetime. I offered a new definition of simultaneity for events at a distance. The maximal nature of the velocity of light for the propagation of causal signals was used to demonstrate that the new definition would never lead us to affirm that a causal signal traveled between events simultaneous relative to any observer, or from an event to an event earlier in time with respect to it with respect to any observer. But if the connection between temporal priority and causal priority is not analytic, why would the fact that the special-relativistic definitions preserved this particular connection between temporal and causal relations have been so important in "justifying" or, better, "legitimating" the relativistic definitions?

The best answer I can offer to this question here is this: In any scientific revolution, in which many antecedently-held propositions are to be rejected as false, there seems to be a principle of scientific conservatism, a principle that tells us that as many as possible of the propositions previously accepted are to be retained in the new theory. The more "fundamental" the propositions of the earlier "total body of accepted science," the greater the priority that is given to holding fast to them through the theoretical revision of some of the body of accepted science. It is not the case that any of these propositions are such that we hold them to be totally inviolable to change. For each, perhaps, there is some observational data that would lead us to drop them as no longer acceptable. But we hold the maximum number that we can of accepted propositions constant in any particular revision, and the more fundamental the proposition to the previous physics, the greater the desire to hold it intact.

In the change from prerelativistic physics to a new theory that can account for the null results of the round-trip experiments, for example, some important propositions of the older physics must go. As we have seen, if the new physics is not to contain arbitrary elements or "unverifiable postulates," then something radical must be done to our views about time and, in particular, to views about simultaneity for events at a distance. But we need not drop any of the prerelativistic notions of genidentical sets of concrete happenings, of a temporal direction for such sets, or of continuity for them. Nor need we drop the prerelativistic assumption that causal effects always propagate forward in time. We can hold on to this even when by "time" we mean the coordinate time of the events relative to a particular observer, and we can certainly maintain this for events that are invariantly temporally related, i.e., in one another's light-cones. What we see is that the new relativistic theory remains compatible with the assumption of prerelativistic physics that there are genidentical causal signals, that they never travel with infinite velocity, and that they can always be taken as propagating from earlier to later events. But this hardly makes any of these propositions analytic or "unrevisable in principle without changing the meanings of the terms," as the possibilities of theories in which these propositions are rejected shows us.

All that we have seen so far is that the adoption of a causal theory of time may have consequences that are unpleasant to some philosophers, and that we can make coherent sense of such scientific developments as the relativistic definition of simultaneity for events at a distance without necessarily adopting a causal theory of time. But to make real progress we will have to examine the causal theory of time at its roots, that is, examine critically the claim that the invariant temporal relations can and must be "defined" in terms of certain causal relations. What we will discover, I believe, is that the proponents of a causal theory of time will find it very difficult, if not impossible, to motivate the "reduction by means of meaning relationships of temporal to causal order." Instead, as we shall see, if there is any reducing of a philosophical kind to be done at all, it is of causal notions to spatiotemporal notions. The philosophical version of the causal theory of time seems incorrect, but there may be a correct philosophical spatiotemporal theory of causation!

What is the usual motivation in a philosophical reductionist program for claiming that all $T$-assertions "translate without loss of meaning into $C$-assertions"? It is usually based on an epistemic claim, the claim that all epistemic access into the existence and nature of $T$-relations is through observational knowledge of $C$-relations and inductive generalization upon it. Thus "material object sentences" translate, for the phenomenalist, into "sense-datum sentences," because all knowledge of the existence of physical objects and their nature comes about from sensory awareness of sense-data and from inductive generalization to the existence of lawlike regularities among these "pure contents of sensation."

From this point of view, a philosophical reduction of *some* temporal assertions to assertions that are *in part* framed in causal language seems motivated. The temporal assertions in question are not those asserting invariant or absolute temporal relations among events but, instead, those attributing temporal relations to events at a distance from one another relative to a particular inertial observer. In special relativity, for example, the "criterion" for simultaneity of events at a distance, relative to an inertial frame, relies upon the propagation of causal signals back and forth between the places, relative to the frame in question, at which the events take place. So part of the definition of '$e_1$ is simultaneous with $e_2$, $e_1$ and $e_2$ spatially separated' involves reference to a genidentical causal signal propagated between the "locations" of $e_1$ and $e_2$. But we must be careful at this point. First, we should note that there are those who would reject definitional status, even in special relativity, for the proposition relating simultaneity for distant events to their causal interrelation, as we saw in the discussion of the conventionality of special relativity in (IV,C,2). More importantly, it is vital to notice that in the definition of simultaneity for events at a distance, there is reference not only to certain relations of causal connectibility among events but to the "purely spatiotemporal" relations of coincidence and noncoincidence of events and of an event being temporally halfway between two other events at the same point. In addition, as we shall see, one can very plausibly argue that the very notion of causal connectibility itself *presupposes* antecedently-understood spatiotemporal notions.

Finally, and most importantly, we should note that this "causal definition of a temporal relation" is a definition of one of the noninvariant or nonabsolute temporal relations, simultaneity for events at a distance. But the causal theory of time has always had as its goal the reduction of the *invariant* temporal relations to causal relations.

The whole schema of reductionist programs usually works by trying to show that what had been taken as an *inference* actually amounts to a *definition*. A representative realist, for example, believes that we *infer* from our observational knowledge of sensory contents to the existence and nature of physical objects, but he has a difficult time justifying his inferences, since the material objects remain ever immune to direct inspection. The phenomenalist obviates this difficulty by denying that any inference need be made, for material objects are nothing but logical constructs out of sensory contents. Similarly, for Einstein for example, we don't infer from causal interrelations among events at a distance to their temporal interrelation, we *define* the temporal order by the causal.

But is there any such inferential step to be "dissolved" by a "meaning reduction" in the knowledge of the *invariant* temporal relationships? Is it plausible to assert something like this: We think that there are such things as invariant temporal relations. But knowledge of the existence and nature of such relations always arises by unjustifiable inference from knowledge of causal relations. So the only meaningful role assertions about invariant temporal relations can have is the one they have as "shorthand" for the assertion of the existence and nature of certain causal relations. Therefore, assertions about invariant temporal relations are meaningful only if they can be translated without loss of meaning into assertions about causal relations.

The answer, I believe, is that it is not plausible. Consider the notions used in the "defining" part of the definitions proposed in a causal theory of time: genidentical set of events, causally continuous set of events, causal order of a set of events. Is it plausible to claim that we directly apprehended the genidentity, causal continuity, and causal order of a set of events and only inferred to the temporal or spatiotemporal structure of the set? Or is it, rather, the other way around: We directly apprehend certain spatiotemporal features of a set of events and on the basis of these features attribute, by inference or by definition, to the set of events certain "causal" features such as the set being a set of genidentical events, being a causally continuous set, and being a causally ordered set.

Curiously enough, some of the proponents of causal theories of temporal betweenness reject a causal theory of temporal priority, just because they realize the absurdity of the claim that we somehow or other directly apprehend the causal priority of events without first establishing the temporal priority of the events. Since finding out which of a causally connected pair of events is "cause" and which "effect" requires discovering the temporal order of the events, they argue, one cannot "define" temporal priority in terms of causal priority. It is just this observation that leads many of them to seek a source for the "direction of time" outside of mere causal interrelation, as I shall discuss at length in Chapter V.

But what these proponents of the causal theory of time fail to realize is that establishing the genidentity or causal continuity of a set of events requires antecedent knowledge of the temporal structure of the set of events, just as discovering its causal priority requires already knowing its invariant temporal priorities. What on earth gives one the motivation for saying that a set of concrete happenings is a segment of the life-history of a single mass point or light ray, except the knowledge that (1) the events in the set are all events in the life-history of *some* particle or light ray of a given kind and (2) the events are "seen" to be spatiotemporally continuous, giving us reason to believe that they are events in the history of *one and the same* particle or light ray. Similarly, why do we think a set of concrete happenings causally continuous, except that the set constitutes an *observably* spatiotemporally continuous set?

There seems to be no grounds whatsoever for looking at the situation as one in which (1) we have direct observational knowledge of genidentity and causal continuity, (2) we seem to infer from these to spatiotemporal continuity, and (3) this shows that properly we should really take the meaning of the spatiotemporal assertions to be reducible to certain complex assertions about genidentical and causally continuous sets of events. The criticism of Reichenbach and Grünbaum leveled against the attempt to define temporal priority in terms of causal priority—the criticism being that we can't "independently" establish causal priorities without first already knowing the temporal priorities—holds against attempts to define invariant simultaneity and invariant temporal betweenness as well. We can't know that a set of events is a set of genidentical events unless we already know that it is spatiotemporally continuous, and we can't know that it is causally continuous or discontinuous without knowing whether it is spatiotemporally continuous or discontinuous, either.

In giving a definition for simultaneity for events at a distance in special relativity we took as "givens" certain spatiotemporal notions, in particular the notions of coincidence and noncoincidence for events and the notion of temporal separation for events at the same place relative to an observer. It now appears that the invocation of "genidentical causal signals" in the definition indicates that we implicitly took some other spatiotemporal notions as givens as well: the notion of a set of events being spatiotemporally continuous and the notion of a continuous set of spatiotemporal locations having, when the set is a set of timelike-separated events, a temporal direction as well.

From this point of view, *some* spatiotemporal relations (for example, simultaneity for events at a distance) can be "defined," but only in terms of other spatiotemporal relations that are taken as the "directly observable" features of events. What we see is that special relativity is motivated by the intuition that only *local* spatiotemporal notions, like coincidence and noncoincidence of events and continuity of a set of events and "temporal directedness" of a "small" set of timelike-separated events, are observationally given. The inference that we wish to replace by a definition is that from *local* to *distant* spatiotemporal relations. But there is no epistemic motivation in special relativity, or from any other source, to back up the claim that *all* of the spatiotemporal notions should be defined in terms of notions not spatiotemporal at all, such as purely causal notions.

Earlier I hinted that it was not only true that the philosophical version of the causal theory of time was false, but that there might even be some reason to accept instead a philosophical version of the spatiotemporal theory of cause. What do I have in mind here? Well, consider for example Hume's analysis of causation. The assertion '$e_1$ caused $e_2$' is analyzed by Hume into a number of components. One component of his analysis is the constant conjunction component: $e_1$ and $e_2$ are causally related only if events of *kind* $e_1$ never occur without events of *kind* $e_2$ and vice versa. This leads to all kinds of interesting questions about scientific laws, their nature, and the justification for our belief in their truth, but we will ignore these. Another component is "psychological": We never affirm that $e_1$ is the cause of $e_2$ unless the mind immediately "leaps to the idea of $e_2$ with an attached feeling of belief" whenever we are presented with the idea of $e_1$. Again we shall ignore this.

The last component, however, is vital for us: $e_1$ is the cause of $e_2$ only if (1) $e_1$ and $e_2$ are contiguous in space and time and (2) $e_1$ is precedent to $e_2$ in time. Now we would want to use spacetime notions, not those of space and time. And since we know spacetime to be a continuum, we know that the notion of contiguity of distinct events will have to be replaced in our analysis by the notion of spatiotemporal continuity of a set of events. Nevertheless, we can see what is going on here in our terms: If Hume is correct, or if an analysis anything like his is correct, then at least a major component of the meaning of any assertion about the causal relationship holding among events will be a component describing the spatiotemporal relations holding among the events. That is, if we are to follow Hume's analysis of causation, we will end up with a theory of causation that is at least *partially* a "spatiotemporal theory of cause."

Here the motivation for the reductionist analysis is clear: We think of causation as a relation holding among events to which we must infer. But, and this is a major component of Hume's argument, we never directly observe any causal connection among events. All we ever observe are constant conjunctions of kinds of events, habits of mind to leap from one kind of idea to another, and *the spatiotemporal relations of contiguity and precedence among events*. We don't infer from these to a causal connection, for having these relations is what it *means* for events to be related by causal connection.

I am not arguing for the correctness of a Humean analysis of causation, but surely there is plausibility in claiming that such notions as genidentity of events and causal continuity among events and causal precedence among events, rather than providing an "observational basis" on which we can *define* spatiotemporal relations among events, are themselves notions plausibly definable in terms of spatiotemporal relations taken as givens, and, perhaps, some other "observational" features of sets of events as well.

The superiority of a spatiotemporal theory of causation to a causal theory of time can be demonstrated another way. Since *actual* temporal relations are associated by our "definitions" with only the *possibility* of the appropriate causal relations, it is plain that events can be temporally related without the associated causal relation actually holding between them. From the point of view of the causal theory of time, this is difficult to understand. It is already clear that epistemically establishing the existence of the spatiotemporal relations is antecedent to establishing the existence of even actual causal relations. Could we really believe that the epistemic route to the belief we have in the existence of certain spatiotemporal relations comes from a *prior* establishment of a merely *possible* causal relation? We have already seen some of the difficulties of a view that bases actual relations on mere possibilities in the discussion of the "pure relationist" account of empty space-time locations and empty spacetimes in (III,B,2).

But from the point of view of the spatiotemporal theory of causation the relationship is clear. Suppose events have a certain spatiotemporal relation. If some other features hold true of the situation, then they have the appropriate causal relation as well, say if the chain of events temporally connecting them can be viewed as occupied by events in the life-history of a genidentical particle or light ray. If these other features are lacking then the events, although remaining spatiotemporally related, fail to have the *actual* associated causal relation. This causal relation remains a mere possibility, but that it is a *possible* relation the events might (but don't) have to one another is now founded upon the *actual* spatiotemporal relation they do possess.

Let me conclude by assuring the reader that although I find the spatiotemporal theory of cause (say, Hume's) more philosophically acceptable than the causal theory of time, I am not proposing it as a *true* philosophical hypothesis. A little reflection will show the reader that on a Humean account both causal action-at-a-distance and causation "backward in time" are a priori prohibited, since causal interaction is defined in these theories as spatiotemporal continuous connection in the "forward" time direction. My own feelings are that such a theory, like many philosophical theories, places overstringent constraints upon the use of scientific concepts, and I for one would be the last to reject a scientific proposal out of hand as "philosophically unacceptable" because it postulated action-at-a-distance, or even causation "backward in time." All that I wished to argue here is that the causal theory of time is implausible, and that if any reduction thesis has any plausibility at all, it is that which alleges the definability of causal notions by temporal notions and not that which maintains the definability of the temporal by the causal.

# BIBLIOGRAPHY FOR CHAPTER IV

## SECTION A

For a comprehensive survey of philosophical views about time, see:

Van Fraassen, B., *An Introduction to the Philosophy of Time and Space,* chaps. 1–3.

A useful anthology of philosophical writings about time is:

Gale, R., *The Philosophy of Time.*

Other selections of general interest are contained in:

Smart, J., *Problems of Space and Time,* especially chaps. 1 and 4.

## SECTION B

### PARTS 1 AND 2

For a survey history of the optical experiments, their unexpected results, and the theoretical response to them prior to relativity, see:

Whittaker, E., *A History of the Theories of Aether and Electricity,* vol. 2.

A survey of the physics at an elementary level is contained in:

Rosser, W., *An Introduction to the Theory of Relativity,* chaps. 1–3.

Two original papers of Lorentz are reprinted in:

Einstein, A. et al., *The Principle of Relativity.*

PART 3

Einstein's development of the theory from the point of view of a "redefinition" of simultaneity is treated in any of the general works cited as references to (II,C,1) and in the work of Rosser cited immediately above.

Einstein's original paper:

"On the Electrodynamics of Moving Bodies," reprinted in Einstein, A. et al., *The Principle of Relativity*,

is as clear a treatment of the approach as any of its successors.

A philosophical treatment is:

Reichenbach, H., *The Philosophy of Space and Time*, chap. 3, part A.

Also useful is:

Van Fraassen, B., *Philosophy of Time and Space*, chap. 5.

PART 4

The references cited to (IV,B,3) above, as well as those to (II,C,1), will serve to introduce the reader to the consequences of adopting the spacetime point of view of special relativity.

PART 5

Two papers arguing that to accept special relativity is to automatically reject such notions as "non-fully-determinate existence of the future" are:

Putnam, H., "Time and Physical Geometry," *Journal of Philosophy* 64 (April 27, 1967).

and:

Rietdijk, C., "A Rigorous Proof of Determinism Derived From the Special Theory of Relativity," *Philosophy of Science* 33 (December 1966).

SECTION C

PART 1

On the Einsteinean redefinition of simultaneity, see the references cited to (IV,B,3).

For a discussion of the method of establishing simultaneity by "slow clock transport," see:

Ellis, B., and Bowman, P., "Conventionality in Distant Simultaneity," *Philosophy of Science* 34 (1967).

A collection of papers on this issue by A. Grünbaum, W. Salmon, B. van Fraassen, and A. Janis is in:

*Philosophy of Science* 36 (March 1969).

There has been much discussion in recent years of the possibility of a relativistic theory compatible with the existence of causal signals whose velocity is greater than that of light in a vacuum. For a survey of the literature on this so-called "tachyon" theory, and a philosophical discussion of it, see:

Earman, J., "Implications of Causal Propagation Outside the Null Cone," *Australasian Journal of Philosophy* 50 (1972).

PART 2

For discussions of the "conventionality" of Einstein's theory, which focus rather narrowly, however, on the choice of $\epsilon = \frac{1}{2}$ and neglect the other conventional alternatives rejected by Einstein in his choice of a definition for simultaneity, see:

Reichenbach, H., *Philosophy of Space and Time,* especially sec. 27,

Grünbaum, A., *Philosophical Problems of Space and Time,* chap. 12. The importance of the other Einsteinean assumptions for a critique of conventionality in special relativity is emphasized in:

Putnam, H., "An Examination of Grünbaum's Philosophy of Geometry," in Baumrin, B., ed., *Delaware Seminar in the Philosophy of Science,* vol. 2.

PART 3

For a careful investigation of an axiomatization of geometry, see:

Reichenbach, H., *Axiomatization of the Theory of Relativity.*
For an alternative approach, see:

Suppes, P., "Axioms for Relativistic Kinematics with or without Parity," reprinted in Suppes, P., *Studies in the Methodology and Foundations of Science.*

SECTION D

PART 1

The fundamental ideas of nonstandard topologies in general relativity, even those with interesting causal consequences, can be developed in an intuitive way—and this despite the fact that their formal, mathematical treatment is one of some sophistication. To this date there is, unfortunately, no "intuitive" exposition of possible nonstandard topologies in general relativity. The reader with some background in mathematics and physics may want to look at:

Kronheimer, E. H., and Penrose, R., "On the Structure of Causal Spaces," *Proceedings of the Cambridge Philosophical Society,* A 63 (1967);[**] and:

Carter, B., "Causal Structure in Space-Time," *General Relativity and Gravitation* 1 (1970).[**]
For a thorough discussion of the problem of singularities in general-relativistic spacetimes, see:

Geroch, R., "Singularities," in Carmelli, M., Fickler, S., and Witten, L., *Relativity.*[**]

PART 2

For a discussion of a realistic cosmological solution of the general-relativistic equations involving closed timelike lines, see:

Gödel, K., "An example of a New Type of Cosmological Solution of Einstein's Field Equations of Gravitation," *Reviews of Modern Physics* 21 (1949).

For a philosophical discussion of closed time see:

Grünbaum, A., *Philosophical Problems,* chap. 7.

PART 3

Reichenbach's extension of his discussion of conventionality to cover theories proposing novel topologies can be found in:

Reichenbach, H., *The Philosophy of Space and Time,* chap. 12.

## SECTION E

PART 1

For Reichenbach's causal theory of time, see:

Reichenbach, H., *The Philosophy of Space and Time,* Part 2.

For further developments, including Reichenbach's own criticisms of his original theory, see:

Reichenbach, H., *The Direction of Time,* Part 2.

Grünbaum continues the discussion in:

Grübaum, A., *Philosophical Problems,* chap. 7.

There are numerous variants of the causal theory of time. They rely upon adopting different notions as "primitive" and utilize different "definitions" connecting propositions about temporal order to propositions about causal order. For an extremely lucid and compact presentation of a variety of these theories, and a discussion of their relation to one another, see:

Van Fraassen, B., *Philosophy of Time and Space,* chap. 6.

PART 2

An extremely enlightening discussion of the causal theory of time, especially with reference to the impact on the plausibility of causal theories of time of cosmologies with nonstandard "causal topologies," see:

Earman, J., "Notes on the Causal Theory of Time," *Synthese,* 24 (July/August 1972).

PART 3

For a critique of the causal theory of time similar to that expressed in this section, see:

Lacey, H., "The Causal Theory of Time: A Critique of Grünbaum's Version," *Philosophy of Science* 35 (December 1968).

CHAPTER V.

*THE DIRECTION OF TIME*

# THE PHILOSOPHICAL BACKGROUND

## 1. *The Intuitive Asymmetry of Past and Future*

Our intuitive world-picture contains the assumption that the time order of the world is "directed." When events are temporally related to one another in any other way than being simultaneous, then one and only one of a pair is *later* than the other, and one and only one (the other one) is *earlier.* The notion that events are not only simultaneous or nonsimultaneous with one another, or between or not between a pair of other events, but have a *priority* ordering in time as well, is a pervasive element of the conceptual schemes of both ordinary and scientific thought, and it deserves a good deal of philosophical attention.

In the discussion of the so-called causal theory of time we have already seen one attempt to try to show that the temporal priority ordering of the world is associated with another kind of ordering, the causal priority ordering, and to justify the claim that this demonstrated "interconnection" of the two kinds of relational structure somehow or other made "philosophically clearer" just what the temporal priority ordering amounted to. We have also seen that many proponents of the causal theory of time as an adequate account of invariant simultaneity and temporal betweenness reject, in fact, the causal theory as a fully adequate account of "what temporal priority is."

In Chapter V I will probe a bit further into some of the issues surrounding the philosophical debate about the nature of the temporal priority ordering of the world and our knowledge of it. Now this issue has moved philosophers to contemplation in many different ways. There has been, for example, much speculation about the temporal priority structure of the world which has taken place in a context of debate in which considerations of the results of physical theory have been assumed irrelevant. The reader will, I hope, not take me as disparaging any such philosophical speculation about the direction of time when I tell him that, for the most part, I shall ignore it. There has also been a great deal of effort expended in attempting to throw philosophical light on the question of "the direction of time," in which careful attention has been paid to the importance of the results of science for resolving the philosophical issues. As usual, it is to these approaches that my attention will be directed, for, as usual, my main concern will be to see in what ways considerations of physical theory and philosophy are, and are not, relevant to each other.

In this section I will simply state, without much discussion or comment, a number of ways in which the "directionality" of time seems to pervade our conceptual scheme. Some of the issues introduced here will, of course, receive more attention later in this chapter. In the next section I will begin to outline the ways philosophers and physicists have thought the results of scientific investigation relevant to the philosophical issue. Four intermediate sections, B, C, D, and E, will explore in more detail some of the features of physical theories which have been claimed to bear on the issue of the direction of time. In Section F I will return to the more philosophical problems, for we will there bring the results about physical theory to bear on the traditional philosophical questions, to see to what extent the results of physics can really help answer philosophical questions and, even more importantly, to see whether the philosophical questions were correctly formulated in the first place.

First, then, let us look at some aspects of the directionality of time as they appear in ordinary conceptual schemes. The past and the future have been considered quite different by us with regard to aspects of knowledge, existence, possibility, causation, and concern, and I will note these aspects in turn.

1. *The knowledge of past and future:* We generally view the accessibility of past and future to knowledge quite differently. Frequently we

feel that, somehow or other, the past is, at least in principle, easier to know than the future. Of course it may be difficult to find out what happened in the past, but this sort of difficulty is frequently thought to be of a different kind from the difficulty we have in knowing the future. These intuitions are tied up with some others we will examine shortly. It is as though the past is "there," although finding out what is "there" may be a matter of no small difficulty. But finding out what the future *will be* is something else again, for here we are dealing with "that which is not yet in existence" and this has made it seem to some that the knowledge of our future requires a projection from the facts of a deeper sort than that required in "merely uncovering" the past. Of course, we can't take this to mean that it is always easier to know anything about the past than it is to know anything about the future. We do know that the sun will rise tomorrow and our knowledge that this will be the case is both easier to come by and more firmly grounded than is our knowledge, say, of the geological history of the world in the Precambrian era.

There is, it is alleged, at least one source of knowledge of the past which is never a source of knowledge of the future. For any given observer, this source of knowledge provides him with knowledge of *his* past, while it is completely impotent by itself to provide him with knowledge of his future. This is the observer's *memory*. If we wish to know about our past, we need only consult our memories, at least in some cases. In other cases our memories aided by external help (for example, the help of a psychoanalyst) will provide us with even more knowledge of our past than could be provided by our unaided memories. But to know what our future will be, we surely cannot simply consult our memories, aided or unaided; we require "prognostication" or inference.

2. *Existence:* In (IV,B,5) above, I noted some intuitions philosophers have had connecting time and existence. Some of these are independent of any notions of directionality, for example, the intuition that objects that don't exist at the present moment cannot be said, properly, to exist at all. But others rely upon an intuitive past/future distinction. There is, for example, the claim that events in the past can be said to have fully-determinate existence, since even if they are not occurring now, they have occurred already; whereas events in the future, being as yet "mere possibilities," cannot be said to have determinate or "fully-actual" existence at all. As we saw in (IV,B,5), the adoption of a relativistic account of time, with the accompanying relativization of time order, for events at spacelike separations to particular inertial systems

certainly requires that these intuitions be modified, making determinate existence at least as relative to an inertial frame as is futurity. But still, one can so modify this view as to make it consistent with a relativistic account of time, and it is one we shall have to keep in mind in reflecting upon the "direction of time."

3. *Possibility:*   The past, it is said, is a realm devoid of "real possibility." What is done is done, there is no use crying over spilled milk, the moving hand having written has written what cannot be erased, etc. But only a fatalist or a strict determinist believes that the future is devoid of real possibility, and the latter denies real possibility to the future only because it is lawlike-determined by facts that, themselves being in the past, are irrevocable. All concepts of freedom of action, of the possibility of altering the world to fit our desires, and, indeed, of there being genuinely alternative real possibilities for the total history of the world rest, it is said, upon an implicit asymmetry of past and future, the latter being the realm of "openness" or "real possibility" and hence markedly distinct from the former which is, at any moment, the realm of the "fixed," actual past.

4. *Causation:*   As I noted in (IV,E,1), at least some proponents of a causal theory of time believe that there is a striking connection between the ordering of events in time and their ordering with regard to causal relations. The connection is between the causal relatability of events and their absolute temporal priority ordering, basically the claim that causal influence always proceeds from an event to an event absolutely temporally subsequent to it in time.

Crudely, if we view the causal relation as a "making" or "determining to be the case" of one state of the world by another, the claim is that the past determines the present and future, and not the other way around. So the asymmetric temporal ordering of events into earlier and later is "matched" by an asymmetric ordering of states with regard to causal determination.

As I noted in (IV,E,2), the possibility of theories that postulate "causation backward in time" makes this intuitive correspondence of causal and temporal order more dubious than we might have thought, but still the intuitive feeling that the making of states by states, and the causally determining of states by states, always proceeds in the "forward" time direction remains strong. So strong, in fact, that theories postulating causation backward in time are frequently alleged a priori objectionable even by those quite reluctant to allow philosophical considerations to rule any scientific theory out of court in advance of experimentation and hypothesis-formation. These are questions I will return to in some detail throughout this chapter.

5. *Concern:* The concern about, or interest in, the future is quite different from that we feel about the past. We can lament, regret, or pleasantly recollect past events, but we don't view them as having the importance for us that events in our future do. After all, today is the first day in the rest of your life, we are told. Past pains we may remember with unpleasantness, but future pains we anticipate with anxiety. For any pain, no matter how intense, in our past, we can at least say, "Thank God that's over!" And, despite the advice of Schopenhauer and others, it does disturb us that there will come a time after which we shall not exist, although we are seemingly indifferent to the fact that the universe existed, perhaps for an infinite time, prior to our birth. We simply *feel* quite differently about what was and what will be, irrespective of the particular quality of the events in question.

Discussing each of these issues in a manner independent of considerations of the relevance of physics would take me much too far afield from the main concern in this book. But it is, after all, these basic intuitions that motivate the original interest in the problem of the direction of time.

## 2. Physics and the Direction of Time

The direction taken by arguments that purport to show that considerations of the structure of physical theories of the world, and of the world described by these theories, are relevant to questions about the direction of time, is one that should be quite familiar to the reader by now. The usual line of attack is to try to find some feature of the material happenings of the world which coordinates in one way or another with temporal priority. That is, to find for any two events that are such that one is earlier than the other some other feature which relates them in an asymmetric way.

The motivation for this search is not always made very clear, but there seem to be at least two distinct implicit motives found in the literature: *the epistemic motivation* and *the metaphysical motivation*. *The epistemic motivation:* Events can be related in temporal priority. But how can we ever know, as we do in fact know, of a given pair of events which is the earlier and which the later? The assumption made is that the ordering of the events in time is not something that can be known immediately or directly. Instead, there must be another relational feature holding between the events that we "observe" and which is the source of our attributions of temporal priority. From here on, the structure of the argument is plain: If the only epistemic ground for attributing temporal priority of one event over another is the observation of this newly discovered relation between the events, shouldn't we take it that what it *means* to say '$e_1$ is earlier than $e_2$' is that $e_1$ bears this newly discovered relation to $e_2$, i.e., shouldn't we be led by our discovered route of epistemic access to relations of temporal priority to adopt a reductionist account of the meaning of assertions attributing a temporal order to events? *The metaphysical motivation:* Events are related by the asymmetric relation of temporal priority. But what gives this asymmetry to the world? On what reality "in the world" is it founded? The assumption here is that it won't do to simply assert, "Well, there is an asymmetric relation, temporal priority, which some events have to others in the world," but that some other, not explicitly temporal, asymmetry must be found to which the temporal asymmetry can be "reduced." It isn't easy to see the plausibility behind this motivation, once one drops the familiar epistemic motivation, but since some authors at least do feel the necessity for "reducing" temporal asymmetry to some other asymmetrical relation, and since they explicitly (sometimes) deny an epistemic motivation to their search, something like this bald "metaphysical assumption of the necessity of a reduction" must be behind their proposals.

The reader who has come this far will certainly realize that I find both motivations dubious and so am doubtful of the whole point of searching for a "physical theory of the direction of time" on a par with, but different in detail from, the causal theory of time discussed in (IV,E). However, since pursuing this issue does provide an opportunity for getting clear about a number of important questions in the philosophy of space and time and another opportunity to rehearse some general arguments that have been important throughout this book, I will pursue this issue in some detail.

For the moment I will only attempt to try to unpack by example the motivations discussed above and then move on to outlining some features of physical theories and the worlds described by them which might be thought relevant to the question of the direction of time. I will reserve an extensive philosophical commentary on the motivation and overall direction of the plan of the proponent of the kind of theory I have in mind until (V,F), below.

To get some insight into the kind of arguments we are going to see presented, we might reflect for a few moments on some features of the causal theory of time discussed in (IV,E). The reader will remember that many proponents of causal theories of absolute simultaneity and absolute temporal betweenness are skeptical about the causal theory of temporal priority that I offered as one possible component of a causal theory of time.

Why? Well first, there is an epistemic-type doubt. Suppose we have two events that are connected by a continuous genidentical set of intervening events. Can we tell which event is first in time and which later? If we had some way of telling, without already knowing the temporal order of the events, which event was cause and which effect, our job would be done—according to the causal theory of time. But can we in general tell, without having antecedent knowledge of temporal priorities, which event is cause and which effect? No, for we can determine this only if certain additional features are present, additional constraints, as it were, on the nature of the causal interaction. Only if there are features of the events, determinable independently of any knowledge about their priorities in time, which we could use to distinguish cause events from effect events, would the causal theory of temporal priority be correct. So it is our job to search for these additional features of events.

Metaphysically, the argument is similar if, as usual, less clear and less persuasive. Causal connectibility, it is claimed, is a symmetrical relation and hence cannot serve as a foundation to which to reduce the asymmetrical relation of temporal priority. What is needed is some additional asymmetrical relation among events to serve as the "ground" for temporal priority. Apparently it is usually not sufficient for these people to simply postulate an asymmetrical *causal* relation '$e_1$ causes $e_2$' to serve this purpose, although why this would be objectionable to them, aside from the epistemic reasons expressed immediately above, is never made clear.

Just what features of the world come to mind as the sought-for foundation for temporal priority? At least two different approaches have been suggester, call them the *lawlike* and the *de facto* approaches:

1. *The lawlike approach:* Suppose it were the case that, as a matter of scientific law, there was an asymmetrical relation that held among events when and only when the events were, intuitively or according to the theory, related by the asymmetric temporal priority relation. So that it was a matter of a lawlike feature of the world that '$e_1$ is earlier in time

than $e_2$' is true when and only when '$e_1$ is $R$ to $e_2$' where $R$ is the other asymmetric relation referred to.

Perhaps, then, $R$ could serve as the asymmetrical relation to which the "earlier than" relation may be reduced.

2. *The de facto approach:* Suppose there were no such asymmetric $R$-relation coextensive with the "earlier than" relation as a matter of scientific law. Still it might be the case that in the *actual* world, although not perhaps in all possible worlds consistent with the *laws* governing the actual world, there is some relation, $R'$, which is as a matter of fact asymmetric and coextensive with the "earlier than" relation. In this case we would still have a relation to which the temporal priority relation could be "reduced," even though the "lawlike approach" had failed.

Now as I have said, there is a rather complicated philosophical argument which one must pursue in order to have any clear idea whether either of these approaches will succeed or even to discover whether there is any point in making the attempt to find such $R$- and $R'$-relations in the first place. But I will postpone this in favor of first pursuing the examining in some detail of the lawlike and the de facto ordering of events in time according to the "best available scientific theories." After this, to which Sections B, C, and D are devoted. I will devote one section, Section E, to discussing an issue almost never noted by the discussants of this problem, the problem of the temporal orientability of spacetime and the connection between the orientability or nonorientability of spacetime and the lawlike nature of the events occurring in it. Then, finally, I will return to the questions at the heart of the issues, the philosophical questions about the *point* of such "physical theories of the direction of time" in the first place.

## LAWLIKE SYMMETRIES IN PHYSICS

### 1. *Geometric Symmetries and Conservation Laws*

The symmetry of the laws governing the behavior of the mass-energy in the spacetime of the world plays an extraordinarily important role in contemporary physics. Originally, the symmetries were discovered as features of known physical laws. In more recent times, the symmetries have been used as devices for discovering new laws. When the form of a law is unknown, one can frequently "home in" on the correct law by assuming, among other things, that whatever the detailed structure of the law is, it obeys one of the previously discovered symmetry conditions found to hold in known laws.

Some of the important symmetries of physical laws are the invariance of laws under translation in space, invariance under translation in time, invariance under spatial rotation, invariance under change of inertial frame, invariance under spatial reflection, and so-called time-reversal invariance. All of these invariance principles are intimately related to the structure of the spacetime in which the material happenings of the world occur, as we shall see. There are other "internal symmetries," expressed in invariance principles, which are not reflections of the space-time structure; I will discuss these very briefly in (V,B,3) below. In this section I will discuss all of the geometric symmetries except time-reversal invariance, for its importance in the present chapter suggests I give it a section of its own.

Some of the importance of invariance, or symmetry, principles for physics can be understood when one realizes that there is an important connection between invariance principles and conservation laws. The connection between these symmetries and the conservation principles familiar to us from mechanics is one which is somewhat hard to demonstrate in the context of classical physics and, surprisingly perhaps, rather easier to show in the context of the quantum theory. I will not "prove" the existence of these connections, even informally, but simply provide references in the bibliography following this chapter which the interested reader can follow up if he wishes.

Prerelativistically, invariance under spatial translation leads to the law of the conservation of linear momentum; invariance under spatial rotation to the law of the conservation of angular momentum; and the invariance of the laws under time translation to the law of the conservation of energy. In special relativity, the principle of invariance under translation leads to a generalized law of conservation of momentum-energy, since in this theory momentum and energy are not separately conserved quantities, but rather components of a four-vector which is conserved as a whole. Invariance under spatial reflection leads to no conservation law at all in classical physics, but in the quantum context it leads to the law of conservation of parity. Interestingly, the invariance of laws under time reversal doesn't lead to a conservation law even in quantum theory. The reasons for this would take us too far afield, but the reader might reflect upon these differences as some indication that although the invariance principles all have a great deal in common, their physical interpretations may in fact vary from principle to principle. We shall see much more of this.

There is an intimate connection between the assertion of the existence of an invariance for physical laws and assertions about the structure of the spacetime arena of physical events. In prerelativistic physics, for example, the principle of invariance under spatial translation can be taken as showing us that the particular position of a physical event in space is irrelevant to the lawlike behavior of the system in question. Picturesquely, this asserts the "homogeneity" of space, i.e., the fact that one point is qualitatively the same as any other. Invariance of the laws under time translation tells us the same thing about the instants of time, that nothing differentiates them in such a way that a physical process will occur differently solely because of the particular time it happens to occur. Relativistically, we are informed by the invariance principle that gives rise to the momentum-energy conservation law that the Minkowski spacetime of the special theory is likewise homogeneous. Qualitatively, one event location in Minkowski spacetime is just like any other.

The principle of invariance under spatial rotation tells us, similarly, that space is isotropic, i.e., the same in all directions. What this means physically is that the particular angular orientation of a physical system is irrelevant, by itself, to the lawlike behavior of the system. Of course, we can "break" the symmetry by imposing an "outside" influence on a formerly isolated system. A system susceptible to magnetic influence will not have its behavior independent of its angular orientation if it is situated in a magnetic field pointing in a particular direction. But this "nonisotropy" is introduced by the interaction of the system in question with the external influence and not by the situation of the system in spacetime itself.

In general relativity it is quite hard, except under certain highly restrictive assumptions, to make any useful sense of the ordinary conservation laws. But this should come as no surprise. It is just in this theory, at least in its curved spacetime interpretation, that the homogeneity and istotropy of spacetime, shared by prerelativistic physics and special relativity, breaks down. About the only cases where conservation laws can be invoked in general relativity is where one assumes that the spacetime is Minkowskian in all directions when one goes out far enough. Restoring the homogeneity and isotropy of the spacetime by imposing these boundary conditions reintroduces the possibility of invariance principles and their associated conservation laws.

The principles of invariance under spatial reflection and under time reversal can be described as principles of invariance under noncontinuous transformations. The others are principles of invariance under continuous transformations. This difference, interestingly, leads to important differences in the physical interpretation of the invariance principles.

Let me start with the principles of invariance under continuous transformations. Suppose we are in a laboratory investigating the behavior in time of an isolated physical system. Let us construct a system at a different point in space in the laboratory whose initial state is the "same" as the initial state of the first system at the same time. We must be careful here, for the initial state of system 2 cannot really be identical to that of system 1, since the systems are at different places. What we want is system 2 started at the zero of time in "the appropriate state" corresponding to the initial state of system 1. As we shall see shortly, the classification of states of "changed" systems into those that are "equivalent relative to their particular systems" is a nontrivial matter. But let us pass over this here.

Suppose our systems obey deterministic laws. Then system 1 will evolve over time, deterministically, through some other states from its initial state. The "active" interpretation of the principle of invariance under spatial translation tells us that system 2 will evolve from its "same initial state" through the same later states as did system 1. Now in quantum theory there is no deterministic evolution of isolated states from initial to later states. Instead, given the initial state, what the theory allows us to compute are certain probabilities of transition to various possible later states. The active interpretation of the principle of invariance under spatial translation tells us, in the quantum-theoretic context, that the transition probabilities for system 2 will be identical to those for system 1.

The "passive" interpretation of the principle of invariance under spatial translation tells us that if one and the same system is described by observers whose positions are different from one another, the two observers will find that the same laws govern the deterministic evolution (or probabilistic evolution) of the system from both points of view. They now characterize the initial and later states differently, for each describes these states from his own "point of view." But if the laws of nature are truly invariant under spatial translation, both observers describe the same *lawlike connection* among the various states.

The situation is similar, for example, in the principle of invariance under spatial rotation. The active interpretation tells us that system 2 which is obtained by rotating system 1 will behave just like system 1, and the passive interpretation tells us that the evolution of a system will be described identically by observers whose reference frames are rotated with respect to one another.

In the case of the principles of invariance under noncontinuous transformations, only the active interpretation makes sense (Fig. 52). Let us leave time-reversal invariance to the next section and focus on spatial-reflection invariance here. Suppose I have a system, system 1. There is no way I can change my orientation as observer such that the system now looks like its mirror image. A right-hand glove cannot be made to look like a left-hand glove if I somehow or other change my position as observer. So for the principle of invariance under spatial reflection, the passive interpretation simply does not make sense. But the active does. Given system 1, I can construct a new system, system 2, which is its

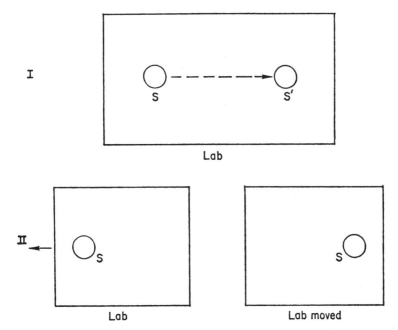

Fig. 52. Active versus passive interpretations of the symmetry principles. In I the active interpretation of translation invariance is illustrated. $S'$ is just like $S$ but translated
to different place in the laboratory. Translation invariance
tells us that $S'$ will evolve from its initial state just as $S$
evolves.

In II the passive interpretation is used. The invariance
principle tells us that a description of the evolution of $S$
by a man in the original laboratory frame will do for the
man in the laboratory that has been translated as well, $S$
left alone.

For the noncontinuous invariance principles, only the
active interpretation makes sense.

mirror image. If I am given a right-hand glove, I can make a glove that
is its left-hand correlate.

Actually, there is some question as to whether the active and passive
interpretations are equivalent, even in the case of the continuous transformations. Since we shall be concerned here only with discontinuous
transformations and their appropriate active interpretations, we need not
pursue this question here. For our purposes it will be satisfactory to
think always of the active interpretation of any symmetry principle, continuous or discontinuous.

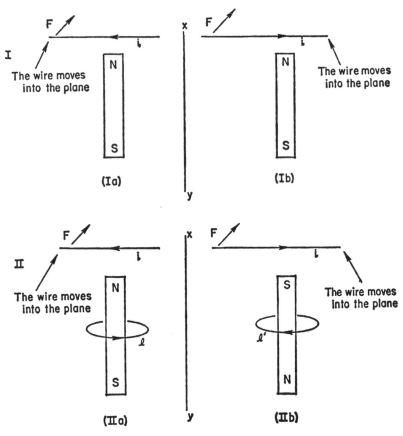

Fig. 53. Electromagnetism and invariance under reflection. In the above illustrations, *xy* represents a mirror on edge to the paper.

Diagram I shows how electromagnetism *seems* to violate reflection invariance. Diagram II shows why it really does not do so.

I*a* is a possible physical situation. With the magnet oriented as shown, and the current flowing through the wire as shown, the wire will suffer a force pushing it into the page. I*b* is the *apparent* mirror image of I*a*. But this is an impossible physical situation, for with the magnet oriented as shown, and the current directed as shown, the wire will actually suffer a force tending to push it out of the page.

II shows what went wrong. A permanent magnet is actually generated by current loops of moving electrons in the magnet. For example, in II*a*, the magnet will be oriented as shown if the current loop has the direction

Let the initial state of system 1 be $S$. Then I start system 2 in state $S'$, where $S'$ is the appropriate "mirror state" of $S$. The principle of invariance under spatial reflection tells us that if at some later time system 1 has evolved to state $S_t$, then system 2 at that time will have evolved to state $S_t'$, the mirror state of $S_t$.

I noted earlier that just what state of the transformed system "corresponds" to a given state of the original system may not be a trivial question, and I can show this here. As the accompanying illustration shows (Fig. 53I), it appears at first glance that the laws of classical electromagnetism are not invariant under spatial reflection. But as the second illustration shows (Fig. 53II), this is illusory. The mirror image of a north-south magnet is not a north-south magnet, but rather a south-north magnet. The invariance principles only make sense, and can only be confirmed or disconfirmed, if one has specified what the transformation of a system is to be and what the proper transformed state of a state of a system is to be—and this may depend upon deep, and even unknown, physical truths.

The other illustrations shows this (Fig. 54). The first shows the reason why physicists reject the principle of spatial-reflection invariance for some physical processes describable in the quantum theory of the basic interactions of elementary particles. The original system evolves in such a way that its apparent mirror image cannot exist in the world. But as the second illustration shows, if we changed our views about what constituted the "proper" mirror image of the first system, and if certain additional laws were in fact correct, then it would turn out, as in the electromagnetic case, that the apparent counterexample to invariance under spatial reflection, found in the study of the weak interactions, was not really a counterexample at all, but only a deception—as was the apparent counterexample to the spatial-reflection invariance of the laws of electromagnetism.

In addition, there are some states, like the thermodynamic states of a system—its temperature and entropy for example—which seem such that the whole notion of "appropriate transformed state of the transformed system" is inapplicable to them.

---

shown by $l$. In the mirror image, now, $l$ is reversed to $l'$, so the orientation of the magnet is reversed. Now with the magnet orientation and current direction as shown in II$b$, the situation is physically possible, for in this case the wire will suffer a force tending to push it into the page.

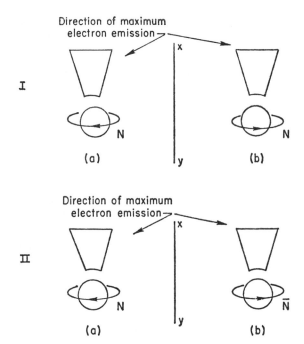

Fig. 54. Weak interactions and reflection noninvariance. In both diagrams $xy$ is the edge of a mirror that is perpendicular to the plane of the paper.

I$a$ represents a spinning nucleus which decays by a weak interaction. The electrons emitted in the decay are emitted preferentially in the direction from which the nucleus appears to be spinning clockwise. I$b$ is the mirror image in $xy$ of this situation. In it, the electrons are being emitted in the direction from which the nucleus seems to be spinning counterclockwise. But this situation is physically impossible if the world really is as in I$a$. So the physical process is not invariant under reflection.

In II we see a way to try to preserve reflection invariance. Suppose we took it that the proper mirror image of a nucleus was an antinucleus (we have no reason to suppose this, in fact, but we are just speculating as to possibilities here). Suppose, in addition, that when we perform the weak decay experiment for an antinucleus, the electrons do seem to be preferentially emitted in the direction from with the antinucleus would be seen to be spinning counterclockwise. Then II$b$ would be just as physically realizable as II$a$, and reflection invariance would be restored.

So we have seen that invariance principles can differ in their physical interpretations. The invariance principles for continuous transformations allow both an active and a passive interpretation, the invariance principles for the noncontinuous transformations allow only the active physical interpretation. And we have seen that in trying to decide whether the invariance principles actually hold, we must first make some important decisions about what is to count as a transformed system under the active interpretation and what are to count as the "correct" transformations of the states of the original system to the states of the transformed system.

We still must ask, "Which of the invariance principles actually hold in the world?" for it is certainly *not* the case that we wish to maintain these principles as "a priori" true. They are supposed to be empirically confirmable or disconfirmable propositions about the world, and it makes perfectly good sense to ask, "Are they true?" I will refrain from this question, however, until (V,B,3) below.

## 2. *Time Reversal as a Symmetry*

What does it mean to assert that the laws of nature are invariant under time reversal or that time reversal is a symmetry of nature?

First, it is important to realize that the time-reversal transformation is a noncontinuous transformation. As a consequence, there is no passive interpretation of the symmetry. That is, given a particular isolated system undergoing a course of evolution according to the laws of nature, there is no standpoint of an observer such that the same system and evolution, seen from this standpoint, can be viewed as the "time-reversed evolution" of the system.

Further, the physical interpretation of the invariance principle does not require that meaning be given to any locution as peculiar as, for example, one that describes a system as 'going backward in time,' or 'time going backward,' or anything of the sort. Rather, there is a perfectly clear active interpretation of the principle.

We start off with a system in state $S$, allowing it to evolve, after time $Dt$ to state $S_1$. At the same time we start off another system, exactly like the first, except that its initial state is the "time-reversed state" of the *final* state of the original system. Call this new state $T(S_1)$. If the laws of nature are time-reversal invariant, then at the end of the interval $Dt$ we will find the second system in the state $T(S)$, the time-reversed version of the *original* state of the first system. Note the "reverse order" of initial and final states for the two systems. This is unique to time-reversal invariance and distinguishes the interpretation of this principle from that of the other symmetry principles.

What is the time reversal of state $S$, i.e., $T(S)$? This will depend upon the particular dynamic parameters characterizing $S$. For our purposes it will be sufficient to deal with the problem of time-reversed states in prerelativistic point mechanics. Here the state of a system at any time is completely characterized by the positions and momenta of the particles at that time. And the momenta, for fixed masses for the particles, depend only upon the velocities of the particles. To get the time-reversed state corresponding to state $S$, imagine the state in which point masses, like those in $S$, are arranged with the same spatial distribution but with each mass having its velocity reversed in direction, though of the same absolute magnitude (speed) as in $S$.

Sometimes the laws of nature are spoken of as time-reversal invariant if "we couldn't tell from watching a strip of film whether it is going forward or backward," i.e., whether the process being shown on the film is going in the direction the actual process filmed went or whether, instead, what we are seeing is a film taken of a process being run backward through the projector. This is correct if by "couldn't tell" we mean "couldn't tell by simply examining the consistency of the process portrayed with the laws of nature." Many processes consistent with the laws of nature never occur, as we shall see, yet their time-reversals are very frequent indeed.

If the laws of nature are time-reversal invariant, then for any process that occurs in the world, the process that consists in starting with the time-reversed final state of the original evolution and ending up with the time-reversed initial state of the original process is equally compossible with the laws of nature. If the laws are not time-reversal invariant, then this is not so. Remember, of course, that in the quantum-theoretic context, it is transition probabilities between reversed states which must equal the probabilities of the unreversed states taken in opposite temporal order for the laws to be time-reversal invariant. The physics of time-reversal invariance, then, is quite simple. What the philosophical implications of all this are will have to wait until (V,F).

### 3. Nongeometric Symmetries in Physics and the Soundness of the Invariance Principles

All of the symmetries mentioned so far are dynamical reflections of underlying features of the spacetime, its homogeneity (as in time and space translation invariance), its isotropy (as in rotation invariance), its symmetry for left- and right-handed systems (as in spatial-reflection invariance), and its symmetry "backward and forward in time" (as in time-reversal invariance). One must not be overanxious to read too much into these notions, as we shall see when I discuss in detail later the "backward and forward in time" symmetry of spacetime implied by time-reversal invariance. Nonetheless, there seems to be a clear association of these invariance principles with various aspects of the underlying spacetime.

But there are other invariance principles, those dealing with "internal" symmetries of systems, whose "source" seems to have nothing to do with the spacetime structure at all. For example, there is charge conjugation symmetry, which tells us that if we take the initial state of a system and replace all the particles in the system with their antiparticles, then the system will evolve into the "charge conjugated" version of the original final state of evolution of the system—or at least, all the transition probabilities among states will remain the same for the values of transition probabilities among the antiparticle states.

Another symmetry, useful in gaining insight into the interrelations among interaction probabilities in various collision reactions of elementary particles, is crossing symmetry. This is founded upon an important, powerful assumption about the functions characterizing the interactions (their analyticity for imaginary values of energy and momentum transfer) and has interesting predictive consequences. But its existence (or nonexistence) as a genuine symmetry of nature seems, like charge conjugation symmetry, to be independent of any structural features of spacetime.

Some other internal symmetries are the gauge invariances, which are related to the conservation of electromagnetic charge, and generalizations of this notion. Again, there are the isotropic spin symmetries and their generalizations. These arise when one considers the various possible interactions the elementary particles can have with one another. A pair of elementary particles can at one and the same time be interacting electromagnetically, by the weak interaction (that responsible for beta-decay for example) and by the strong interaction (that which is responsible, among other things, for holding the nucleus of atoms together against the electrostatic repulsion of the protons in it).

The strong interaction seems to be totally independent of the charge on the particles, neutrons and protons interacting by it indiscriminately. If one follows up this principle of charge independence and generalizes from it, it is possible to come up with a general symmetry principle of the internal kind (SU(3) symmetry) which is as useful in classifying the various elementary particles into "families" as are the geometric principles of symmetry in classifying, along with certain other rules, the possible electron energy states of an atom. But there is no need for me to follow up these notions.

One question I haven't yet discussed is this: How true are the various symmetry principles? Are the laws of nature translation invariant, rotation invariant, reflection invariant, etc.?

We have seen that once gravitation is brought into the picture, i.e., once we have found it necessary to move from the special to the general theory of relativity, then even the oldest established symmetry principles (translation invariance and rotation invariance) and such familiar associated conservation laws as the conservation of energy and momentum (or, in special relativity, energy-momentum) become doubtful. If we neglect this, however, the basic translational and rotational symmetries —summed up in the relativistic version by speaking of invariance under the transformations of the proper Lorentz group, i.e., translations and/or rotations and/or velocity changes to new inertial frames, but no reflections and no time reversals—seem well established, indeed.

With regard to reflection invariance and time-reversal invariance the situation isn't so clear. There is strong evidence that the weak interactions among elementary particles lead to processes that are *not* reflection invariant, at least if we have been correctly interpreting what the proper state to call the "reflection image" of a given state is. Time-reversal invariance has also been found by some experimenters to be violated in some elementary particle interactions, but here the experimental evidence is much less clear cut than it is in the "violation of the conservation of parity" experiments that seem to force us to reject reflection invariance.

To what extent *should* the invariances hold? That is, do we have any fundamental theoretical reasons for expecting that the invariances, in particular the noncontinuous invariances, will correctly describe nature? The only solid theoretical result is the so-called CPT theorem of relativistic quantum field theory. This tells us that if one makes the basic assumptions of relativistic quantum field theory—which include (1) assuming the correctness of the quantum-field-theoretic approach in general, (2) assuming the Lorentz invariance of any correct physical theory, and (3) assuming the principle of microcausality, i.e., that the field operators have zero commutators for points at spacelike separation—then the invariance principle one gets by combining charge conjugation, space reflection, and time reversal must hold.

To see what this means, start with a system in state $S$. Assume it has a certain transition probability over an interval of time to state $S_1$. Then the $CPT$ principle tells us that if we start with a system in state $CPT(S_1)$, obtained from $S_1$ by reversing velocities and spins, reflecting the result in a mirror, and then replacing particles by antiparticles, then the transition probability to $CPT(S)$ over the same time interval will equal the original transition probability from $S$ to $S_1$.

The $CPT$ theorem gives us no reason whatever to expect, theoretically, that reflection invariance or time-reversal invariance will hold. All that it assures us is that if one of the three symmetries, the two mentioned plus charge conjugation symmetry, fails then at least one other must fail as well.

# SOME DE FACTO IRREVERSIBLE PROCESSES

## 1. *The Expansion of the Universe*

Even if the laws of nature are time-reversal invariant, it is possible for the physical world to be characterized by rather gross asymmetry in time. The fact that the time reversal of every process that occurs is possible, i.e., not inconsistent with the physical laws, does not mean that such processes ever in fact occur. As I have had occasion to note before, what happens in the world depends not simply upon which *laws* hold in the world, but upon which initial conditions and boundary conditions hold as well. And these latter may very well be such as to introduce into the physical world as we know it a pattern of processes characterized by a very plain "asymmetry in time direction." I will examine some of these in this section, the next, and in Section D below. Once again, at this point I will pretty much stick to the exposition of the physics, bringing philosophy to bear on it in Section F.

There is one asymmetric feature of the temporal behavior of the world which is of striking importance. Curiously enough, it is one scientists have been aware of only for a rather short time. If we perform astronomical observations on galaxies other than our own, we discover a pervasive shifting to the red side of the spectrum of the light emitted by various processes in these galaxies. The normal explanation of the red shift is the Doppler effect, accounting for the change in the frequency of the light in terms of a velocity of recession between the galaxy in question and our own. As far as one can tell, the relative velocity of recession of remote galaxies from our own seems to increase in linear proportion to the distance of these galaxies from our own. Unless we adopt a highly anthropocentric view of the universe, in which our galaxy is for some reason chosen as a distinctive center of expansion, the most reasonable explanation of this proportionality of distance and velocity of recession is that the entire galactic universe is in uniform expansion, with each galactic observer seeing all other galaxies receding from him with a velocity of recession proportional to their distance from his own galaxy.

Now this expansion is certainly a process strikingly asymmetric in time. But it is never alleged that the presence of expansion rather than contraction is somehow a result of asymmetries of the laws of nature in time, for it is usually thought that a uniformly contracting universe is as lawlike-possible as one in uniform expansion, at least up to that singular point of a contracting universe where "everything comes together."

For the moment, draw no "philosophical" inferences about the expansion of the known galactic universe "defining" the direction of time, or anything of the sort. Rather, simply notice that as a process *in* time, the expansion is a striking example of a process whose states arranged in temporal priority will be arranged in another important relational order as well. That is, the earlier states of the universe are the more contracted states; the later a state the more expanded it appears.

Interestingly, there have been notable speculative attempts to tie up this global, astronomical asymmetry in the behavior of the known physical world with other, smaller scale, asymmetries as well. For example, there has been, as we shall see, some speculation to the effect that the asymmetries observed in radiation phenomena and in thermal phenomena can somehow or other be "explained" in terms of their connection with the overriding asymmetry of galactic expansion.

## 2. Radiation

A somewhat less spectacular, but perhaps far more important, asymmetry of the physical world in time can be found in the study of radiation phenomena. Now the laws governing electromagnetic radiation, and its interaction with charged matter, are certainly time-reversal invariant, as a fairly easy study of the Maxwell equations shows. Yet we observe matter-of-fact asymmetries in time of radiation phenomena of a striking nature. There is more than one possible theory to account for radiation.

There are, for example, the orthodox field-theoretic account and the less widely held, but conceptually quite important, action-at-a-distance theory of Wheeler and Feynman. Both of these theories postulate laws that are time-reversal invariant, yet both come to terms with the observed asymmetry in time of radiation phenomena. And both account for this asymmetry in an interestingly similar way: in terms of the initial conditions or boundary conditions correctly describing the actual world.

Let us consider the field-theoretic approach first. If we consider a charged particle in accelerated motion, the theory of the interaction of charges with the field allows the possibility that what we will observe will be energy taken from the motion of the particle and converted into the energy of electromagnetic radiation that is emitted, the net result being radiation "propagated off to infinity" (some may be absorbed by matter it runs into, but this is considered accidental and irrelevant) and a net force of deceleration exerted on the particle. And this is in fact what we observe.

But the equations allow another solution as well. One in which radiation "comes in from infinity," "collapsing" onto the charged particle, which absorbs the energy of the radiation field and picks up energy thereby. But this phenomenon, a spontaneous acceleration of particles accounted for by their absorption of a "coherent" incoming radiation, we never observe. Why not?

The answer given is usually in terms of initial conditions. For the latter phenomenon to occur (and it is possible as far as the laws are concerned, for they are time-reversal invariant) one would have to imagine a radiation shell of great radius which as far back in time as we go, is always seen at any point in time as "coming in from infinity and heading for collapse at a particular spacetime location." If the radiation doesn't come in from infinity then still we would have to imagine, at some time in the past, a coherent motion of charged particles at a great distance from one another, all moving so as to emit the appropriate collapsing radiation field. But the assumption of such initial conditions violates the intuition that unless there is some causal reason for order in the world, disorder or randomness is the general phenomenon to expect. Such coherent states, requiring a high degree of order among events at great distances from one another, which would be required in order that we see the time-reversed process of radiation *from* an accelerated charged particle, are states that, in fact, never exist in our world. There is nothing lawlike-impossible about them, it is just that as a matter of fact such a "pre-established harmony" does not exist in the actual world as it is.

So the asymmetry of the world in time with regard to radiation phenomena is, in this account, a result of a matter-of-fact randomness in the distribution of initial conditions characterizing the world at any time. Insofar as the state of the world is ordered, the order is the causal result of a process that can be traced back in time to some local happening. Notice that in this field-theoretic account, there is no talk about causation backward in time, or anything of the sort. The theory presumes a direction of causation forward in time, and this would still be assumed even if the time-reversed processes were seen to exist. There is nothing about them which has "backward causation"; the only thing about them which is peculiar is the required existence of "unexplained" highly-ordered initial conditions at any given time. It is useful to note here that time-reversal invariance of the laws of a theory by no means implies that the theory postulates causation acting backward in time, for nothing in the physical interpretation of the meaning of the invariance principle makes reference to causal influence of any kind other than the ordinary determining of the future by present and past.

The action-at-a-distance theory of electromagnetism is interesting in that it does postulate backward causation. As I noted in (IV,E), we should be very reluctant to label it an a priori absurdity just because it has this feature, for there is no reason to believe that such theories are inconsistent. It is, of course, true that if we were to adopt some such theory, we might have to radically revise our conceptual scheme in many ways, but this would not be the first time that a revolution in physics has upset conceptual orthodoxy.

Why does the action-at-a-distance theory of electromagnetism require causation backward in time? The greatest difficulty for any action-at-a-distance theory of electromagnetism is to account for *radiation*. In the field-theoretic account matter in motion can give rise to radiation—radiation possessing energy and momentum, and carrying them off to infinity.

If an action-at-a-distance theory is to retain the conservation laws of energy and momentum—and yet account for all electromagnetic phenomena in terms of the causal interaction at a distance and over a time interval of material particles, dropping the postulation of fields from its ontology altogether—it will have to assume that, in field-theoretic language, all radiation sooner or later gets absorbed by some particle. None "escapes to infinity" leaving us with vanishing energy and momentum to account for.

So the action-at-a-distance theory accounts for one of its difficulties with radiation by simply asserting that what in the field-theoretic account is the possibility of radiation going off to infinity (or radiation coming in from infinity, for that matter) is not a real possibility at all. If we wish to talk in field-theoretic language, then we must realize that all radiation has a source in some charged particle and all radiation is, sooner or later, absorbed by some material object.

But radiation presents the action-at-a-distance theory with an even greater difficulty, requiring a more drastic revision of old concepts for the theory to work. When, in the field-theoretic account, a particle radiates, there is an immediate interaction between the particle and its emitted field. It is this interaction that accounts for the "radiation reaction" force on the particle which tends to counteract the forces leading to its acceleration in the first place. This radiation reaction is the reason why one must use linear accelerators to accelerate electrons to high energies. The radiation reaction due to accelerating the electron around a circular path is so great that one cannot feed in the energy necessary to overcome it.

So field theory offers an "explanation in principle" of the force of radiation reaction. I say an explanation in principle only, since field theory has never adequately accounted for the details of the radiation reaction. The reason is clear. Usually one treats particles in the theory as charged point masses. But, according to field theory, the field at the origin of the point mass diverges to infinity and all the laws break down here. This is one reason why proponents of action-at-a-distance theories have been anxious to eliminate the field in the first place.

How does an action-at-a-distance theory account for radiation reaction? Well, if it is a force acting on the particle electromagnetically, then it must be a force exerted on the particle by the other charged particles of the universe, for electromagnetic theory is the description of such actions-at-a-distance. There is no field for the particle to interact with, and self-interactions of point particles are denied. But now we become perplexed. The action-at-a-distance theorists are well aware that their theory must allow a finite velocity of propagation of causal influence. A particle here influences (forces) a particle there by a change in its state only at a time later than that of the change of state of the influencing particle. In fact, the causal influences propagate with the velocity of light. But when we accelerate a charged particle, the force of radiation reaction appears *instantaneously!* How did the distant particles know, in advance, that the test particle was going to accelerate at time

$t = 0$, so that they could "move" in appropriate ways prior to $t = 0$ so that the appropriate force would be exerted by them on the test particle at $t = 0$, allowing for its time of propagation from the distant to the test particle?

The answer of the action-at-a-distance theorist is astonishing. When a particle is moved, it "propagates" causal influences affecting other particles at a distance. But it propagates the causal influences both forward and backward in time! The influence does take a time $t = d/c$ to "reach" a particle at a distance $d$ from the moved particle, but the influence affects the motions of particles at time $t = +d/c$ and at time $t = -d/c$ when the particle is moved at time $t = 0$. The backward portion of the causal influence so moves particles at a distance *earlier* in time than the time of the motion, that their return forward causal influence, arriving at the test particle at $t = 0$ just when the test particle begins its acceleration, is just the right amount to induce the radiation reaction force on the test particle.

But now, the action-at-a-distance theorist has a new problem to solve. If causal actions really propagate backward as well as forward in time, why does it appear to us that they propagate only forward in time? When we move a particle, we expect an associated motion of another particle at a distance $d$ at time $t = +d/c$ caused by the motion of the original particle. But we never "see" the motion of the second particle at $t = -d/c$, which we account for according to the action-at-a-distance theorist as the result of the "backward-propagating causal influence" of the motion of the first particle at $t = 0$.

The action-at-a-distance answer again rests upon the invocation of initial conditions. The basic state of the universe is one of randomness, with the particles having their initial positions and velocities at a time distributed at random. The net effect of this randomness, combined with the forward and backward in time causal influences being propagated from particle to particle, is to give the world the appearance of one in which the correlations of motion all act forward in time. I can't pursue the details of this argument here; the reader will find them in the references cited. I might note that as we might expect, this theory, along with all other theories allowing causation backward in time, imposes more stringent constraints upon allowable possible states of the world at a time than do normal physical theories with causation propagating only forward in time, for the states of the universe at any time must obey severe self-consistency conditions. "Real future possibilities" seem out of place in a universe governed by laws allowing causal influence to affect the past, since any "change" in a future state must be reflected in changes in states in the past as well.

For our purposes the following points are of interest: (1) The action-at-a-distance theory, unlike the field theory, postulates genuine causal influences propagating backward in time. (2) The action-at-a-distance theory, like the field theory, postulates laws that are time-reversal invariant. (3) The action-at-a-distance theory, like the field theory, does postulate a world that is not symmetrical in its behavior in the forward and backward time direction; and, like the field theory, it accounts for this asymmetry in time in terms of the initial conditions which, in fact, describe the world as it is. For both of these theories, the world's behavior in time could look very different indeed and yet still be governed by the same time-reversal invariant physical laws. The field theory tolerates the lawlike-possibility of worlds in which we do see coherent radiation coming in from infinity and setting charged particles into accelerated motion. And the action-at-a-distance theory allows worlds in which, contrary to the structure of our world, the causal influences *appear* to propagate only in the backward time direction. In both accounts, then, the world's "direction in time" is a feature not of any asymmetry of the laws of nature in time, but of the actual initial conditions that, at any time, characterize the state of the world at that time.

## ASPECTS OF STATISTICAL PHYSICS

### 1. *The Second Law of Thermodynamics and Statistical Physics*

The most pervasive irreversible phenomena we observe, that is, processes that take place but whose time reversals are "never" observed to occur, are those treated by thermodynamics and the statistical mechanics to which it is usually alleged to "reduce." It is these processes and the theories describing them which have received the greatest attention in the literature devoted to the problem of the direction of time.

Now thermodynamics and statistical mechanics are theories dealing with a remarkably diverse range of phenomena. The equilibrium and nonequilibrium behavior of gases, diffusion processes, the magnetization of materials in imposed magnetic fields, and phase transitions of materials subject to variable external constraints are some of the physical phenomena encompassable by these theories. For simplicity's sake, I will confine my attention to the treatment of the equilibrium and non-equilibrium behavior of gases, although all of the remarks I make can easily be generalized to handle many other situations as well.

What is a typically irreversible process of a gas which thermodynamics and statistical theory attempt to account for? The examples are well known and we need treat only one or two cases here. Suppose we have a gas isolated from the external world by a container impervious to the transfer of heat. The gas is held in the container by a movable piston upon which a specific force is exerted. If we place the piston at a particular position, the gas soon settles down into an unchanging state characterized by a few parameters—its volume, pressure, and temperature—the equilibrium state of the gas for the specified constraints of volume and pressure. We discover that for all the equilibrium states of the gas, although the various parameters take on different values, they remain functionally interrelated by a specific equation in every such equilibrium state, the functional relation being expressed in the ideal gas law or one of its modifications.

Suppose we change the constraints on the gas by moving the piston, i.e., by changing the volume of the container of the gas. After we have finished moving the piston, the gas will soon reach a new equilibrium state. If we move the piston very slowly, this change from one equilibrium state to another is reversible in the sense that we can restore the gas to its original equilibrium state by once again moving the piston, requiring just as much work put into (taken out of) the new transition as came out of (went into) the original.

But there are *irreversible* transitions between equilibrium states as well. Suppose we have a gas confined to one-half of a box by a membrane. We remove the membrane and the gas moves to a new equilibrium state in which it fills the entire box, no work being done on the environment external to the box nor heat transferred from or to it in the process. But there is no way of getting the gas back into its original half of the box without supplying energy into the system from an outside source. Once the membrane was removed, the transition of the gas from the state of occupying half the box to the equilibrium state of occupying the whole box was spontaneous. But the gas never spontaneously removes itself back into its original half-box state. The isolated system has undergone an irreversible thermodynamic change.

In the moving piston case, changes induced by rapid motions of the piston are likewise irreversible, in that a return to the original equilibrium state from the new one will require a net input of energy into the system from the outside world.

It is found that one can systematically treat reversible and irreversible transitions in thermodynamics by introducing one new concept—a novel "thermodynamic parameter," the *entropy* of the gas. Entropy can be defined, for gases, in terms of the other thermodynamic parameters. It is a number assignable to a gas only when it is in an equilibrium state; but this is not a novelty, for only a gas in equilibrium can properly be said to have a temperature. Any two equilibrium states of an isolated system connectible by a reversible thermodynamic transition have the same entropy. If equilibrium state $S_2$ can be reached from equilibrium state $S_1$ by an irreversible thermodynamic transition, then state $S_2$ has a higher entropy value than state $S_1$. Given the definition of entropy, the time asymmetry of thermodynamic processes can be summarized in the famous second law of thermodynamics: In any transition of an isolated system from equilibrium state to equilibrium state, the entropy cannot decrease.

In reversible transitions, of course, the entropy doesn't increase either. But in irreversible transitions, like the filling up of the newly available space in the box by the gas, once the membrane is removed the entropy of the system actually increases from the initial to the final equilibrium state. The second law of thermodynamics is plainly not a time-reversal invariant law, since it allows processes forward in time, the expansion of the gas to fill the box, but prohibits their time reversals, say the spontaneous retreat of the gas back to its original half of the box. Naturally we have assumed here that a gas characterized by a certain set of thermodynamic parameters has its "time-reversed state" characterized by the same parameters. This is a curious assumption but, as we shall see, we need not pursue it, for the whole status of the laws of thermodynamics, and most particularly of the Second Law, looks quite different when we examine the statistical mechanics to which thermodynamics soon becomes "reduced."

Statistical mechanics is an attempt to account for the macroscopically observable thermodynamic behavior of gases (and other systems) in terms of their microscopic atomic constitution. The original attempts to reduce thermodynamics to an atomic theory could not properly be called statistical mechanics at all. A better designation would be to call the program a reduction to a kinetic-atomic theory of gases. We cannot investigate the problems and changes in the atomic theory of thermodynamic features of the world in anything like the detail it deserves, for the evolution of the theory from its original kinetic theory beginnings to the contemporary full-blooded statistical theory is one of the most complex chapters in the history of physical theory. Further, it is one of very great interest in its own right philosophically, since the notions of probability, of explanation, and of reduction which are invoked at various stages, cry out for philosophical clarification. I will only go into as much detail as we require to get reasonably straight about those features of the microscopic theory which bear upon the question of the asymmetrical behavior of physical systems in time and hence, allegedly, upon the question of the direction of time itself. In Section F we will see that ultimately we must come to the conclusion that the results of statistical physics are a good deal less philosophically important than some have made them out to be. But it will be helpful nonetheless to have our understanding reasonably clear of what the physical theory actually does and does not say.

Boltzmann, developing his work upon the foundation of the kinetic theory of Clausius and Maxwell, attempted the construction of a microscopic theory that would account both for the functional interrelation of the thermodynamic parameters for gases in equilibrium (ideal gas law, etc.) and for the well-known "approach to equilibrium" of any gas originally not in an equilibrium state, as summarized in the second law of thermodynamics. His original equilibrium theory met with spectacular success, as did, apparently, his demonstration of the inevitable approach to equilibrium of nonequilibrium gases. But this latter, the original kinetic theory of irreversible processes, met with stringent objections in principle. Getting around those objections, a task originally taken on by Boltzmann himself and pursued in a somewhat different manner by Gibbs and his successors, threw important light upon the very foundations of the microscopic approach, eventually putting the equilibrium theory itself in a new light and radically changing our understanding of the traditional Second Law.

Let me start with Boltzmann's equilibrium theory. We begin with certain assumptions about the constitution of the gas out of numerous molecules, and with additional assumptions about the laws of nature governing the interaction of the molecules with one another, the laws governing their interactive forces, and the general principles of dynamics, which we will assume to be classical mechanics. We look at the "phase space" for a single molecule. Each point in this space represents a particular position and momentum for a molecule. If the molecule has no internal degrees of freedom, for example, this is a six-dimensional space. We divide up the molecular phase space into small boxes. A molecule in a given box is one whose position and momentum are within certain small, specified ranges.

Now, the $n$ molecules of the gas can be divided up among the boxes in many ways, each "way" constituting a *permutation* of the molecules among the boxes. There are certain constraints on these distributions, first that the total number of molecules be a fixed constant, $n$, and second that the total energy of the molecules be also a fixed, known constant.

(I am restricting attention here to the case where the gas in equilibrium is kept isolated from the surrounding environment.)

We can speak of *arrangements* as being specified by the *number* of molecules in each box, irrespective of which particular molecules are in which box. Some arrangements are obtained by more permutations than others. In particular, there is one arrangement reachable by a vastly larger number of permutations than any other; we assume that when a gas is in equilibrium the arrangement of its molecules is this "most probable" arrangement.

Next, we need a "reduction" step, and identification of each macroscopic thermodynamic parameter characterizing a gas with some function of the arrangement of its microscopic constituents. For example, we "identify" pressure with the mean rate of transfer of momentum to a wall per unit time per unit area by means of the collisions of the molecules with the wall; temperature with the mean kinetic energy of the molecules, etc. When this is done we discover the following extremely interesting fact: When the molecules of the gas are distributed among the boxes of molecular phase space according to the most probable arrangement, the functional interrelation of the microscopic features of the gas identified with the macroscopic thermodynamic parameters is just that interrelation among the macroscopic parameters known to hold at equilibrium from the macroscopic theory of gases. So we have now, Boltzmann claims, *explained* the well-known laws governing the equilibrium behavior of gases.

What about the irreversible approach to equilibrium of a gas originally not in equilibrium? Boltzmann attacks the problem like this: Let us assume that the molecules of the gas are distributed according to any arrangement among the boxes other than the "most probable" equilibrium arrangement. Can we then show that they will, by means of their interaction with one another, approach the equilibrium arrangement and then stay there? We can if we make certain assumptions about the detailed nature of their interaction and an additional nonmechanical assumption as well. The additional assumption is Boltzmann's famous *Stosszahlansatz,* or hypothesis with regard to collision numbers. What it assumes is that a molecule of specified momentum will meet, in a given time, a number of molecules of any other momentum depending only upon (1) the volume swept out in the time by the first molecule, (2) the density of the gas, and (3) the fraction of the molecules in the gas at the time which have the second momentum. It is, essentially, an assumption to the effect that there is no "correlation" among the molecules so that the probabilities of collision are fundamentally independent of the relative momenta of the molecules to one another. With the *Stosszahlansatz* Boltzmann proves that a gas whose molecules are originally distributed in any but the most probable arrangement will approach the most probable arrangement, i.e., approach equilibrium, monotonically, and once having gotten close to equilibrium will stay there (Fig. 55).

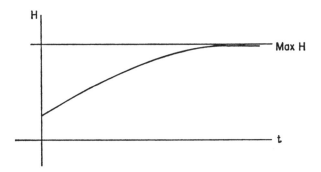

Fig. 55. Boltzmann's original nonequilibrium theory. *H* is the entropy of an isolated piece of gas, and *t* is the time.

In Boltzmann's original nonequilibrium theory, as in pre-statistical thermodynamics, a gas that starts off with less than maximum entropy, i.e., in nonequilibrium, will monotonically approach a state of maximum entropy, i.e., will monotonically approach equilibrium. This is supposed to be true of *every* nonequilibrium gas.

It is at this point, the derivation of the Second Law as a law by the microscopic theory, that the theory comes under severe attack (Fig. 56). Here are two objections to it: (1) *The objection from reversibility:* Boltzmann "proves" that any gas in a nonequilibrium microscopic state will go to a state closer to equilibrium at any later time. Now take the closer-to-equilibrium state. Consider the new state obtained by reversing the velocities of all the molecules in the original close-to-equilibrium microscopic state. By the time-reversal invariance of the underlying mechanical laws, a gas in this state, $S_2$, reversed, will evolve into a gas in state $S_1$, reversed, this being the time-reversal of the original far-from-equilibrium state $S_1$. But if $S_2$ was closer to equilibrium than $S_1$, then $S_2$ reversed will be closer to equilibrium than $S_1$ reversed, and so we have described a gas evolving from one state to another further from equilibrium, contrary to Boltzmann's *lawlike* assertion of the inevitable ap-

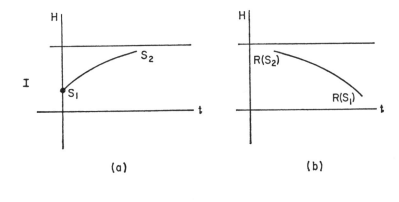

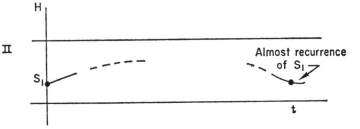

Fig. 56. Objections to Boltzmann's original theory. Diagram
I illustrates the objection from reversability. If I*a* represents
a gas in nonequilibrium state $S_1$ approaching equilibrium
state $S_2$, then I*b* represents a gas in possible near-equil-
ibrium state, the reverse of $S_2$, which must, by time reversal
of the underlying laws, approach nonequilibrium state,
R ($S_1$) the reverse of $S_1$.
    Diagram II represents the objection from recurrence. If
the gas is isolated, and originally in $S_1$, a state far from
equilibrium, then after some finite time the gas will be in
a state very close to $S_1$ and, hence, almost as far from
equilibrium as was $S_1$.

proach to equilibrium. So the *Stosszahlansatz*, from which Boltzmann's
kinetic second law was derived, must be false. (2) *Objection from recur-
rence*: Imagine a gas in a far-from-equilibrium microscopic state $S_1$. A
theorem of Poincaré, on almost-periodic motion, tells us that for each
such state, with the possible exception of a "very small" class of states,
if we wait long enough the gas will eventually evolve into a state as close
to $S_1$ as we like. Hence, the gas will eventually evolve into a state as
close to being as far from equilibrium as was $S_1$ as we like, despite Boltzmann's assurances that the gas in $S_1$ will go to states closer to equilibrium

than $S_1$ and never reverse its approach in the equilibrium direction. Once again, Boltzmann's assumptions must simply be false.

Boltzmann's response to these objections was to make a far deeper analysis of the conceptual foundations of the microscopic theory, changes sufficiently deep to be called the transition from the kinetic theory to a genuine probabilistic or statistical account. The modifications are important for our purposes in two ways: (1) the manner in which the importance of initial conditions and probabilistic considerations is brought into the theory, and (2) the view of "approach to equilibrium in time" as the new version treats it, with particular regard for the symmetry in time of the modified theory.

I cannot pursue the modified theory in close detail, although to do so would be to perform a conceptual analysis of the development of a scientific theory in one of its most interesting phases. A few crucial features of the account are these, however:

1. *The change from a lawlike to a statistical theory:* In the original account, the "approach to equilibrium" behavior of an isolated gas was viewed, as in macroscopic thermodynamics, as a lawlike feature of the world. *Every* nonequilibrium gas *had* to monotonically approach equilibrium, or at least never move further from it, just as is the gas in a world in which the second law of thermodynamics is true. In the new theory, the approach to equilibrium of an isolated gas, or better, the "pervasiveness of equilibrium" of such a gas that I will describe in (2), is a feature that may or may not hold of a gas, depending upon the particular situation of the microscopic constituents of the gas at any given time. The overwhelming majority of isolated gases will behave as the picture describes them, but exceptions wildly deviant in their behavior from the expected behavior are possible, and may in fact occur. For example, the theory is perfectly compatible with there being an isolated gas which, starting in a state very far from equilibrium, passes only through states equally far from equilibrium and never gets any closer at all to the equilibrium distribution. Such gases will be very rare indeed, but not impossible.

2. *The change from a "monotonic approach to equilibrium" to a "pervasiveness of equilibrium" view of gases:* In the original theory, a particular isolated gas in a nonequilibrium state is viewed as going to states ever closer to equilibrium as time goes on. The picture is plainly

not time-reversal invariant, for as we move back in time we see the gas getting further and further from equilibrium, monotonically.

The new picture is symmetrical in time. If we view an isolated gas over a very long time span, what we expect in the new viewpoint is this: Most of the time the gas will be very close to equilibrium. Sometimes it will be further from equilibrium, but the further from equilibrium it gets, the less frequently we observe the gas in such a state. If we pick a state of the gas when it is very far from equilibrium and examine its neighboring states in time, the most likely situation we encounter is that both the state antecedent in time and that subsequent in time are closer to equilibrium. Less often we observe, with equal frequency, two situations: (a) the earlier states are closer to equilibrium and the later further away, and (b) the earlier states are further from equilibrium and the later closer to it. Least frequently, and as we get quite far from equilibrium very infrequently indeed, what we see is both the neighboring earlier and later states being further from equilibrium.

The picture is perfectly symmetrical in time (Fig. 57). If we come

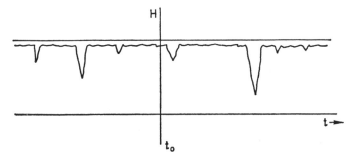

Fig. 57. The revised Boltzmann picture. $H$ is the entropy of an isloated gas and $t$ is the time.

In the revised Boltzmann picture a gas is viewed, if isolated throughout time, as almost always at its state of maximum entropy, i.e., almost always near equilibrium. There are fluctuations away from equilibrium which occur spontaneously, but the greater the fluctuation away from equilibrium, the less frequently it occurs. The picture is clearly symmetric in time.

In the revised Boltzmann picture, even this picture is supposed to be true only of the typical isolated gas. There will be "statistical exceptions" that behave in quite different ways, always remaining, for example, at highly nonequilibrium, low entropy states.

upon a gas which we know to have been isolated and know will remain isolated and if the gas is very far from equilibrium, then we infer that, with a high probability, the gas was earlier in a state closer to equilibrium and will soon be again in a state closer to equilibrium.

If this picture is correct, why then do we macroscopically infer to a second law of thermodynamics that is so baldly asymmetrical in time? The answer goes something like this: To be sure, if we came upon a gas known to have been isolated, and which we know will remain isolated, and if it is far from equilibrium, then it is likely that its earlier and later states are both closer to equilibrium than the state we see. But in the world we inhabit, we very rarely "prepare" isolated systems far from equilibrium by isolating an equilibrium gas and waiting the long time necessary for one of its fluctuations which will take it to a far-from-equilibrium state. If we want to prepare a gas in isolation and far from equilibrium we have a much more efficient way of doing so. The world in which we live, or at least the portion of it known to us, is very far from equilibrium, as the existence of the brightly shining stars assures us. If we wish to prepare an isolated piece of gas far from equilibrium, we have an easy way to do so; we simply break off a chunk of our far-from-equilibrium world and isolate it. Once it is isolated it behaves, with possible rare exceptions of course, as expected. That is, it proceeds to later states closer to equilibrium.

So the asymmetry in time indicated by the Second Law now appears in a very new light. First of all, it is not a lawlike feature of the world at all. Even the expected behavior of gases with regard to approaches toward equilibrium depends upon the gas having one of the likely microscopic states, rather than one of the more unusual but certainly possible nonequilibrium states, which evolves into states even further from equilibrium. Secondly, the "tendency toward equilibrium" should be viewed not as a "directedness toward equilibrium in the forward time direction," but rather as a time-symmetrical "prevalence of states near equilibrium over time."

## 2. The Statistical Theory of Irreversible Processes

It will be worthwhile to pay a bit more attention to the treatment of the "approach to equilibrium" in the theory of statistical mechanics, if only to emphasize the fact that in trying to wield scientific results for philosophical purposes, it is sometimes not all that easy to get the science right. For the scientific results themselves may not be the "transparently clear givens" the philosophers frequently make them out to be.

First some remarks on the Boltzmann approach. Boltzmann's theory has, of course, not remained static since his replies to his critics which established the new probabilistic and time-symmetrical approach. The work of Chapman, Enskog, and the developers of what the physicists call the *B-B-G-K-Y* approach have carried Boltzmann's work further to the point where, by making appropriate *Stosszahlansatz*-like assumptions, the "approach to equilibrium" not only can be shown to occur, but the details of the process can be predicted. Such phenomena as transfer phenomena, by which nonequilibrium irregularities in the system are smoothed out so that the gas becomes ever more equilibrium-like in its state, can be calculated in detail. In all these developments, however, what is being done is finding the assumptions needed to replace the rejected *Stosszahlansatz* to give a nonequilibrium theory not subject to the criticisms of Boltzmann's original theory and finding out the detailed consequences of these assumptions for the statistically inferable behavior of a nonequilibrium gas. The underlying Boltzmann interpretation, with its "merely probabilistic" approach to the regularities and its fundamental assumption of time-symmetric behavior for gases in isolation, remains constant.

In the Boltzmann approach, the "approach to equilibrium" can be viewed as due at least in part to the way we look at the gas macroscopically. If the gas is truly in isolation, then the evolution of the gas must remain, truly, perfectly reversible in time. In a crude sense, the information about the original microscopic state of the gas in its initial nonequilibrium configuration is never lost. If the original highly-ordered state of the system (being, say, in the left-hand side of the box at the start) appears to vanish in the approach to equilibrium (when we examine the final equilibrium state of the gas, we have no way of inferring from this state that the gas was originally in the left-hand side of the box, for no matter what the original state of the gas, it would end up in the same final equilibrium state), then this "loss of information" is only appearance. If we could *know* the exact microscopic state of the gas, even when the gas has reached equilibrium, we could infer from this the initial state, by following its course backward in time using the underlying time-reversal invariant dynamical laws. So the asymmetry, that we can infer to what equilibrium state a gas in nonequilibrium will go but cannot infer from the equilibrium state the nonequilibrium state from which it came, is simply a result of the "coarseness" of the macroscopic description of the gas. If the "fine" microscopic description were available, we could make the inferences go both ways.

An alternative approach to the problem of evolution of gases from nonequilibirum to equilibrium works somewhat differently. We have taken gases to be isolated. But, in reality, no gas ever is *really* isolated. There is always some interaction between the gas and the external environment. If nothing else, the gravitational interaction of the molecules of the gas with the external matter of the universe exists and is uneliminatable. If we take this uneliminatable nonisolation into account, and if we confine our microscopic knowledge of the gas to knowledge of *its* microscopic state, ignoring the microscopic state of the whole remainder of the universe, then a sort of "microscopic irreversibility" occurs. If we know the microscopic state of the gas when it has reached equilibrium, we still cannot infer its original nonequilibrium state. For in the evolution from the nonequilibrium to the equilibrium state, information has been lost through the walls of the container by the interaction of the gas with the external environment. To be able to follow the evolution of the microscopic state backward in time, we would need not only the microscopic state of the gas when it has obtained equilibrium, but the microscopic state of the whole universe at that time. An astonishingly weak interaction between the gas and the external environment can sufficiently mess up the internal correlation among the states of the molecules of the gas, so as to make the "inference backward in time" from final to initial state impossible even over very short time intervals. This, of course, shows no time-reversal noninvariance of the laws of nature at all. It only shows that if one throws information away, dissipating it throughout parts of the universe one will not examine, then even microscopic inference from present to past states quickly becomes blocked.

There is a somewhat different approach to statistical mechanics, that of the American physicist J. W. Gibbs, to which I might devote a little attention; for the "statistical Second Law" in this account looks rather different from the probabilistic, time-symmetrical version we saw Boltzmann ultimately come up with.

First let us look at the Gibbs equilibrium theory, confining our attention to his treatment of a gas kept in energetic isolation from its surrounding environment.

Gibbs introduces a new kind of phase-space. In this space, $2mn$-dimensional for a gas with $n$ molecules of $2m$ degrees of freedom each, each point represents a total microscopic state of the gas, i.e., each point in this space specifies the position and momentum to be assigned to every molecule of the gas at a time. The constraints on the gas, its volume and its total energy, confine the points describing possible states of the gas to a specified subregion of this phase-space. The Gibbs equilibrium theory works by picking a probability distribution over the points in this "allowable" region of phase-space, i.e., by assigning to each possible microstate of the gas (or, more precisely, each collection of possible microstates) a certain probability. For the study of equilibrium a probability distribution is picked which is constant in time, in the sense that if a probability is assigned to a given collection of possible microstates at a given time, and if probabilities are assigned to collections of microstates at later times by tracing the evolution of microstates into new microstates according to the underlying deterministic behavior of the molecules, the same probability will be assigned to the same set of microstates at any later time.

Next one makes an identification of some quantities defined in the statistical theory with the ordinary macroscopic thermodynamic parameters. In this theory, one identifies the macroscopic parameters with some "average" values of certain functions of the microstates, where the average is computed by means of the chosen probability distribution. One then shows that the functional interrelations of these averages of microscopic states is the same as the well-known equilibrium functional interrelations of the thermodynamic parameters familiar from the macroscopic theory.

The logic of this explanation of thermodynamic laws requires a good deal of philosophical thought and attention, attention I will not be able to give it here. The reader might note, for example, that whereas equilibrium values are associated with "most probable" values of certain functions of microscopic states in the Boltzmann theory, in the Gibbs theory they are associated with "average" values over all possible microstates. Only the large numbers of molecules contained in any macroscopic piece of gas makes the numbers come out the same in the two theories.

In the Gibbs theory, there is an average over a function of the microstates which can be associated with the thermodynamic parameter of entropy. This suggests a Gibbsian approach to nonequilibrium which might be interesting to follow up. Suppose we change the constraints on a blob of gas. What this amounts to is changing the region of phase-space allowable to points representing possible microstates of the gas. Suppose we have such a set of points confined to a region. Given the selected probability distribution, there is an entropy assignable to this set of points. Now enlarge the allowable phase-space for the system. As time goes on, a system in any particular microstate will evolve in a deterministic fashion characterized by the underlying dynamical laws of interaction of the molecules. In the phase-space this is represented by a one-dimensional path proceeding from each point in the phase-space, the "trajectory" for the microstate of a system started off in the designated microstate. Can we show, once we have "opened up" the allowable regions of phase-space for the points representing the microstates of the system, that the evolution in time of this set of points will lead to a new set with higher statistical entropy? If so, we might identify this "increase of statistical entropy" for the set of points with the "approach to equilibrium" of particular pieces of gas.

As a matter of fact, we can show that the original statistical entropy chosen is constant in time, no matter how we change the allowable phase-space for the points. This is provable from the laws of dynamics governing the behavior of the system. To get a Gibbsian portrayal of the "approach to equilibrium" we must pick a new parameter to characterize the evolution of the gas from nonequilibrium to equilibrium states. This can be done, and the new concept is called the *Gibbs coarse-grained entropy*. If we start with a set of points with a specific coarse-grained entropy and then change the allowable phase-space for the points, the coarse-grained entropy will increase as time goes on. Interestingly, to show that it will requires adding to the original "statistical" assumption of the Gibbs theory, which was the adoption of an initial probability distribution over the points in phase-space representing possible microstates of the system, a new assumption, a sort of Gibbsian *Stosszahlansatz*. Getting this assumption right and showing that it has the intended result of guaranteeing a monotonic increase of coarse-grained entropy is no easy task. In fact, Gibbs merely speculated about this new assumption, and only now have workers in statistical mechanics become fairly clear about its nature.

So there is a Gibbsian "correlate" to the second law of thermodynamics. It is the result that tells us that the statistical coarse-grained entropy of a collection of points representing possible microstates of a system subject to specified constraints will increase over time if we at some point modify the constraints on the system in an appropriate way. The result is provable, as I have said, only relative to two assumptions: (1) The assumption of a particular probability distribution over the initial microstates at the initial time and (2) the additional assumption necessary to guarantee the increase of coarse-grained entropy over time once we have modified the constraints.

One must be extremely cautious in attempting to apply these results to an analysis of the actual behavior of actual pieces of isolated gas over time. To see this, we need only make a few observations about the differences between the Boltzmann statistical correlate of entropy and the Gibbs correlate. Suppose we have a gas subject to certain fixed constraints. How does its Boltzmann entropy behave? If Boltzmann is right, the statistical entropy defined in his way usually has a value close to the equilibrium value. Sometimes it fluctuates from this value, but if a given state of the gas is one with a Boltzmann entropy far from the equilibrium value, then with a high statistical probability (frequency) the antecedent and subsequent states of the gas in time will be states with a Boltzmann entropy closer to the equilibrium value.

The Gibbs entropy (the original or the coarse-grained) for this gas is constant in time. It is constant by definition, for the Gibbs entropies are functions only of the constraints placed upon the gas. If they are constant, it is constant, no matter how the gas actually behaves.

This difference in the role played by Boltzmann and Gibbs entropies holds in the nonequilibrium situation as well. Suppose we have a gas subject to certain constraints and in equilibrium and we then change the constraints, say by removing a partition confining a blob of gas to one-half of a container. What happens to the entropy of the gas? According to the thermodynamic Second Law, the gas *must,* by a law of nature, go to a new equilibrium state with a higher entropy. According to Boltzmann, what the gas will do depends upon the particular microstate characterizing it when the constraints were changed. Almost always the gas will evolve to a new state with a higher Boltzmann entropy, but this is the result of its initial condition, not simply of the *laws* governing its molecular dynamics. There may very well be astonishing exceptions, where the Boltzmann entropy of the gas remains constant, or even decreases. The Gibbs fine-grained entropy of the gas will remain constant, for it must by its definition and by the underlying laws of dynamics. The Gibbs coarse-grained entropy *must* increase, no matter what the gas actually does—at least if we make the additional Gibbsian statistical assumption necessary to show increase in coarse-grained entropy in any situation. For the change in the coarse-grained entropy, relative of course to making the Gibbsian assumptions, depends by its definition only upon the change in the constraints imposed on the gas and then changed. It is a parameter totally independent of the actual microscopic states of the gas, their actual evolution, or the observable evolution in the macroscopic thermodynamic parameters due to the changing microscopic states.

By this time the reader should be convinced that the study of the "logic" and explanatory role of the statistical-mechanical replacements for the old Second Law is no simple task. Fascinating as the pursuit of these issues might be, I will forgo any further examination of them at this point. We have enough in the way of resources now, however, to be able, in Section F, to look closely at the allegations of philosophers that the "directedness" of physical systems in time has something to do with "the direction of time itself." As we shall see there, both the alleged directedness of systems in time, and the relevance of this to the direction of time, have frequently been badly misconstrued by philosophers.

## TIME ORIENTABILITY AND THE
## DIRECTION OF TIME

Philosophical discussions about the direction of time have usually taken place in the context of simply assuming that spacetime is time-orientable. In fact, it is usually not even noticed that such an assumption is being made at all. I shall for the most part continue this tradition, and in Section F, where the philosophical issues will be pursued in depth, I shall assume throughout that the underlying spacetime is in fact time-orientable. I might say a few words in advance, however, about just what a time-nonorientable spacetime would be like, and about the important connections between the orientability and nonorientability of the spacetime and the possible lawlike behavior of material processes in such spacetime worlds.

Let me start by reflecting on some differences between a three-space that is spatially orientable and one that is not. In an orientable three-space there will be possible objects in a region which, although similar in shape and size, cannot be made congruent to one another by any continuous motion of the objects. And this will hold globally as well, for no continuous motion of any kind will bring the objects into congruence, despite the fact that they are mirror images of one another. If we call one such object left-handed, and its mirror object right-handed, we will be able to classify similar objects anywhere in the space into left- and right-handed. That is, if we consider the class of all objects that can be made congruent to one another either by continuous motions or by reflections, it can be unequivocally split into two distinct classes. Each class is such that any object in it can be brought into congruence with any other object in it by a continuous motion. But no object in one class can be brought into congruence with an object in the other class by any continuous motion whatsoever.

In a nonorientable space, such an unequivocal partitioning is impossible. We may find two possible objects that are mirror images of one another. It may be impossible to bring one of these objects into congruence with the other by a continuous motion confined to sufficiently small region of the space. But any object can be brought into congruence with an object that is its mirror image by some continuous motion or other. As a consequence, the class of all objects that are bringable into congruence with each other either by mirror imaging or by continuous motion is simply identical to the class of objects that can be brought into congruence by continuous motion alone.

In a time-orientable spacetime, as I noted above in (IV,D,1), there is a parallel partitionability of the light-cones relative to each event into forward light-cones and backward light-cones. In such a world, one could continuously transform any "forward-pointing" timelike vector at a point into another forward-pointing timelike vector at another point, keeping the vector timelike while "moving" it from one event to the other.

In a time-nonorientable spacetime, such a globally unequivocal partitioning of light-cones and their interiors into forward and backward regions is impossible. For in such a world we can take a timelike vector connecting two events and, by a continuous motion from one event to another keeping the vector timelike at each stage in its transformation, take a vector that points "forward in time" from $e_1$ to $e_2$ and transform it into a vector that goes from $e_2$ to $e_1$ in the forward time direction. It is still true in a time-nonorientable world that at any given event there will be timelike vectors that cannot be transformed continuously the one into the other, always keeping the vector timelike, if the path of motion of the vector is confined to a sufficiently small region of spacetime; just as in a spatially-nonorientable space there will still be mirror-image objects "at a point" which cannot be continuously transformed into one another if they are kept in a sufficiently small region about the point. But a *global* partitioning of directed timelike vectors into "forward time pointing" and "backward time pointing" is impossible.

I will not pursue a detailed philosophical investigation of what a time-nonorientable spacetime would be like. It should be noted, however, that the description of such worlds is perfectly consistent, and that a superficial a priori philosophical rejection of such worlds would be wrong-headed. I might note, as a matter of fact, that some spacetime structures which have actually arisen in the study of possible "cosmological" solutions of the general-relativistic field equations are, curiously, time-nonorientable. For example, the elliptic interpretation of de Sitter spacetime, a modified version of a standard cosmological solution to the field equations, is time-nonorientable.

Would certain laws of nature be excludable as laws governing the behavior of material processes occurring in a time-nonorientable spacetime? The answer is, "Yes." If we assume, crudely, that the laws of nature can be made to take the same form at every point in spacetime, then certain kinds of laws are excludable in nonorientable spacetimes of certain sorts. For example, suppose the weak interactions really do violate parity conservation, so that there is a way, at each point, of discriminating right-handed physical systems from their left-handed mirror images that makes no use of the right-left distinction itself. Now if this association of the spatial orientation of a system with the other feature, not requiring specification of the orientation in its definition, holds at every point in spacetime, then we can use this feature associated with the orientation of the system to define a globally consistent right-left distinction for the orientation of physical systems. This implies that the spacetime is spatially orientable, for otherwise such a globally consistent right-left orientation distinction could not be drawn. Similarly, if the laws of nature are of the same form everywhere, and if they are time-reversal *non*invariant, then the spacetime must be time-orientable. For the feature of systems associated with the time ordering of their states by the laws of nature, an association that tells us that one state is later than another if and only if it differs from the other in a manner that requires no reference to the temporal ordering of the states, allows us to establish the globally consistent notion of "forwardness" or "backwardness" in time, which distinction is impossible in a time-nonorientable world.

But from this point on we will assume that the spacetime of the world is, indeed, time-orientable, as most philosophers have long implicitly assumed. Still we must ask the important philosophical question we have been putting off throughout this chapter: What is the relevance of the symmetry or asymmetry of the processes of the world in time, as a matter of law or as a matter of fact, to the existence of a direction of time itself?

## ASYMMETRY IN TIME AND THE
## ASYMMETRY OF TIME

### 1. *Laws and the Direction of Time*

From this point on in this chapter I will be attempting to bring to bear on the philosophical issues some of the physical resources we have just been accumulating. My central concern is to try to determine the real relevance, or irrelevance, to questions about the direction of time of possible lawlike or de facto features of the physical processes of the world. Generally, the conclusions will be that the various features of the physical processes in the world, which have been alleged either epistemologically or metaphysically crucial to an understanding of the direction of time, are not really anywhere near as relevant to the interesting philosophical issues as they have been made out to be. But many different approaches to this question have been tried and the grounds for skepticism vary from case to case, so the full treatment of this issue will take some time.

Let me begin with a discussion of the relevance of the time-reversal invariance or noninvariance of physical laws to the problem of the direction of time. Suppose we start with a bold hypothesis and see how far it goes under critical scrutiny: If the laws of nature are time-reversal invariant, then there can be no asymmetric relation of temporal priority. If the laws of nature are time-reversal noninvariant, there can be such an asymmetric relation of temporal priority, and it will be founded epistemically and metaphysically on the irreversibility of the laws.

First the claim that time-reversal noninvariance is *necessary* for temporal asymmetry in the world. The first reason for being doubtful of this is a very obvious one: Although some recent experimental evidence does give us some, though not very conclusive, reason for believing that the time-reversal invariance of the laws of nature does fail at least a little on the submicroscopic level, to believe that the world does or does not have a real and knowable temporal priority ordering depending on the results of these experiments seems absurd. Do the proponents of this position really wish to allege that if the laws of nature all do turn out to be time-reversal invariant, our whole impression that the world of events is a world in which an asymmetric temporal priority relates event to event is an illusion? Surely there is something implausible about a philosophical analysis that makes the existence and knowability of an obviously-present feature of the world hinge upon the existence of other features that may, in fact, not exist and that, even if they do exist, have observational consequences only under the most unusual of situations.

We can see something of what is going on here if we compare time-reversal invariance to invariance of the laws of nature under spatial reflection. It now appears that the laws governing the weak interactions are not invariant under spatial reflection transformations. But suppose that weak interactions, like strong and electromagnetic interactions, were invariant under the spatial reflection transformation. Would this mean that the distinction between right- and left-handed gloves, for example, had broken down? Or that we could not tell of a glove whether it was right- or left-handed? Surely this is absurd. And it would be equally absurd to believe that just because the laws of nature are time-reversal invariant (as they may very well be for all we really know) we could not have pairs of events related by one being later in time than the other, or we could not tell empirically which of the pair was the later event.

All of this suggests that the true nature of the case is that there is in the world an asymmetric relation holding among events, the temporal priority relation, and that we can know when this relation holds or fails to hold, at least sometimes, without relying upon any features of the lawlike nature of the world in time. We will see that this conclusion is supported by other arguments as well.

Now suppose that the laws of nature are time-reversal noninvariant.

Does this provide us with any real help in getting clear about what the direction of time itself amounts to?

The reader will remember from (V,A,2) that one ambition of reductionists about the direction of time was to try to find states of isolated systems which were such that either as a matter of law, or as a matter of fact, two states of the system were related by a temporal priority relationship when and only when the states had features of some nontemporal kind which were properly asymmetrically related to one another. A little attention will show that the connection between time-reversal invariance or noninvariance of the laws of nature and the presence or absence of such lawlike or de facto states related coextensively with the asymmetrical temporal priority relation is not as simple as it might first appear.

First, assume that the laws of nature are time-reversal invariant. Could it be the case that for every isolated system in the world, or for the universe as a whole, that at each time there was a state of the world such that when time $t_1$ was before time $t_2$, then the state at $t_1$, $s(t_1)$, was related to the state at $t_2$, $s(t_2)$, in such a way that as a matter of physical law, $s(t_1)$-type states had to be earlier states of the system than $s(t_2)$-type states? Perhaps surprisingly, the answer is, "Yes." Suppose we have the system at $t_1$ and $t_2$ with its appropriate states $s(t_1)$, $s(t_2)$. Does time-reversal invariance of the laws tell us that there is a possible system which evolves from $s(t_2)$ at one time to $s(t_1)$ at a later time? Of course not. What it tells us is that a system in the *time reversal* of $s(t_2)$ will evolve, as a matter of law, into the *time reversal* of $s(t_1)$. Time-reversal invariance doesn't tell us that the universe could, consistently with the laws, occur with its states in reverse order, but only that it could occur with its *reversed* states in reverse order. But let this go, although there are some interesting questions to be pursued in this direction. I have already noted questions tied up with that issue, that symmetry principles take on full testable meaning only when we have fully characterized what the appropriate transformation of a given state is to be.

Suppose that the laws of nature are time-reversal noninvariant. Does this mean that for any isolated system, or even for the universe as a whole, at each time there must be a state such that, as a matter of law, the earlier states are related asymmetrically by some nontemporal relation to the later? To see how little really does follow from time-reversal noninvariance we need only note the following: (1) It is compatible with the time-reversal noninvariance of the laws of nature that the whole universe be such that if one views the order of its states in time, the time-reversed order of time reversals of the states are equally possible in the light of the laws of nature. (2) It is compatible with the time-reversal noninvariance of the laws that the universe be such that the universe that consists of just its states at a time in reverse temporal order, without even time-reversing the states themselves, is a possible physical universe.

In order for the laws of nature as a whole to be time-reversal noninvariant, what is required is only that there be some possible pair of states of an isolated system such that the situation "$s(t_1)$ earlier than $s(t_2)$" is possible but "time reverse of $s(t_2)$ earlier than time reverse of $s(t_1)$" isn't. But this certainly does not imply that this pair of states is ever realized in fact. It is perfectly possible for the universe to be such that, despite the time-reversal invariance of the laws, the time-reversed arrangement of *its states'* time reversals is compatible with the time-reversal noninvariant laws. It is even possible for the universe to be such that the time-reversed arrangement of its states—not their time reversal—in reversed temporal order is possible.

For an example of a universe of the first kind, simply consider a universe governed by a set of time-reversal invariant laws and one time-reversal noninvariant law. Imagine that the reaction governed by the noninvariant law simply never occurs. To imagine the second kind of universe, just imagine a universe totally "static" in nature. Of course, there are interesting questions about how we would know that the laws were time-reversal noninvariant if we "lived" in such universes, but that isn't the issue here.

Suppose we give as much away to the proponent of the time-reversal noninvariance of laws as "fixing" the direction of time as we can. We grant him that for every isolated system, and for the universe as a whole, times related to one another by temporal priority are times at which the systems have states that are asymmetrically related to one another by a nontemporal relation and are such that their time reversals can never, by the laws of nature, appear in the reversed time order in an isolated system. Do we still have any more insight into the direction of time?

I doubt it. Certainly the following is true about such a world: The world in its actual time order is compatible with the laws of nature. The world consisting of the time-reversed states of the actual world in the reversed time order is not compatible with the laws of nature. But so what? In this other "possible world," possible in the sense of logically consistent, there are still true universal generalizations, laws of nature, describing the interrelations of physical processes. And the laws are naturally time-reversal noninvariant. Of course, we know these lawlike assertions to be *false*, since they are incompatible with the generalizations we know to be true lawlike descriptions of the world.

But how do we know the real laws to be true and the others false? The naïve answer, and I believe the correct one, is that we know which set of generalizations truly describes the world because we know, independently of our knowledge of the lawlike behavior of physical processes in time, what the actual time order of events really is. Only this "independent" knowledge of temporal order would allow us to decide which of the lawlike descriptions is, in fact, the true lawlike description of the world.

It seems, at this point, that there is a fundamental confusion on the part of the reductionists. Once we know which laws are the true time-reversal noninvariant laws of the world, if there are any such, then we can indeed use such knowledge of the laws to determine if a record of the occurrence of some process is a record of the process as it actually occurred, or the time-reversed record of the process. If we run a film of a process that has features that can only occur in one temporal order because of the existence of time-reversal noninvariant laws, then we can use the laws to tell us whether the film is being run forward or backward. But telling whether a record of a process is going in the correct order is *not* determining the order of the process itself in time. We can do that by simply observing the process and seeing which parts come later than others. And being able to tell from the noninvariant laws that a particular recording of a particular process is going in the right time direction is hardly using antecedently-given laws to tell the direction of the universe in time, or the direction of time itself.

Now there is a reply that the reductionist can give to this last line of argument. But this reply is itself infected with difficulties. The reply is this: The argument just given assumes that we establish empirically the lawlike association between the nontemporal asymmetry among the states and their temporal order. Of course, if that were the case, we would first have to independently know the temporal order of the states to get the associating off the ground. But what this argument fails to understand is that the association is merely a *definition* of the temporal ordering relation. What it *means* to speak of one state as being an earlier state of an isolated system than some other state is that the first bears to the second the appropriate asymmetric "not prima facie temporal" relation.

How should we reply to this "definitional" retort? Here are a few comments:

1. Does this analysis really do justice to what we mean by 'earlier,' for example? To throw some doubt on this we might argue as follows: For the moment we have given the reductionist many benefits of the doubt, allowing him, in fact, a lawlike association of some other asymmetrical relation among states with the temporal relation. But don't we understand the *meaning* of 'earlier' and 'later' independently of the existence and nature of this other feature of the world? It seems that we must, for it is doubtful whether such another asymmetrical relation lawlike associated with the temporal priority relation actually exists and yet we certainly know the meanings of the temporal terms. It is one thing to grant a falsehood for the purposes of asking what the world would be like were it other than it is. It is another thing entirely to maintain that meaningful words in our language can be said to get their meaning from a "definition" which is, as a matter of fact, not well formulatable at all given the world the way it is.

2. The adoption of the definitional approach gives rise to a now familiar and dangerous situation, the philosophical postulation of an "overabundance of analyticity." If the definitional relation between the temporal and other relation really holds, then it is impossible in principle for scientific change to ever lead us to conclude that, as a matter of fact, we were wrong in assuming a lawlike association of the two relations. But do we really wish to be put in a situation such that we simply could not discover that what we took to be the time-reversal noninvariant laws of nature really were not the correct laws or, indeed, such that we could not discover that, contrary to our original belief, the real laws of nature really are time-reversal invariant?

3. If we define '$e_2$ is later than $e_1$' in terms of '$e_2$ is R to $e_1$' where R is the "other" asymmetric defining relation, couldn't we equally well have defined the later-than relation by the converse asymmetric relation to R? But is this "arbitrariness" of which events are taken to be later than other events really what we take the situation to be? We could, of course, trivially decide to interchange the *words* 'is earlier than' and 'is later than' in all contexts, but is it the case that what we *mean* when we say that one event is, for example, later than another, is arbitrarily specifiable in terms of any other asymmetric relation the events have to one another as a matter of law? Isn't the situation rather that if we *find out* that there is some asymmetric relation lawlike-associated with a particular temporal relation, that on the basis of this *empirical dis-*

*covery* we can from then on "test" for the temporal relationship by looking for the other relation? But if this is correct, the lawlike association of the two relations is an empirical discovery, requiring an independent ascertainability of an independently-existing temporal relationship, and not a *definition* of the temporal relationship at all.

## 2. De Facto Asymmetries and the Direction of Time

If the attempt to found a theory of the direction of time on lawlike asymmetries in nature is as dubious as we have just seen, what are the prospects of a theory that attempts to "define" temporal order in terms of de facto asymmetries in time of physical processes in the world? Well, prima facie they are that much worse. But since a great deal of attention has been devoted to the relation between temporal order and some important de facto asymmetries of physical processes in time, it seems reasonable for me to devote some attention to this issue. Throughout this section I will focus on the theory that associates the direction of time with those asymmetrical features of the world described by the increase of entropy of isolated systems "explained" and described in statistical mechanics. Other de facto theories are possible, but I will be able to probe all of the interesting philosophical issues by concentrating on this one approach.

All de facto reductionist positions rest either the foundations of our knowledge of temporal priorities or the very existence of temporal priorities on some alleged "matter of fact" association of the temporal relations among events with some other relationship holding among them. Not surprisingly, the most fundamental difficulties with the lawlike reductionist's position carry over in this case, but with additional aspects which make the "de facto reductionist" position even less plausible than that of the "lawlike reductionist."

Is the association of temporal priority relationships with the other "nontemporal" asymmetrical relationship supposed to be one that is empirically discoverable or one that "defines" the temporal relationship? If the former then, as before, there must be an existing temporal relationship, independently knowable to us, in order that any evidence for the association of this relationship with the other asymmetrical relationship can be forthcoming. If we couldn't independently determine that two events had a certain temporal-priority relation to one another why would we ever think that, as a matter of fact, there is a close association in the world between states of an isolated system having a temporal relationship to one another and the same states having some other asymmetrical physical relationship to one another?

But if the association is supposed to be definitional of the temporal relation, then the difficulties of "too much analyticity" reappear in a magnified form. It is bad enough to transform a discovered *lawlike* association of two relations into one that is analytic and hence not modifiable by any new scientific observation without "changing the meanings of the words." But it is the height of the unjustifiable philosophical practice of "converting the true to the a priori" to promote a merely de facto regularity *discovered* to hold in the world into an irrefutable analytic proposition. Suppose scientists do tell us, as we have seen they do not, 'that in any isolated system, as a matter of fact, the entropy of later states is always higher than that of earlier states. And suppose they take this to be a result not of some thermodynamic Second *Law* but of the "matter of fact" distribution of initial conditions of isolated systems' microstates in the world. Would we really wish to argue that the scientific change that would come about when someone plausibly maintained that he had found a system in which entropy decreased in time could only be the result of his changing the meanings of earlier and later? We must be very wary indeed of promoting de facto regularities into definitions.

But the de facto approach suffers not only from these difficulties, difficulties it has in common with the lawlike approach, but also from significant difficulties of detail as well. Let me begin by assuming that the de facto theory will attempt, somehow or other, to associate the direction of time with the relation between states which is that of one state being of higher entropy than another.

What kind of entropy is intended here? It seems clear that it is not the entropy of the old thermodynamics which is being referred to, for that is connected with the temporal priority of states by means of a time-reversal noninvariant law, the second law of thermodynamics, and the de facto theorists all agree that there is in fact no such true lawlike assertion and that the thermodynamic concept of entropy is vitiated in its use for describing the world by being a concept definable only in a theory known to be incorrect. So it is some statistical notion of entropy which is in mind here. It can't be any of the Gibbsian entropies, for these, as we saw in (V,D,2) are not, properly speaking, properties of individual systems which depend upon their changes in microstate at all. The Gibbsian entropies of a system are fixed by the constraints upon it, not by the actual state the gas takes on while so constrained. It is plainly the Boltzmann entropy that the proponents of the de facto entropy theory of the direction of time have in mind. For a system subject to given constraints, this quantity of the system can indeed change over time with the changing micro-states of the system, and there is, thus, at least some initial plausi-bility in looking for interesting associations of the changes in the Boltzmann entropy of a system with other temporal changes in its state.

Once we realize that it is the Boltzmann statistical entropy with which we are dealing, some reflections on the final Boltzmann position, adopted in response to the criticisms of his earlier presentation and presented in (V,D,1) above, will make us skeptical from the very beginning about any theory of the direction of time which rests the asymmetric temporal relations upon any notions about the statistical entropy of isolated systems.

How will the Boltzmann entropy of an isolated system behave in time? First of all, that depends upon the system. For a given system one can choose appropriate initial conditions of the microstates at a time such that the Boltzmann entropy will, after that time, monotonically increase, monotonically decrease, or fluctuate up and down in any way we like. This won't disturb the de facto theorist though, since he is, after all, associating the direction of time with changes in entropy of isolated systems *not* fixed by lawlike considerations, but fixed only by the "matter of fact" entropic behavior of systems in the world.

What kind of entropic behavior of isolated systems in time should we normally expect, according to the corrected Boltzmann theory? If we consider a system isolated from minus to plus infinity of time, its entropic behavior according to Boltzmann should normally be expected to be most of the time in a state close to the equilibrium state, i.e., a state of maximal Boltzmann entropy, with sporadic fluctuations away from the pervasive state to states with much lower entropies. The lower the entropy of a state, the less frequently it will occur. The curve depicting the entropic behavior of the "expectable" system will look perfectly symmetrical in time, as I have noted.

It begins to look pretty implausible to try to define the asymmetric temporal-priority relations in terms of relations among states definable in terms of their entropy, for the latter don't have, in the Boltzmann account, any plausible "preferential time ordering" associated with them. Now one could "bite the bullet" and say something like: "The direction of time changes as the entropic direction system happens to be taking changes." But if one begins to say things that are *that* peculiar, one begins to wonder if the proponent has abandoned entirely any connection between what he means by "the direction of time" and what we ordinarily mean by it at all. When we are giving analyses of concepts that pervade ordinary discourse and scientific theorizing, we can't simply choose to mean by expressions anything we like. Of course scientific (and perhaps philosophical) results can lead us to believe that previous views about the nature of time were wrong. Presumably that was the major accomplishment of the change from prerelativistic to relativistic physics. But do we really wish to make the very radical assertion that "time is continually changing its direction" simply because some systems, or even many systems, have some particular feature of them, like their Boltzmann entropy, fluctuating up and down as time goes on? As we pursue the details of various possible "entropic theories of the direction of time," we shall see this same peculiar difficulty with them occur again and again.

Just which system or systems should we consider when attempting to associate the changing entropy of the system or systems with the direction of time? A natural proposal would be to take the relevant system to be the universe as a whole, identifying the forward direction of time with the time direction in which the total entropy of the whole universe is increasing. But if the universe is infinite, and contains an infinite number of particles, then the notion of its total entropy is simply not well defined, since the Boltzmann statistical notion is applicable only to finite systems.

And if the whole universe does have a total entropy, we simply don't know, as a matter of fact, whether it is increasing, decreasing, or remaining stationary in what we intuitively identify as the forward time direction. If Boltzmann is correct, it is in fact just as likely to be decreasing as increasing. In any case, if the universe had an entropy that fluctuated in time, in the ordinary sense of the forward time direction and as Boltzmann would predict, the definition proposed would lead us to the conclusion that the time direction kept changing. Now, once again, we could talk that way if we chose. But to do so would be to radically revise almost everything else we would normally be inclined to say about the world. For example, every time the entropy curve for the universe changed direction in our ordinary sense, we would be compelled to assert that the direction of time changed, and consequently that future states began determining past states, we began to remember the future instead of the past, etc. Such a revision of our way of speaking

THE DIRECTION OF TIME

would be, truly, a trivial and silly "mere semantic change," for we could far more easily describe what had happened by simply asserting that the entropy of the total universe had started to decrease instead of increase in the forward time direction. What we see here is this: The "directionality" of time, whether it exists in its own right and is knowable by us directly or not, is tied into our whole network of physical laws. To maintain that time changed in direction just because one particular system (even the whole universe) began to change differently than it had in the past in time with respect to one particular feature (entropy)—and not even a feature whose changes in time are, as we previously thought, lawlike-connected to the increase of time—seems grossly to inflate a particular feature of the world in time into a defining characteristic of time itself.

If we choose small isolated systems in the universe instead of the whole universe for the systems whose entropic change "defines" the direction of time, the situation becomes even worse. One proposal would be to attribute to each isolated system a direction of time which is the time direction in which its entropy increases. This would mean that even spatially-very-nearby systems would be going in different directions in time. This could have some very upsetting consequences. For example, we would end up saying of two states of system 1, $s_1$ and $s_2$, that $s_1$ was earlier than $s_2$, and of two states of system 2 nearby, $s_1'$ and $s_2'$, that $s_2'$ was earlier than $s_1'$, despite the fact that a timelike vector from $s_1$ to $s_2$ could be transformed continuously, always being kept timelike, into a timelike vector from $s_1'$ to $s_2'$.

An analogy will make this clear: Suppose that right-handed gloves are usually red, and left-handed ones black. We come upon two gloves near to one another. The first is red and the second black. So we call the first right-handed and the second left-handed, despite the fact that we can bring the gloves into congruence by a local, continuous motion of one of the gloves! To adopt the version of the entropic theory of time which identifies the time direction of a given system with its increase of entropy is to sacrifice the most fundamental things we want to say about the direction of time in a time-orientable world simply to make a priori true propositions which previously we took to be sometimes true and sometimes false, propositions to the effect that the entropy of an isolated system increases as time goes on.

To try to define the direction of time as the direction in which entropy is changing in a majority of isolated systems won't do either. First of all, this notion won't be well defined unless the number of systems we consider is finite. And if we pick a finite number of systems, the entropic behavior of the majority will behave in time with the same time-symmetric fluctuational behavior Boltzmann would have us expect of a single system.

We could pick a single system and choose the positive direction of time as the direction of its entropy increase, using continuous transport of timelike vectors to fix time directions among events in other systems. But surely this is the *reductio ad absurdum* of the whole approach. This particular system will have its entropy fluctuate in time (in the ordinary sense) and we now will have to project changing directions of time onto the whole universe every time the microstates of the system are such that the entropy changes its direction of change. In addition, surely it is illegitimate to carry the de facto approach out to the point where we try to define so important a relation among events as their temporal priority in terms of the merely contingent behavior of some particular system picked at random.

Of course there are connections between the direction of time and the entropic change of isolated systems. In our local region of spacetime, filled with mass-energy in a state far removed from equilibrium and hence with quite low entropy, if we branch off from the whole system a system of low entropy and wait awhile it usually will be found to have higher entropy later. In fact, if we pick several such systems their change in entropy is usually consistent in the sense that they are found to all (or nearly all) increase in entropy as we watch them in time.

But so what? If we watch a typical diner in our world who is eating oysters and steak at the same meal, then, usually, the oyster-eating event will come before the steak-eating event. But what does this particular feature of familiar systems in time have to do with the direction of time itself. Nothing. But neither does the contingent fact about the increase of entropy of branched systems of low entropy in our local region of the universe.

The Boltzmann theory provides a time-symmetrical statistical theory about the changes of entropy in time to be expected of isolated systems. The particular contingent facts about the mass-energy we are familiar with in our local region of spacetime gives a reasonable expectation about how the entropy of systems we branch off from the main body of mass-energy in our region will change in time. If the system branched-off is initially at low entropy, i.e., far from equilibrium, then usually we will find that later it is at a higher entropy, i.e., closer to the equilibrium state.

The Boltzmann theory is a theory about the probabilities of initial conditions of isolated systems in general. The well-known Second Law behavior of the systems with which we are familiar is an attribute of the particular distributions of initial conditions in the systems with which we deal. Both are theories about the behavior of systems in time, but neither is relevant to any question defining the direction of time itself.

### 3. *Should There* Be *a Theory of the Direction of Time?*

The difficulties we have seen infect theories of the direction of time founded upon alleged time-reversal noninvariance of physical laws, and the even more profound difficulties that detract from the plausibility of the de facto entropy theory, make one pause to wonder whether the pursuit of a philosophical *theory* of the direction of time was a reasonable undertaking in the first place.

The basic assumption of all the traditional approaches I have been criticizing is that a reduction of the relations of temporal priority to some other relations, not prima facie temporal and not requiring temporal-priority notions in *their* definitions, is necessary. As I have noted, allegations of ontological reducibility seem to have two possible sources: (1) A scientific identification, like the reduction of light waves to a kind of electromagnetic radiation, and (2) an epistemologically-motivated philosophical reduction like the alleged reducibility of talk about space and time to talk about the spatial and temporal relations among concrete happenings.

Now the theories of the direction of time we have been examining are plainly supposed to be of the philosophical sort. But when we examine their motivation, we find that the usual epistemological warrant for such a philosophical reduction seems to be lacking. If we can determine "by observation" when events have a given temporal priority relation to them, or, rather, if we can determine this directly for at least some events and then project the relation onto others by means of continuous transport of timelike vectors and the like, then there is no reason to search for a reduction of the temporal-priority relation to some other relation among events in the first place—and no reason to "translate" the meaning of assertions about temporal priorities into assertions about some other relation among the events in question.

If we are to have any theory of the world at all, then we must believe in the existence of some entities and in the existence of some relations among them. And if we are to have any epistemic access into the world at all, then at least some of these relations must be knowable to us "directly" and not in terms of inferability from other "directly observable" relations. Just as the theory of special relativity assumes that the coincidence and noncoincidence of events is directly apprehensible by us, and the continuity of sets of events as well, as we have seen, so we may suppose that at least some relations of temporal priority are also among the directly inspectable features of events.

Notice that in claiming this we are not opting for a substantival as opposed to relational theory of spacetime. Temporal priority may be taken simply to be a relation that holds or fails to hold among concrete happenings, although the substantivalist will believe it to hold or not hold among spacetime locations themselves as well. In rejecting the goal of a theory of temporal priority, we are not opting for a substantival theory of spacetime, but simply rejecting the claim that temporal-priority relations are among the "theoretical fictions" of the world which must be shown to be either reducible to some other kind of relation or not to exist at all.

It may be that there are some relations among events we will wish to hold that are a priori coextensive with the relations of temporal precedence or succession. But they are not likely to be the relations speculated upon as the foundation for the direction of time by any of the theorists who find a foundation for the direction of time in the lawlike or de facto irreversibility of physical processes in time.

# BIBLIOGRAPHY FOR CHAPTER V

## SECTION A

PART 1

A good place to start in examining the philosophers' discussion of the features of time rehearsed in this section would be:

Swinburne, R., *Space and Time*, chaps. 8 and 9, and the references cited to these chapters.

PART 2

For an introduction to this subject, see:

Van Fraassen, B., *An Introduction to the Philosophy of Time and Space*, chap. 3, sec. 2;

or:

Swinburne, R., *Space and Time*, chap. 11.

## SECTION B

PARTS 1, 2, AND 3

An excellent introduction to the role of symmetry principles in physics and their association with conservation laws is:

Wigner, E., "The Conceptual Basis and Use of Geometric Invariances," *Reviews of Modern Physics* 37 (1965).*

See also the articles on this subject collected in the same author's:

Symmetries and Reflections.

For additional material on the interpretations of symmetry principles, including a discussion of the relation between the active and passive interpretations, see:

Earman, J., "Einstein and Mach: Covariance, Invariance and the Equivalence of Frames, and the Special and General Principles of Relativity," (mimeographed).

## SECTION C

### PART 1

A collection of papers dealing with such de facto asymmetries in time of physical processes as those discussed here and in (V,C,2) and (V,D), including a discussion of whether some of these asymmetries are dependent, physically, on others, is:

Gold, T., The Nature of Time.

### PART 2

In addition to the items on electromagnetic radiation, retarded and advanced potentials, and the connection of these with the expansion of the universe contained in the item cited immediately above, the reader might look at:

Wheeler, J., and Feynman, R., "Classical Electrodynamics in Terms of Direct Interparticle Action," Reviews of Modern Physics 21 (1949); and:

Wheeler, J., and Feynman, R., "Interaction with the Absorber as the Mechanism of Radiation," Reviews of Modern Physics 17 (1945).

## SECTION D

### PART 1

The reader interested in the development of Boltzmann's thought from the original kinetic theory to the statistical theory of irreversible processes might start with:

Brush, S., Kinetic Theory, 2 vols.,

which contains the most important seminal papers in the history of this development.

A very careful discussion of the crucial issues is contained in:

Ehrenfest, P., and Ehrenfest, T., The Conceptual Foundations of the Statistical Approach in Mechanics.*

### PART 2

For an introduction to Gibbs's approach to statistical mechanics, the original work:

Gibbs, J., *Elementary Principles in Statistical Mechanics,* *
is still a clear and authoritative source. Chapter 12 of this work contains
Gibbs's rather sketchy proposals for a nonequilibrium theory based on his
approach.

A good contemporary treatment of nonequilibrium statistical mechanics
from a Gibbsian point of view is:

Penrose, O., *Foundations of Statistical Mechanics.* **

For the approach that "blames" the approach to equilibrium of nonequilibrium systems on their being not really isolated (as opposed to Boltzmann
and Gibbs who hope to account for the approach to equilibrium of genuinely
isolated nonequilibrium systems), see:

Blatt, J., "An Alternative Approach to the Ergodic Problem," *Progress in Theoretical Physics* 22 (1959).

## SECTION E

For an introduction to the notion of temporal orientability and nonorientability and its relevance to the philosophical issues, see:

Earman, J., "Sense and Nonsense about Entropy and Time" (mimeographed).

## SECTION F

### PART 1

For a very insightful discussion of these issues, see:

Earman, J., "Sense and Nonsense."

### PARTS 2 AND 3

For introductions to the thesis that there is an important connection between
de facto irreversibility and the notion of the direction of time, see:

Van Fraassen, B., *Philosophy of Time and Space,* chap. 3, sec. 3;

or:

Swinburne, R., *Space and Time,* chap. 13.

The thesis is developed at great length and in great detail in:

Reichenbach, H., *The Direction of Time,* especially chaps. 3 and 4.

Another detailed investigation in the Reichenbachian direction is:

Grünbaum, A., *Philosophical Problems of Space and Time,* chap. 8.

A careful critique of the Reichenbach-Grünbaum approach, both physically
and philosophically, is:

Earman, J., "Sense and Nonsense."

CHAPTER **VI.**

## *EPILOGUE*

I have examined quite an array of philosophical problems concerning space and time in the course of this book. If any general thread can be said to tie them together, it is the issue of reductionism. In Chapter II we saw how the reductionist attempts to undercut the difficulties in the epistemology of geometry by attempting to "reduce" propositions about geometrical structure to propositions about the local spatiotemporal relations among material objects. In Chapter III we saw how the "pure relationist" attempts to reduce all talk about spacetime as a substantival entity in itself into spatiotemporal relational talk about material things or events.

In Chapter IV we saw how a reduction of distant temporal relations to local spatiotemporal relations of coincidence and continuity was at the heart of the special-relativistic revisions in our conception of time. Also in Chapter IV we saw, in the various causal theories of time, a program to reduce talk about spatiotemporal relations in general to talk about causal relations among events. Finally, in Chapter V we reviewed the attempts to reduce assertions about the temporal priority among events to assertions about some other asymmetrical relations among them.

How successful are these reductionist programs? Where the programs consist in the reduction of one kind of spatiotemporal language to another, as in the local, relationist reduction of spatiotemporal language in general to talk about local spatiotemporal relations among concrete events, we saw that the issues were very problematic indeed. Both the positions adopted by the reductionist and by his antireductionist opponents seem fraught with philosophical difficulties when examined in detail.

In the case of the reductionist programs where the attempt is made to reduce a branch of spatiotemporal talk to language that is not prima facie spatiotemporal at all, the programs seem a good deal less plausible; for, as we have seen, such attempts as the causal theories of time or the asymmetrical physical process theories of the direction of time seem to rest upon extremely implausible epistemological and metaphysical assumptions. It seems that if there is going to be a reductionist "purification" of spatiotemporal discourse, it will be a reduction of all spatiotemporal talk to some spatiotemporal fragment of the original language and not a reduction of spatiotemporal discourse to some other kind of discourse entirely.

We might reflect upon some general features of reductionist programs at this point. Above, I have made the reductionist program seem to be a program about the elimination of one kind of language in favor of another, or at least the reduction of one kind of discourse to another. But, as we have seen, linguistic considerations of this kind are inseparable from metaphysical and ontological considerations. The pure relationist does not only believe that talk about spacetime as an entity can always be translated without loss into talk about spatiotemporal relations among concrete happenings. He believes, as well, that there is no such thing as substantival spacetime. So the reductionist programs can be viewed on the one hand as semantic, making claims about the meanings of propositions familiar from everyday or scientific discourse, or as metaphysical or ontological, making claims about just what constitutes the genuine real nature of the world.

The most interesting feature of reductionist programs we have uncovered is their reliance, when they are plausible at all, upon epistemological considerations. The usual course of a reductionist argument is to claim that since all the evidential basis for making $A$-type assertions is contained in the class of $B$-type assertions, $A$-type assertions must be translatable into $B$-type assertions without loss of meaning. And since all knowledge of the existence and nature of $A$-type entities comes from observation of the existence and nature of $B$-type entities, $A$-type entities must be reducible to $B$-type entities in any correct ontology.

As we have seen in detail in the study of geometric conventionalism and as a recurring theme throughout the book, the critical arguments for and against reductionist positions are usually twofold: (1) Is the reductionist's epistemological thesis correct? (2) If it is correct, is he justified in drawing the semantic and ontological conclusions he draws from the epistemological result? As we have seen, the answers to both these questions are rarely simple.

In some portions of this book, my own position has been one of a skeptical withholding of judgment. I have presented the reader with a number of alternatives, and some of the best arguments I know for and against them, and left the situation at that. In other places, I have come out for a definite stand, most clearly in my rejection of the causal theories of time and the various physical asymmetry theories of the direction of time.

I would not be particularly upset to realize that many, or even all, of my readers disagree either with my hesitation on some issues or with the conclusions I draw on others. My primary aim in this book has been to try to convince the reader of the truth a number of "metaphilosophical" and "metascientific" propositions. These I hope he will agree have been pretty conclusively demonsrated.

Here are a few of them: (1) To try to carry out a philosophical discussion about such a problem as the nature of space and time without relying upon the results of physical theorizing is to walk blindfolded when one is equipped with extraordinarily useful eyes. (2) When one makes use of scientific results in philosophical analyses, one had better be sure that he has got the science right, for philosophy resting on badly understood science is almost as bad as philosophy that ignores science altogether. (3) On the other hand, one must be extremely wary of adopting the position that the scientific results are a "given" which the philosopher can simply accept as true. As we have seen, the adoption of one scientific theory rather than another, sometimes and in very crucial cases indeed, rests as much upon the philosophical presuppositions of the scientists as it does upon the hard data of the laboratory. You can't do very good philosophy unless you get your science right. But you can't do science in full self-conscious understanding, unless you realize how much it depends upon philosophical modes of reasoning as well.

# INDEX